PRACTICAL PHYSICS

Ernest Zebrowski, Jr.
Associate Professor of Physics
Community College of Beaver County

Gregg Division
McGraw-Hill Book Company
New York   Dallas   St. Louis   San Francisco   Auckland
Bogotá   Düsseldorf   Johannesburg   London   Madrid
Mexico   Montreal   New Delhi   Panama   Paris
São Paulo   Singapore   Sydney   Tokyo   Toronto

Library of Congress Cataloging in Publication Data

Zebrowski, Ernest.
   Practical physics.

   Includes index.
   1. Physics.   I. Title.
QC23.Z382   530   79-18549
ISBN 0-07-072788-0

1234567890 DODO 8876543210

Sponsoring Editor: Myrna Breskin & Cary Baker
Editing Supervisor: Karen Sekiguchi
Design Supervisor: Eileen Kramer & Nancy Ovedovitz
Art Supervisor: George T. Resch
Production Supervisor: Frank Bellantoni

Cover Designer: Dave Thurston
Text Designer: Blaise Zito Associates

Contents

To the Instructor

It has been ten years since I first taught a one-semester applied physics course to a class of auto mechanics and welding students. Since then, this course's student mix has expanded to include prospective machinists, electricians, HVAC technicians, and automotive body specialists. The distinguishing feature of these student programs is their manual skills orientation. Various other construction and industrial trades (carpentry, masonry, plumbing, airframe mechanics, etc.) would also fit the category.

While certainly it is not the role of a physics course to teach manual skills, the educational interests of students cannot be honestly ignored either. Students have the right to ask why I have included a particular topic and in what practical connection they might encounter it again. But in teaching the course, I quickly found existing textbooks to be woefully inadequate in providing answers to such questions. It was only after some first-hand excursions into the fascinating worlds of the mechanic, machinist, welder, builder, electrician, and plumber that the answers began to leap at me. Physics, it turns out, is quite relevant.

This book is an attempt to combine a series of trades applications with a sequence of basic physical principles in such a way that each justifies the other. Topics irrelevant to the trades (e.g., the kinematics of falling bodies) are summarily omitted. Although much of the material is qualitative, in those instances where measurements and calculations form the basis of the application, a detailed quantitative treatment is presented. Throughout, my intent has been to free the instructor of much of the responsibility of justifying the material's relevance. Along the way, I hope I have also succeeded in making the text interesting to student and instructor alike.

In the interest of avoiding any misinterpretation of the method behind my madness, I present the following comments on some of this book's features.

Application Sections

Many, if not most, trades applications are simultaneously based on more than one physical principle. At the same time, the details of many applications will not be familiar to the physics instructor. The inclusion of separate applications sections in each chapter permits the instructor to point them out without claiming expertise and permits the student to see how physical principles appear in practice.

Some of the applications sections are historical accounts (e.g., the atmospheric engine in Chap. 6 and the attempts at perpetual motion in Chap. 9). These are intended not only to place technology in a broad social context but also to spark student interest. As much as trades students may claim to dislike history, I have found very few who were not excited by well-chosen stories from the history of technology. The instructor may exercise his/her own judgment on whether to hold students responsible for such material.

Computational Exercises

Few mathematical prerequisites are assumed. Students are expected to perform arithmetical calculations with the help of a calculator and should know enough about scientific notation to be able to read the tables of unit conversion factors. Previous exposure to basic algebra is useful but not essential.

Several numerical exercises which include quantitative material are listed following each section. The calculations are straightforward and are intended to reinforce the definitions and familiarize students with the units. Students should be encouraged to complete the exercises as they encounter them in the reading. Answers are listed within each exercise section for easy reference.

The end-of-chapter problems, while never requiring complicated mathematical procedures, are nevertheless more challenging than the exercises. It is expected that many students will have difficulty with some of these. The detailed solutions and answers are listed separately in the *Instructor's Guide*. Throughout, I have attempted to establish practical incentives for each calculation and to avoid "fun with figures" exercises.

Graphs   Because many equipment operating manuals and other trades publications make liberal use of graphs, many of the quantitative descriptions are given graphically. This also permits the inclusion of some nontraditional topics (e.g., the variation of friction coefficients with stress and temperature) and areas that would otherwise present algebraic complications (e.g., centripetal force). Repeated emphasis is made in such cases on the effects of scaling (e.g., the wind-load stress quadruples when the windspeed is doubled). The approach encourages the student to view a graph as a description of physical behavior rather than simply as a way of presenting data values.

Units   This book uses predominantly metric units, although some of these are not proper International System of Units (SI) units—the liter, hectare, and kilogram-force, for instance. The student is explicitly made aware of the distinction.

Some scientific purists will be upset with my use of the kilogram-force in the first half of the book. Certainly such a drastic deviation from convention deserves some justification. In fact, this decision was not made hastily or without a great deal of soul searching. Use of the kilogram-force, I am convinced, eliminates many bewildering and unnecessary computational complications and consequently allows the most straightforward approach to a great deal of physics.

I hasten to point out that the kilogram-force (kg_f) is in no way a phony unit. Its size has been precisely established by international agreement: 1 kg_f is *exactly* 9.806 65 N, and 1 pound-force lb_f is *exactly* 0.453 592 37 kg_f. These factors have even been programmed into some unit-converting calculators. But the beauty of the kilogram-force is that it represents the weight of a 1-kg mass, to an accuracy of ±0.5 percent, at all points on or near the earth's surface. In fact, at 40° latitude and elevations up to 1000 meters (m), the error is no more than 0.05 percent, or 1 part in 2000.

The simplifications wrought by the kilogram-force are many. Mass-weight calculations are eliminated from equilibrium problems involving dead loads. Weight density and mass density are numerically the same, which leads to a straightforward calculation of pressure from depth, or vice versa. The buoyant force on a body in kilogram-force becomes numerically equal to the mass of fluid displaced in kilograms. And a mass of 1 kg acted on by a net force of 1 kg_f accelerates at the gravitational acceleration. In short, a large number of the intermediate problem-solving steps that confuse beginning students are eliminated.

In fact, current industrial and commercial practice makes use of a similar relationship with U.S. Customary System (USCS) units. Although physicists have long told their students that the slug is *the* USCS unit of mass, the pound-mass (lb_m) is actually everywhere around us. By law, goods are ordinarily priced by their mass, not their weight. Yet who among us has ever bought a slug of potatoes? And even very good physicists express latent heats in units of British thermal units per pound (Btu/lb). Certainly this is not intended to imply that a substance's latent heat would be different on the moon! The pound here is really the pound-mass, whether we have thought about it or not. In the same way, pounds per cubic foot (lb/ft^3) as a unit of density is often taken by physicists to be the pound-mass per cubic foot (lb_m/ft^3). Otherwise, they would require the standard density tables to list the latitude and elevation (or the value of g) for which the entries are accurate. When fluid pressure is calculated from such density values, it is done subject to the approximation that 1 lb_m of fluid weighs about 1 lb_f under normal conditions. So the use of conjugate mass and force units is, in fact, a firmly established practice. The correspondence of the kilogram and kilogram-force parallels that of the pound-mass and pound.

My contention, based on at least some experience in industry prior to teaching, is that industrial personnel will have little patience with a system that forces them to multiply the mass unit by 9.81 to get the force unit. After all, they never did adopt the slug-pound system. A kilogram-newton system is a totally unnecessary complication in industry. While accuracy in mass measurement is essential, a small fraction of 1 percent uncertainty in force (including weight) calculation is in all common instances quite tolerable in industry and commerce.

The kilogram-force per centimeter squared (kg_f/cm^2) is a particularly nice unit for describing pressures in industrial applications. To within an accuracy of 3 percent, 1 kg_f/cm^2 is the same as 1 standard atmosphere (atm). The actual atmospheric pressure is frequently even closer to 1 kg_f/cm^2. Since most hydraulic devices have dimensions conveniently measured in centimeters, the kilogram-force per centimeter squared easily lends itself to calculating force from pressure, and vice versa. In contrast, the SI pressure unit (the pascal, abbreviated Pa) amounts to the rather small force of 1 newton (N) spread over the rather large area of 1 m^2. Nhis is roughly the pressure developed on a mattress when a bedsheet is spread over it. Normal atmospheric pressure becomes 101.325 kPa, and many industrial pressures must be expressed in megapascals. With this SI pressure unit, calculation of force from pressure or vice versa inevitably involves annoying unit conversions.

Lest I be accused of trying to start a crusade, let me stress that these comments represent more than my own personal judgments. In fact, the engineering community has long ignored the physicists in choosing a practical working unit of force. This is reflected in the numerous torque wrenches whose scales read in kilogram-force times meters, pressure gauges that read in kilogram-force per centimeter squared, and even spring scales that are calibrated in gram-force and kilogram-force. Of course, many of these abbreviate the kilogram-force as simply kg, which leads to an ambiguity that I have been careful to avoid in this book.

The previous considerations notwithstanding, the newton cannot be completely ignored. Energy is conventionally expressed in joules (J), where 1 J = 1 N·m. And power is expressed in watts (W), where 1 W = 1 J/s. The kilogram-force times meters as an energy unit is, in fact, becoming scarce, and the Common Market countries are currently eliminating the metric horsepower (*cheval vapeur*, or *Pferdestärke*), equal to 75 kg_f·m/s, in favor of the kilowatt. In purely scientific activities, of course, the newton has long reigned dominant over the kilogram-force. For these reasons, the student is told about the newton from the beginning, and this force unit also appears in some of the early exercises.

Throughout this book, I have stressed that no quantitative description is valid without a consideration of the measurement units. Units are introduced along with each new physical quantity, and detailed tables of conversion factors are always included nearby, within the chapter itself rather than in an appendix. (These tables are meant to be studied, not just used for reference.) But in no case has erroneous information intentionally been included under the banner of simplicity—particularly with regard to the units of force and mass.

Ernest Zebrowski, Jr.

To the Student

When I was a student, I always skipped over this section of my textbooks. I'd like to attract your attention with a flashing light that says "read me," but unfortunately that isn't possible. So I'll just have to take my chances.

Right now you are probably going to school to acquire some skills that will get you a good job. You wonder why you are being told to study physics. I think it's a fair question.

My answer is that a knowledge of physics will help you a great deal in the years ahead. Technology seldom stands still very long. New inventions appear, new manufacturing processes are developed, and changing consumer demands create new markets for products and services. But you can't go back to school every time there is a new innovation in your field. To be successful in the long run, you will have to get used to doing some learning on your own. From time to time, you will want to keep up to date by reading equipment or construction manuals, trades magazines, and other technical publications.

Nobody can predict what technical innovations will appear during your working life. But we *can* predict that these future innovations will be based on the established principles of physics. If you have a basic knowledge of physics, you will find it easy to understand many new developments as they occur.

Of course, this book contains only a small fraction of what is known about physics. (Otherwise, it would be the size of an encyclopedia.) So

I've had to do quite a bit of picking and choosing. The material that follows was chosen because it forms the basis of a broad variety of today's important trades applications. Since new developments often grow out of old ways of doing things, I expect this material to continue to be useful to you for a long time.

Now I've been talking as if you were going to work in the same job for a very long time. But studies show that there is a good chance you will change jobs many times, especially if you are young. If you know something about trades applications outside your primary field, your options increase. Knowing the material in this book will help you adapt to new trades-related jobs and new employers. A welder, for instance, who understands the principles of hydraulics and simple machines is more flexible and more valuable to an employer than one who doesn't.

Physics is important in helping us understand how things work. But it is also important in helping us *predict* how things will work, before we actually try them out. For instance, no one wants to install a heating system in a building only to find out later that it is inadequate. It makes more sense to first predict the heating requirement and then choose the heating system. Such predictions will usually be numerical.

For this reason, some of the material in this book involves calculation. You will save yourself a great deal of trouble by using a hand-held calculator. When you encounter numerical exercises in the text, you should take the time to pick up your calculator and work them out immediately. The answers have been listed right along with the exercises so you can see how you're doing.

At the end of each chapter is a short list of additional problems. You will find most of these more difficult than the in-chapter exercises. Since they are intended to make you think about how the principles apply, you shouldn't be too discouraged if you find yourself thinking a great deal before solving some of them. If you are unsuccessful, your instructor has all the detailed solutions and answers.

As you read this book, you should also pay close attention to any terms that may be new to you. At the end of each chapter is a list of terms you should know; if you missed any, go back and find them and see what they mean. Having a good technical vocabulary will someday be very useful.

Although many years have passed since I was a student myself, I haven't completely forgotten what it was like. I suspect that you'll find some parts of this book easy and other parts difficult. I'm hoping that you'll find most of it interesting. But most of all, I hope my efforts will have played a role in leading you to a productive and personally rewarding life. Good luck.

Ernest Zebrowski, Jr.

Measurement

Modern industry, commerce, and technology could not function without measurements. Dies have to be set, rolling mills have to be aligned, and the humidity in textile factories has to be controlled. Welders regulate their gas and oxygen pressure by reading pressure gauges; machinists check dimensions with micrometer calipers; auto mechanics use dwell meters, tachometers, feeler gauges, and other measuring instruments. We all read the speedometer and fuel gauges on our cars, and we buy materials according to their measured weight, volume, or length. Our utility bills are based on measurements from electric, water, and gas meters. So *measurements* are all around us—both on and off the job.

In this book we talk about many physical quantities—things like volume, weight, pressure, temperature, velocity, voltage, electric current, and so on. These quantities are useful to us only because they can be measured. We begin, then, by looking at some of the basic features of all measurements.

1-1 Measurement Units

Suppose that we phone a shipper and ask for the rate to ship a package a certain distance. The shipper will certainly want to know the package's weight, so we tell him that it weighs 12. Does this tell him what he needs to know? Of course not. Twelve could mean 12 ounces (oz), or 12 pounds (lb), or 12 kilograms (kg), or maybe even 12 tons if we are shipping a hydraulic press. The number 12 by itself doesn't mean very much. To make any sense, we have to give a *unit* with the number.

To fully appreciate what the measurement unit represents, we need to think about the measurement process itself. Let's say that you have to measure the distance between two towns. If you use the odometer on the instrument panel of your car, you may get a result of 10.3 miles (mi). But if your odometer indicates kilometers (km) instead of miles, the numerical result will be different: 16.6 km. Expressed in feet, this same distance is 54 400 ft, or in meters (m) it becomes 16 600 m. There are still other possibilities. You could measure the distance in furlongs, leagues, rods, perches, or any of nearly 100 other distance units. The numerical result obviously depends on which unit you choose to use. Yet in all cases, it is exactly the same distance that you've measured.

Now the point is this: The measurement itself is a *comparison* of the unknown distance with the size of the distance unit you use. Compared to a unit called the *mile*, the distance between towns is 10.3 times as great. Compared to a unit called the *kilometer*, it is 16.6 times as great. And so on.

Any time we make a measurement, we are really making a comparison. The basis of this comparison is the measurement unit, which must be written as a part of the result.

measurement unit   Definition: A *measurement unit* is the quantity used as the basis of comparison in a measurement.

In ancient times, measurement units were very crude. Carpenters measured boards by the length of a forearm—a distance called the *cubit.* Longer distances were measured in *paces,* which was the length of a double step. Then 1000 paces was called a mile. A *hand* was the width of a palm. Standardized as 4 inches (in), it is sometimes used today for measuring the shoulder height of horses. The inch was based on a knuckle length. In the middle ages, it was redefined as three barley seeds placed end to end. The yard was the distance between a person's nose and the longest fingertip on the outstretched arm. We could go on and name many other units which sound strange by today's standards. The problem with most of these units was that they varied considerably from time to time and place to place. As technology progressed, standardization became crucial. If the inch were based on barley seeds today, it would be impossible to mate machine screws made at one factory with nuts made at another.

1-2 The International System of Units (SI)

We have all heard of the metric system, and most of us have noticed the increased number of metric units being used in the United States. Cereal boxes list the mass in grams as well as pounds, some brands of wine and soda are sold by the liter instead of the quart, and road signs are beginning to list distances in kilometers instead of miles. Metric units like the gram, the liter, and the kilometer are used routinely in most parts of the world today. In fact, only 12 countries* (as of this printing) do not yet use metric units.

The metric system was originally developed by the French during their revolution of 1789. This was the first serious attempt at a completely standardized system of weights and measures. This original metric system was gradually improved through the years, and in 1960 an international conference renamed the updated system "Le Système International des Unités." The English translation is "International System of Units." Ordinarily, it is referred to as the *International System,* or *SI* for short.

More recently, Congress passed into law a bill known as the Metric Conversion Act of 1975. This law began a nationwide program for conversion to SI units of weights and measures. It also continued the use of the term "metric system" to mean the same thing as the SI. Many large industries are now training their workers to use metric measurements (Fig. 1-1).

The SI conference of 1960 did several important things. One was that it established that only seven fundamentally different quantities can be measured. We can measure distance, which is basically the same thing as length, height, thickness, depth, and so on. We can measure time, which is fundamentally different. The other five are mass, electric current, temperature, luminous intensity, and chemical amount of sub-

Fig. 1-1 Many U.S. industries are now training their workers to use metric units. (Courtesy United States Steel International, Inc.)

*Barbados, Burma, Gambia, Ghana, Jamaica, Liberia, Oman, Nauru, Sierra Leone, Tonga, Trinidad and Tobago, and the United States.

stance. All other physical quantities are some combination of these basic seven. The units for the seven fundamental quantities are called the SI *base units.* They are listed in Table 1-1.

Of these seven quantities and their base units, we deal with the first five in some detail in this book. The last two are used mainly in optics and chemistry, which we won't get into.

But are there really only seven kinds of things that can be measured? We can measure the speed of a car with a speedometer, yet speed isn't listed in Table 1-1. Shouldn't this quantity be added to the list? The answer is no. The quantity *speed* amounts to length divided by time [miles per hour (mph), kilometers per hour (kmph), meters per second (m/s), etc.]. Length is one of the fundamental quantities, and so is time. So by measuring speed, we are referring to two of the fundamental quantities in Table 1-1. The point is that everything we can possibly measure is some combination of these seven base quantities. We return to this idea later.

TABLE 1-1   THE SEVEN FUNDAMENTAL PHYSICAL QUANTITIES AND THEIR SI BASE UNITS

Quantity	SI unit	SI symbol	SI primary standard
Length	Meter	m	Based on the wavelength of light from a special krypton lamp
Mass	Kilogram	kg	A cylinder of platinum alloy kept at the national standards laboratory in France
Time	Second	s	Based on the frequency of the radiation from a special cesium oscillator
Electric current	Ampere	A	Based on the magnetic force between two wires carrying this much current
Temperature	Kelvin	K	Based on the temperature at which water will simultaneously boil and freeze if the pressure is right
Luminous intensity	Candela	cd	Based on the radiation from a specially prepared sample of molten platinum
Amount of substance	Mole	mol	Based on the properties of carbon-12

1-3 Measurement Standards

We said that measurement is the process of comparing a physical quantity with a measurement unit. This brings us to a practical problem: How do we keep track of the size of our units? Since a unit has to have some relation to physical things, we can't just keep track of its size on paper. What we need is a *physical* way of recording the size of the units we use. Such physical records are called standards.

standard

Definition: A *standard* is a permanent or readily repro-duced physical record of the size of a unit of measure-ment.

Scientists have carefully established a standard for each of the SI base units. These are described in Table 1-1. Most are based on the physical behavior of certain substances—a special light source using the element krypton forms the standard for the meter, for instance. Every-day measurements are based on standards that have been copied from these *primary standards*. Figure 1-2 shows one example—a set of gauge blocks that are used as a standard for precision machine work. Machin-ists can compare their work with this standard by using a caliper.

Many measuring instruments carry their own standards internally. For instance, a surveyor's transit (Fig. 1-3a) compares the positions of

Fig. 1-2 Gauge blocks are used as a working length standard for precision machine work.

Fig. 1-3 Measuring instruments are used to compare physical quantities with their standards. If a standard is not part of the instrument itself, one must be used in calibrating the instrument.

landmarks in the field with the angular gradations on a finely ruled scale in the instrument. Other instruments, such as the aircraft altimeter (Fig. 1-3*b*), must be *calibrated* against a standard. The pilot does this before takeoff by manually setting the altimeter to read the known altitude of the airport runway. This altitude then becomes the standard for later altitude measurements.

So a standard might be an actual part of the instrument, or it may be used in the instrument's calibration. The thing to remember is that the standard can never be eliminated. Somewhere along the line, a standard must figure into every measurement.

1-4 Conversion of Units

We often encounter cases where a measurement is made in terms of one unit, but we need to express it in terms of another. For instance, we may know the volume of a swimming pool in cubic feet when we are going to pay for the water by the gallon. In such cases, we need to convert the measurement from one unit to another.

Application: Interchangeable Parts

Prior to 1800, every gun, clock, and other mechanical device was built start to finish by a single craftsperson who made one piece at a time and fitted them together along the way. Any pioneer with two broken guns was out of luck: even if the guns were built by the same gunsmith, the parts could not be interchanged. Moreover, this manufacturing technique was slow and expensive and required skill and experience.

In 1798 Eli Whitney* accepted a government contract to deliver 10 000 muskets within 15 months. To manufacture on this scale required a completely new approach—each worker turning out a large number of the same part and then passing these parts onto someone else for the actual assembling. But if this was to work, all samples of the same part had to be identical. Whitney designed filing jigs, stencils for drilling holes in the right places, mechanical lathe stops, and dies and molds to make the component parts in large numbers. Although he missed the deadline, the government was lenient. Soon the practice was adopted by clockmakers and other manufacturers. In 1908 Henry Ford applied the idea of interchangeable parts in the mass production of the Model T, and the world has never been the same since.

But it's not all as easy as this might sound. It turns out that no two manufactured items can ever really be identical. Since drill bits tend to wear with use, the 1000th hole will not be as big as the first hole drilled. And different-sized cutters and drills wear at different rates. Temperature changes cause the blanks to expand or contract when they are worked—possibly by different amounts on different days. There are also other factors, such as machine vibrations or variations in the metallurgy or chemistry of the raw materials. To allow for these effects, manufacturers specify a certain *tolerance,* or allowable error, in each part's dimensions. In-tolerance parts always mate, but some samples fit together better than others. In fact, if all left-side body parts of a new car are at the upper tolerance limit, and the right-body parts are all at the negative limit, the left side of the car may be 2 cm (nearly 1 in) longer than the right side! In pumps, engines, or valves, the tolerances are usually much more critical. The manufacture of interchangeable parts always requires accurate inspection and measurement procedures.

*The same Whitney invented the cotton gin a few years earlier.

Let's begin with the units of length or distance. As we have seen, the SI unit for this quantity is the meter. All other units of length have been defined in terms of this base unit. Table 1-2 lists the numerical relationships between the meter and some other common length units. We see, for instance, that 1 foot (ft) is the same as 30.480 centimeters (cm) and 1 m is 39.370 in. Some of the very large or very small numbers are written in powers-of-10 notation: thus 1 ft is equivalent to $3.048\ 0 \times 10^{-4}$ km. If this notation is new or confusing to you, you should turn to Appendix B for a complete explanation.

The procedure for converting units is fairly simple, once we get used to including the unit with each number we write. Example 1-1 shows how this is done.

Example 1-1 Converting Meters to Feet.

A certain drawing calls for a set of ridge rafters to measure 8.32 m in length. The only available tape measure reads in feet. We therefore need to find the number of feet corresponding to 8.32 m.

From Table 1-2, we see that the relationship between feet and meters is

$$1\ \text{ft} = 0.304\ 80\ \text{m}$$

where we have used abbreviations for the units. This means that

$$\frac{1\ \text{ft}}{0.304\ 80\ \text{m}} = 1$$

In other words, if we write a fraction whose numerator and denominator are equal, the fraction must have a value of 1. We can now multiply the metric measurement by this fraction whose value is 1:

$$8.32\ \text{m} \left(\frac{1\ \text{ft}}{0.304\ 80\ \text{m}} \right)$$

This makes the unit "m" cancel out, top and bottom, and leaves us with a division problem:

$$8.32\ \cancel{\text{m}} \left(\frac{1\ \text{ft}}{0.304\ 80\ \cancel{\text{m}}} \right) = \frac{8.32}{0.304\ 80}\ \text{ft}$$

$$= 27.3\ \text{ft}$$

TABLE 1-2 CONVERSION FACTORS FOR COMMON UNITS OF LENGTH

	cm	m	km	in	ft	mi
1 centimeter =	1	0.01	10^{-5}	0.393 70	0.032 808	$6.213\ 7 \times 10^{-6}$
1 meter =	100	1	0.001	39.370	3.280 8	$6.213\ 7 \times 10^{-4}$
1 kilometer =	10^5	1000	1	39 370	3280.8	0.621 37
1 inch =	2.540 0	0.025 400	$2.540\ 0 \times 10^{-5}$	1	0.083 333	$1.578\ 3 \times 10^{-5}$
1 foot =	30.480	0.304 80	$3.048\ 0 \times 10^{-4}$	12.000	1	$1.893\ 9 \times 10^{-4}$
1 statute mile =	$1.609\ 3 \times 10^5$	1609.3	1.609 3	63 360	5280.0	1

1 angstrom (Å) = 10^{-10} m
1 fathom = 6 ft = 1.828 8 m
1 furlong = 220 yd = 201.168 m
1 international nautical league = 3 international nautical miles (nmi)
1 international nautical mile (nmi) = 6076.12 ft = 1852 m
1 mil = 0.001 in = 2.540×10^{-5} m
1 rod = 16.5 ft = 5.029 21 m
1 yard (yd) = 3 ft = 0.914 40 m

Our answer, then, is that the rafters should measure 27.3 ft in length.

But what if we had written the fraction the other way, with 0.304 80 m on top and 1 ft on the bottom? The fraction still would have a value of 1, since numerator and denominator are still equal. But now the unwanted unit doesn't cancel:

$$8.32 \text{ m} \left(\frac{0.304\ 80 \text{ m}}{1 \text{ ft}} \right)$$

This immediately tells us that we should turn the fraction upside down before doing the arithmetic.

We could also have gotten the right answer by starting with

$$1 \text{ m} = 3.280\ 8 \text{ ft}$$

Application: The Micrometer Caliper

The metric version of this familiar instrument can measure dimensions as small as 0.01 mm, which is much finer than can be read by eyeballing a conventional rule. The caliper achieves this great sensitivity through a screw turning in a threaded sleeve. One complete turn of the thimble advances the screw and spindle a given amount—usually 0.50 mm. Multiples of this distance are indicated on a fixed scale on the sleeve (or barrel) of the instrument. These scale divisions are uncovered as the thimble is unscrewed. The thimble's circumference is further divided into 50 divisions, so a rotation through one of these divisions moves the spindle by 1/50 of 0.50 mm, or 0.01 mm. To read a micrometer, first we read the fixed scale to the last 0.50-mm division, and then we read the rotating scale to get the number of hundredths of a millimeter beyond the last multiple of 0.50 mm.

Using the caliper requires some skill. Many are equipped with a ratchet so they cannot be overtightened, but those that are not require a "feel" that comes only with experience. In any case, the anvil and spindle must contact the work at precise right angles to get a true reading.

The calipers shown here measure outside dimensions. Specially designed micrometer calipers can measure inside diameters (of pipes, for instance) or the depth of holes. Micrometer dials may also be found on precision gas metering valves, certain optical devices, and other instruments.

(This relationship is also read from Table 1-2.) The solution is then set up this way:

$$8.32 \text{ m} \left(\frac{3.280\ 8 \text{ ft}}{1 \text{ m}} \right) = 8.32\,(3.280\ 8)\,\text{ft}$$

$$= 27.3 \text{ ft}$$

The answer is still the same. ◀

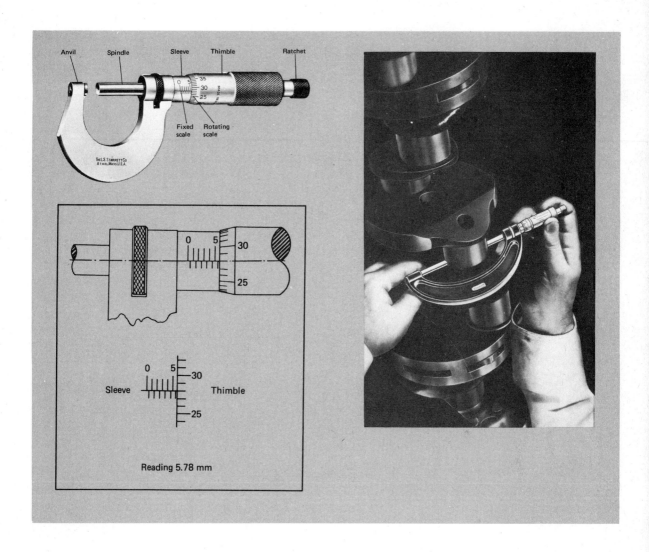

Reading 5.78 mm

Let's summarize this procedure for converting units:

1. Make sure you know the unit of the original measurement.
2. Consult a table of conversions to get the relationship between this original unit and the unit you want.
3. Write the conversion relation as a fraction whose value is 1.
4. Multiply the original measurement by this fraction, making sure that the unwanted unit cancels between numerator and denominator.

Unfortunately, many students feel that writing the converting fraction and the units is a waste of time. Can't we simply multiply or divide by the number we find in the table? The answer is yes—the conversion finally amounts to either a multiplication or a division problem. But even the best of us sometimes multiply when we should have divided. The procedure outlined here is designed to prevent this kind of mistake. If the unwanted unit cancels, we can be sure that we've set up the conversion correctly.

We get an added bonus by using this procedure. Later in this book, we see that certain problems are made very simple by carrying the units through the calculation and using converting fractions. You are strongly encouraged to get used to the procedure now.

Exercises

1. Convert a measurement of 32.62 ft to meters.
    Answer: 9.943 m
2. Find the number of statute miles in 134.2 km.
    Answer: 83.39 mi
3. How many centimeters are there in 18.31 in?
    Answer: 46.51 cm
4. Convert a measurement of 47 231 ft to international nautical miles.
    Answer: 7.773 2 nmi
5. How many meters are there in 100.000 furlongs?
    Answer: 20 116.8 m
6. Convert 0.067 in to mils.
    Answer: 67 mils
7. The depth of a certain channel in the ocean is 32.0 m. What is it in fathoms?
    Answer: 17.5 fathoms
8. What is the length of a 100-yard (yd) football field in meters?
    Answer: 91.44 m

1-5 The SI Prefixes

One of the main advantages of the International System of Units is that the large units are related to the small units in a consistent way. In comparison, the old U.S. Customary System (USCS) units are simply a hodgepodge. For very fine machine work, we measure length in mils, where 1000 mils adds up to 1 in. But then there is 12 in to 1 ft, 3 ft to 1 yd, and 1760 yd to 1 mi (a statute mile). Ocean depths are measured in fathoms (equal to 6 ft), and there are furlongs (220 yd), rods (16.5 ft), chains (4 rods), leagues (3 nmi), and so on. To complicate things, a nautical mile is different from a statute mile, and there are four slightly different varieties of the foot. All these are just units for measuring length. If we need to talk about volume, or speed, or weight, or pressure, the units really get chaotic. The International System simplifies things considerably.

In SI, there is just one unit for length—the meter. If this is an inconvenient size for our purposes, we create multiples or submultiples of this unit by following a simple procedure. Prefixing the unit with *kilo* gives us a *kilometer,* which is exactly 1000 times as large as a meter. Or prefixing it with *milli* gives us a *millimeter,* which is exactly 1/1000 m. Standard *SI prefixes* are listed in Table 1-3. Example 1-2 shows how these prefixes are used.

TABLE 1-3 STANDARD SI PREFIXES

SI prefix	SI symbol	Factor	
tera	T	1 000 000 000 000	or 10^{12}
giga	G	1 000 000 000	or 10^{9}
mega	M	1 000 000	or 10^{6}
kilo	k	1 000	or 10^{3}
milli	m	0.001	or 10^{-3}
micro	μ	0.000 001	or 10^{-6}
nano	n	0.000 000 001	or 10^{-9}
pico	p	0.000 000 000 001	or 10^{-12}
femto	f	0.000 000 000 000 001	or 10^{-15}
atto	a	0.000 000 000 000 000 001	or 10^{-18}

Example 1-2 Using SI Prefixes.

A very small electric current is measured, with a result of 0.000 173 amperes (A). Let's write this in terms of a more conveniently sized unit.

Since the result is a decimal fraction, we need a fairly small unit. Looking at Table 1-3, we see that the prefix "milli" means a factor of 0.001. We may therefore write

$$1 \text{ mA} = 0.001 \text{ A}$$

Notice that we have abbreviated amperes by A, in accordance with Table 1-1, and milliamperes by mA. We may now use the standard procedure for converting units:

$$0.000\ 173\ A = 0.000\ 173\,\cancel{A}\ \left(\frac{1\ mA}{0.001\,\cancel{A}}\right)$$

$$= \frac{0.000\ 173}{0.001}\ mA$$

$$= 0.173\ mA$$

Note that the unit in this result is read "milliamperes."

This is an acceptable result, but we could just as well have done it another way. From Table 1-3, we see that the prefix "micro" means a factor of 0.000 001. We can then write

$$1\ \mu A = 0.000\ 001\ A \qquad \text{or} \qquad 1\ \mu A = 10^{-6}\ A$$

Although the symbol μ is the Greek letter mu, we read it here as "micro." Then

$$0.000\ 173\ A = 0.000\ 173\,\cancel{A}\ \left(\frac{1\ \mu A}{10^{-6}\,\cancel{A}}\right)$$

$$= \frac{0.000\ 173}{10^{-6}}\ \mu A$$

$$= 173\ \mu A$$

or 173 microamperes (μA).

Is this answer different from the first one? No. A current of 0.173 mA is exactly the same as 173 μA. Which one we use is a matter of choice, just as we can buy salami by the pound or by the ounce. Whether we ask for 0.5 lb or 8 oz, we are getting just as much. ◀

Because the SI prefixes are so easy to use, tables of conversion factors do not list all the SI multiple units separately. In Table 1-2, for instance, we don't see units like millimeters and micrometers. Example 1-3 shows how to handle conversions involving such units.

Example 1-3 Converting Inches to Millimeters.

The cylinder of a certain engine is supposed to be 3.128 in in diameter. We want to check it for wear, but the only available micrometer caliper is metric and therefore measures in millimeters (mm). What is the proper diameter in millimeters?

Table 1-3 tells us that the prefix "milli" means a factor of 0.001. Therefore,

$$1 \text{ mm} = 0.001 \text{ m}$$

Consulting Table 1-2, we see that

$$1 \text{ m} = 39.370 \text{ in}$$

Together, these two conversion relations take us from millimeters to inches. We can speed up the calculation by multiplying by both converting fractions at once:

$$3.128 \text{ in} = 3.128 \,\cancel{\text{in}} \left(\frac{1 \,\cancel{\text{m}}}{39.370 \,\cancel{\text{in}}} \right) \left(\frac{1 \text{ mm}}{0.001 \,\cancel{\text{m}}} \right)$$

Notice again that these fractions were written so each unwanted unit cancels. Then

$$3.128 \text{ in} = \frac{3.128}{39.370 \,(0.001)} \text{ mm}$$

$$= 79.45 \text{ mm}$$

There is no limit to the number of converting fractions that can be strung together in this way. Using a hand-held calculator, you can do the arithmetic without writing down any intermediate products, so there is a definite advantage in setting up the calculation like this. ◀

We need to mention one other thing about the SI prefixes. The original metric system included four prefixes that are not an official part of the new International System. These are listed in Table 1-4. Although use of these four prefixes is being discouraged, there is one that is so

TABLE 1-4 METRIC PREFIXES THAT ARE NOT AN OFFICIAL PART OF THE SI

Prefix	Symbol	Factor
hecto	h	100
deca	da	10
deci	d	0.1
centi	c	0.01

entrenched that we can expect to see it for some time. This is the prefix "centi," which is commonly combined with the meter to give a convenient unit for length, the *centimeter*:

$$1 \text{ cm} = 0.01 \text{ m}$$

Other than this example, the use of these four prefixes is not very common.

Exercises

9. Express the following measurements in millimeters (mm):
   (*a*) 0.14 m
   (*b*) 542 μm
   (*c*) 2.67 cm
   (*d*) 1.03 m
   (*e*) 0.613 in
   (*f*) 0.084 1 ft
   Answers: (*a*) 140 mm, (*b*) 0.542 mm, (*c*) 26.7 mm, (*d*) 1030 mm, (*e*) 15.6 mm, (*f*) 25.6 mm
10. The distance between the earth and the sun is 93 000 000 mi. Express this in (*a*) meters, (*b*) kilometers, (*c*) megameters (Mm), (*d*) gigameters (Gm).
    Answers: (*a*) 1.5×10^{11} m, (*b*) 1.5×10^8 km, (*c*) 1.5×10^5 Mm, (*d*) 150 Gm

1-6 Measurement Uncertainty

We can count the number of bricks delivered to a construction site, and we can claim that the result is exact. But if we try to measure the weight of the same bricks, nothing we do will give an exact answer. This is because measurement is fundamentally different from counting. Measurement is a comparison with a standard, and in practice such comparisons are never perfect whole numbers, or even perfect fractions. Furthermore, the measuring instruments are never perfect. As a result, there is always some *uncertainty* in any measurement.

We usually try to keep measurement uncertainty as small as possible, and this is what we mean by an *accurate* measurement. But if a public right-of-way is off by a few centimeters, no one is really going to care very much. Or if a spark plug gap is off by a mil or two, the engine hardly performs any differently. So the fact that measurements aren't perfect is not really a hindrance. As long as the measurement uncertainty is small, the measurement can still be useful.

Usually a number of sources work together to contribute to the measurement uncertainty. For one thing, the standard is probably a copy

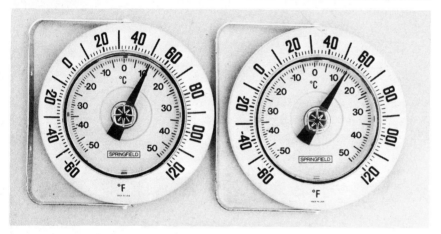

Fig. 1-4 Uncertainty in measurements: a demonstration you should try.

of another standard that itself may be many steps removed from the SI primary standards listed in Table 1-1. Thus the standard itself is probably not perfectly accurate. In addition, the measuring instrument was probably subject to manufacturing tolerances on the assembly line, so no two instruments measure exactly the same. You can verify this for yourself by trying the demonstration shown in Fig. 1-4. Instruments are also affected by age and wear. Steel tapes contract in the cold and give inaccurate readings. Electronic instruments may be affected by temperature and humidity. Changes in barometric pressure can affect certain kinds of flowmeters. And so on.

Can we estimate the amount of uncertainty in a measurement? The answer is yes, in a crude way. The first step is to check on whether the measurement is reliable.

Definition: A *reliable measurement* gives the same result when it is repeated.   *reliable measurement*

Application: Fuel Economy

With the rising cost of gasoline, we have all become conscious of our cars' fuel economy. In old USCS units, distance is measured in miles and fuel is sold by the gallon. By dividing the distance by the fuel consumed, we get *miles per gallon* (sometimes abbreviated mpg) as the index of fuel economy. With metric units, distance is expressed in kilometers and gasoline is sold by the liter (1 U.S. liquid gallon is 3.785 L). The index of fuel economy is then *kilometers per liter*. For comparison, 20.0 mpg is equivalent to 8.50 km/L.

Although fuel economy varies considerably with driving conditions, we can easily measure an average value over a distance of several hundred kilometers. Here's how:

1. Fill up the fuel tank.
2. Set the trip odometer to zero. If there is no trip odometer, write down the actual odometer reading.
3. Drive until the tank is nearly empty.
4. Fill the tank to the same level as you did in step 1. The station pump has a flowmeter that registers the number of gallons or liters of fuel you've bought. This is also the amount of fuel you've used since the last fill-up. Write down this figure.
5. Divide the number of gallons of fuel by the mileage driven, or the number of liters of fuel by the kilometers driven. The result is your average rate of fuel consumption.

With extreme care, it is possible for this procedure to give a result accurate to three significant digits. But if you've driven at high speeds or there have been drastic changes in weather, the result is no more accurate than two significant digits.

Now obviously we can't tell if a measurement is reliable unless we try it a second time. This gives rise to a cardinal rule of measurement: *We never report or use a measurement that is the result of just a single trial.* In fact, it is a good idea to triple-check each measurement we make.

Now just because a measurement is reliable, we shouldn't think that it is exact. Remember that no measurement is ever exact. But we can usually assume that a reliable measurement made with a quality instrument is as accurate as the instrument's smallest scale division. Let's look at an example.

Example 1-4 Uncertainty in a Voltage Measurement.

A voltmeter is used to measure a battery's voltage. The smallest scale division is 0.1 volt (V). The measurement is made and repeated, giving the same result: 12.2 V. What is the uncertainty?

  If the voltmeter was properly calibrated, we can expect that the voltage is very close to 12.2 V. If it had been higher than 12.25 V, we would have read 12.3 on the scale (to the nearest scale division). If it had been lower than 12.15 V, we would have read 12.1 V from the scale. This means that the measurement uncertainty is plus or minus 0.05 V. We may write the result as

$$12.2 \pm 0.05 \text{ V} \quad \blacktriangleleft$$

  In the last example, we began with a reliable measurement. In such cases, we always take the uncertainty to be plus or minus half of the instrument's smallest scale divison. Let's now look at an example where the measurement is unreliable.

Example 1-5 Uncertainty in a Distance Measurement.

Suppose that we want to measure the distance between two towns. We can do this with the odometer on our car's instrument panel. The odometer should first be checked against a measured mile. Let's assume that we've done this and that the instrument is found to be accurate.

  We then drive between the city limits of the first town and the second town. The odometer totals 24.3 mi. But on the return trip, the mileage is only 23.9 mi. The next time we make the trip, we get results of 23.8 and 24.4 mi. So what do we report as the distance?

  If we say it is 24.1 mi, for instance, we are implying that the measurement was much better than it really was. The actual uncertainty is *greater* than half of the smallest scale division. The only fair way to report this result is to say that it is 24 mi. This implies that the result is within at least 0.5 mi of the correct distance.

  Why might we get unreliable results like these? Do they mean that the odometer is malfunctioning? Probably not. Chances are, the car did not follow the exact same path each time. There are lane changes and if the road was curvy, there could be quite a difference in distance depending on the lane. Furthermore, the tires expand slightly as a result of heating and inertial effects at high speeds, so making the trip first at a slow speed (perhaps in heavy traffic) and then at a fast speed gives different odometer readings. Finally, changes in weather have a slight effect because of changes in barometric pressure that affect the tire size. ◄

There are mathematical procedures for analyzing unreliable measurements, but these are beyond the scope of this book. The point here is that unreliable measurements do exist. When we make such measurements, we should not quote the results to the accuracy of the instrument's smallest scale division. Instead, we omit the last digit or two of the measurement.

1-7 Significant Digits

Throughout this book, we describe many things numerically. We assume that these numbers have been measured, or at least that they *can* be measured. This means that each of our numbers has an uncertainty, as discussed in the last section.

For instance, the torque specification for the cylinder bolts of a certain engine is 13 kilogram-meters ($kg_f \cdot m$). This automatically means that an uncertainty of 0.5 $kg_f \cdot m$ is allowable. If we use a torque wrench that can be read to the nearest 1 $kg_f \cdot m$, we can torque the bolts to this specification. But if the specification had read 13.0 $kg_f \cdot m$, this would imply a much smaller uncertainty—no more than 0.05 $kg_f \cdot m$. We would then need a more accurate instrument.

The point is that the numbers 13 and 13.0 do not mean the same thing. The second has a smaller uncertainty and is therefore more accurate. We say that 13 has two significant digits while 13.0 has three significant digits.

significant digits

Definition: The *significant digits* in a number are those that are measured, or at least measurable. A zero is not *significant* unless it is between two other digits, or unless it is to the right of the decimal part of a number. (The zero in 1.30 is significant while the zeros in 0.04 are not.)

We should always pay attention to the number of significant digits in the measurements we use. If we are writing our own measurements, we should include only as many significant digits as we really have. If we are using someone else's measurements, we should consider them to be no more accurate than the number of significant digits tells us. As for the uncertainty, it is always one-half of the place value of the last significant digit. Some examples are given in Table 1-5. Notice that the number 2300 is considered to have just two significant digits and an uncertainty of ±50. If the measurement was more accurate than this, we can place

TABLE 1-5 EXAMPLES OF THE SIGNIFICANT DIGITS AND UNCERTAINTIES IN SOME NUMBERS

Number	Significant digits	Uncertainty
6	1	±0.5
0.8	1	±0.05
90	1	±5
94	2	±0.5
94.1	3	±0.05
2300	2	±50
23$\bar{0}$0	3	±5
2300.0	5	±0.05
0.604 1	4	±0.000 05

a bar over a zero to show that it is significant: thus 23$\bar{0}$0 has three significant digits and an uncertainty of ±5.

We need to be particularly careful about significant digits when we do calculations. Suppose that we drive 151 km and it takes us 2.30 hours (h). We want to know our average speed, so we divide the distance by the time. Using a calculator, we get

$$\text{Speed} = 65.652\,173\,91 \ \frac{\text{km}}{\text{h}}$$

Does this make any sense? We start out with two measurements with three significant digits each, then calculate a result with 10! Of course, the result can't be this accurate, because doing the arithmetic cannot possibly improve the original measurements. The only sensible thing is to round off the result to the same accuracy as the measurements we started with:

$$\text{Speed} = 65.7 \ \frac{\text{km}}{\text{h}}$$

This brings us to an important rule:

Significant-Digit Rule: The result of a multiplication or division can never have more significant digits than the least accurate quantity used in the calculation.

significant-digit rule

Let's look at an example where we use the significant-digit rule.

Application: Spark Plug Sizes

Many people who complain about the upcoming U.S. metrication program do not know that U.S. spark plugs have *always* had metric threads. Before 1900 the only available spark plugs were made in Europe. So, early U.S. engine manufacturers had little choice but to put metric threads in their plug holes to use the plugs they could get. This started the ball rolling, and it never stopped.

The two common spark plug thread diameters are 14 and 18 mm: their hex sections usually have a 16- or 21-mm diameter. By slightly adjusting the manufacturing tolerances, U.S. plug-makers have gotten the 16-mm hex section to fit a ⅝-in socket wrench and the 21-mm section to fit a 13/16-in wrench.

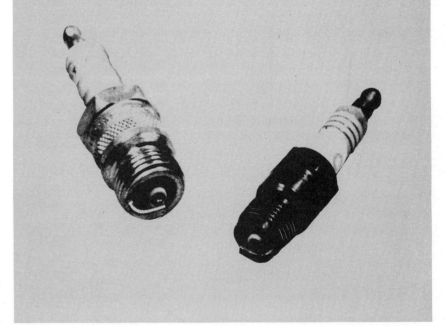

Example 1-6 Significant Digits in a Calculation.

The electric current required to operate a device can be found from the device's power and voltage rating:

$$\text{Current} = \frac{\text{power}}{\text{voltage}}$$

The power is specified in watts (W), the voltage in volts, and the current in amperes.

A certain spotlight is rated at 175W. The voltage is 110V. This gives a current of

$$\text{Current} = \frac{175 \text{ W}}{110 \text{ V}}$$

$$= 1.590\,909\,091 \text{ A}$$

How many digits do we keep?

According to the significant-digit rule, we keep only as many as the least accurate quantity used in the calculation. This is the 110 V, which has just two significant digits. The result is therefore

$$\text{Current} = 1.6 \text{ A} \qquad \blacktriangleleft$$

Exercises

11. Give the number of significant digits in the following measurements:
    (a) 56.5 m
    (b) 0.403 1 kg
    (c) 38 470 pascals (Pa)
    (d) 29.0 A
    (e) 2000 V
    (f) 31$\bar{0}$ seconds (s)
    Answers: (a) 3, (b) 4, (c) 4, (d) 3, (e) 1, (f) 3
12. Give the measurement uncertainty in each of the measurements in Exercise 11.
    Answers: (a) ±0.05 m, (b) ±0.000 05 kg, (c) ±5 Pa, (d) ±0.05 A, (e) ±500 V, (f) ±0.5 s
13. A car is driven 472.8 km on 57.2 liters (L) of gasoline. Calculate the fuel economy in (a) kilometers per liter (Km/L) and (b) miles per gallon (mpg).
    Answers: (a) 8.27 km/L, (b) 19.4 mpg

Summary   Every measurement is a comparison of a physical quantity with a unit. The size of a unit is found from a physical record called a standard. Throughout history, there have been many strange units with strange standards that led to a great deal of confusion in commerce and technology. But since the development of the International System (SI), most of the world has adopted the same units and standards.

No measurement is ever exact or perfectly accurate. Consequently, all physical quantities have some uncertainty. This uncertainty is usually taken to be one-half of the place value of the rightmost significant digit. No number should ever be written with more significant digits than the accuracy of the measurements it is based on.

Terms You Should Know

measurement	SI prefixes
measurement unit	uncertainty
standard	accuracy
International System (SI)	reliable measurement
base units	significant digits

Problems

1. A low bridge carries the sign "Clearance— 11 ft 5 in." If the sign is to be metricated, what should it read?

2. A certain lathe can hold work 48 in long between centers. What is this in metric units?

3. An oval race track is 2.82 mi around. How many laps should be run to make a 200-km race?

4. A piece of lumber 2 × 4 actually measures 1.5 in by 3.5 in. Express these dimensions in metric units.

5. Construction specifications for a building call for using 2250 ft of electrical cable. The cable is available in 100-m coils. How many coils are needed?

6. A certain airplane lands in 985 m of runway if the wind is calm. For each 10 kmph of head wind, the landing roll is shortened by 50 m. The plane approaches a small airport runway 3000 ft long in a head wind of 20 mph. Determine if the plane can land, and if so, how much runway it has to spare.

7. A transmitting antenna rises 236 ft above ground level. Its height is to be indicated on a metricated map. What should the legend read?

8. A micrometer caliper reads to the nearest mil. A measurement of 32 mils is made. What is the uncertainty in (a) USCS units, (b) metric units?

2 Physical Description

In this book we look into the physical principles behind many machines, engines, and industrial processes. Our descriptions are reasonably technical, so that there can be little confusion about exactly what is meant. In this chapter, we lay out the ground rules for such physical description. These rules are not limited to this book. Equipment operating manuals, schematics and flowcharts, technical handbooks, engineering reports, and other technical publications use the same basic methods we discuss here. Although some of this material may be a review to the reader, it is still important to take some time to carefully recognize the advantages and limitations of these methods of description.

Introduction

When something is described with words rather than numbers, the description is said to be *qualitative*. If we say that a building is *big*, or a temperature is *high*, or a truck is *heavy*, or a bridge is *weak*, this is qualitative description. Such description obviously leaves many questions unanswered. The building may be big, but is it capable of storing 20 truckloads of fiber glass? The temperature is high, but will it melt a solder joint? The truck is heavy, but will its tires support its weight? The bridge is weak, but can we cross in a car? To answer questions like these, we need numbers to compare. And, as we discussed in Chap. 1, such numbers are based on measurement.

2-1 Qualitative Description

Yet qualitative description isn't totally useless in technical applications. If the air smells like rotten eggs, we can be sure that hydrogen sulfide is leaking somewhere. This is qualitative, and no measurement has been made. If we dip litmus paper into a liquid and the paper turns red, we know that the liquid is acid. This is qualitative. It doesn't tell us how strong the acid is, but it does say that it is acid and not alkaline. We do a road-test on a car and notice an ignition miss. We don't need to know how often the engine is misfiring; if it is enough to be noticeable, then it is enough to justify a tune-up.

Thus qualitative description does play an important role in technical work. We use it in many parts of this book.

2-2 Quantitative Description

Many technical applications require that we make comparisons. These are best done *quantitatively*; that is, with numbers. If an arc-welding application requires a certain amount of electric current, it is difficult to describe this in words. But it is easy to say that the current must be 80 A, for instance. The comparison is then made by reading the actual current from the ammeter built into the power supply. The advantage of such quantitative description is that it is more accurate than qualitative description.

Are there any limitations to quantitative description? Are there any cases where we must do things qualitatively rather than quantitatively? Yes. Smell, for instance, can be described only in a qualitative way: it is next to impossible to put a number on it. Taste is the same. Color is so difficult to describe through numbers that in most practical cases we stick with words like "red," "dark green," "lavender," and so on.

For most other physical quantities, we have a choice. Table 2-1 lists the names of some physical quantities and the corresponding qualitative terms. If accuracy is not important, we can be satisfied with saying

TABLE 2-1   *QUANTITATIVE AND QUALITATIVE DESCRIPTION*

Quantitative (measurement)	Qualitative (judgment)
Weight	Heavy or light
Length	Long or short
Volume	Big or little
Friction coefficient	Rough or smooth
Temperature	Hot or cold
Tensile strength	Strong or weak
Velocity	Fast or slow
Modulus of elasticity	Stiff or flexible
Sound intensity	Loud or quiet
Light intensity	Bright or dark

that something is "hot," for instance. But if we need to be more accurate than this, we have to make a temperature measurement and get a numerical value.

Quantitative description is numerical, and it must be based on measurements. Qualitative description is usually based on judgment. If we have quantitative information to start with, we can restate it qualitatively. Thus boiling water at 100°C can be described as hot, while boiling oxygen at −183°C is cold. But going the other way—qualitative to quantitative—is not possible without additional information. With this in mind, let's look at some of the methods of quantitative description.

2-3 Data Tables

A single measurement is described by a single number, or possibly by a number followed by its tolerance. But when an entire set of numbers (data) is to be reported, the numbers need to be arranged in some organized way to prevent confusion. An accepted method is to use *data tables* such as Table 2-2. Here we see a list of the carbon contents of various carbon-steel tools. Such a table may list the items alphabetically or, as in this case, in order of high to low numerical values.

Of course, we already used another kind of data table in Chap. 1. Table 1-2 allowed us to look up the conversion factor between two length units. By having this information arranged in tabular form, we could easily find any of 30 different relationships.

This is the idea behind all data tables: The numbers are grouped in one place in a form that makes them easy to look up. We should keep this in mind if we ever have to make a table ourselves.

TABLE 2-2 CARBON CONTENT IN CARBON-STEEL TOOLS

Tool	Carbon %
Razor	1.25
File	1.25
Twist drill	1.15
Pipe cutter	1.15
Tap and die	1.10
Circular saw	0.85
Cold chisel	0.85
Pliers	0.75
Crowbar	0.75
Wrench	0.75
Hammer	0.75
Screwdriver	0.65

Steel is an alloy of iron and small amounts of other substances—principally carbon. Steels with carbon content greater than 0.75 percent are easy to harden, but difficult to weld. With lower carbon content, welding is easy but hardening is difficult.

2-4 Graphs

Graphs use circles, bars, lines, or some other geometrical form to represent numerical information. One of the simplest types is the *circle graph*, an example of which is shown in Fig. 2-1. Here we see what happens as gasoline is burned in a conventional car engine. The circle represents 100 percent of the energy in the unburned gasoline. The size of each pie slice shows us how much of this energy goes in each of five directions. Since more than one-third is lost in the exhaust, for instance, this section is slightly more than one-third of the entire circle.

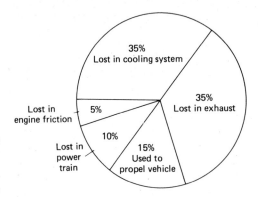

Fig. 2-1 A circle graph: Where the energy goes in a typical gasoline-powered car engine. The circle represents 100 percent of the energy in the fuel.

Now certainly we could list this same information in a table. But the advantage of the graph is this: In one glance, we can take in all this information and quickly see how it is related. The point is made very dramatically that only a small part of the gasoline's energy goes where we want it—to propel the vehicle. More than twice as much goes into the cooling system, for instance.

The most common type of graph is a line drawn on a rectangular grid. The *line graph* in Fig. 2-2 describes the performance of a certain car as it accelerates from rest. From this graph, we can read the car's speed at any instant in time. In 10 s, for instance, the car reaches a speed of 88 kmph [about 55 mph]. In 20 s, it is traveling at 126 kmph [79 mph].

Again, this same information could have been listed in a table. The advantage of the graph is that it is more compact and allows us to make comparisons more easily. We quickly see, for instance, that the acceleration levels off at higher speeds. And we can draw this conclusion without sifting through a long list of numbers.

Notice that the graph contains all the labels we need to read it. The horizontal axis is labeled with both the quantity being represented (time) and its unit of measure (seconds). The same is done with the vertical axis. Each axis is also labeled with a series of numbers which are equally spaced. In other words, the distance between 20 and 30 kmph on the graph is the same as the distance between 90 and 100 kmph. On the

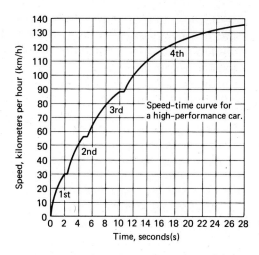

Fig. 2-2 A line graph showing the performance of a vehicle with a four-speed transmission.

horizontal axis, the distance between 0 and 2 s is the same as the distance between 22 and 24 s, and so on. Finally, the graph has a brief *legend*, which tells us what is being represented. Without this legend, we wouldn't know if the description applied to a car, an airplane, or a sailfish. The legend and all the labels are considered to be a part of the graph itself.

We often find it convenient to combine two or more graphs on the same grid. Figure 2-3 is an example. It is found that temperature affects the strength of a piece of metal, but that some metals behave differently from others. This graph shows how the strength of copper decreases with increasing temperatures, and how structural steel actually gets stronger for awhile before the temperature gets high enough to weaken it. In other

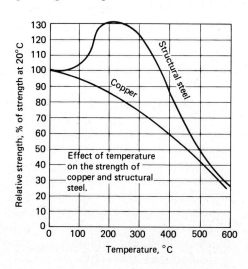

Fig. 2-3 Sometimes more than one graph is drawn on the same grid. This makes it easy to make comparisons.

words, the graph tells us not only how copper and structural steel behave, but also how the behavior of these two metals is different. We often
combine graphs like this when we want to compare the behavior of
different substances.

Exercises

1. Use Fig. 2-1 to answer the following questions. (a) How much of the
   energy in the gasoline is lost in the power train of a typical car? (b)
   Compared to the energy lost in engine friction, how much of the
   energy in the gasoline is lost in an engine's exhaust?
   Answers: (a) 10 percent, (b) 7 times as much
2. Use Fig. 2-2 to answer the following questions. (a) How long was the
   car in first gear? (b) How long was the car in third gear? (c) What was
   the car's top speed in first gear? (d) What gear was the car in when
   it hit 70 kmph? (e) How long did it take the car to reach a speed of
   100 kmph?
   Answers: (a) 2.0 s, (b) 4.6 s, (c) 30 kmph, (d) third gear, (e) 12.2 s
3. Use Fig. 2-3 to answer the following questions. (a) At what temperature does structural steel attain its greatest strength? (b) At what
   temperature does structural steel have the same strength it has at
   20°C? (c) At what temperature will copper be half as strong as it is
   at 20°C? (d) At what temperature will structural steel have half its peak
   strength? (e) At what temperature will the strength difference between
   copper and structural steel be greatest?
   Answers: (a) 215°C, (b) 370°C, (c) 450°C, (d) 450°C, (e) 275°C

2-5 Logarithmic Graphs

In the last section, we said that the numbers on the axis of a graph
should be equally spaced. When this is done, we say that the graph has
a linear scale.

linear scale

Definition: A *linear scale* relates a physical quantity to
distance on a graph in such a way that the same distance
always represents the same change in the quantity.

For instance, if one division represents 2 s on the horizontal axis of Fig.
2-2, then each additional division represents an additional 2 s.
    But sometimes it is impossible to draw a graph on a linear scale.
In Table 2-3 we see some data taken on samples of medium-carbon
steel. The samples were subjected to a stress that was alternately applied and removed a large number of times. This was found to weaken the
steel—a phenomenon known as "fatigue." The table lists the number of
times the load was applied and removed, as well as the corresponding
measured tensile strength. If we want to graph these data, we run into a

TABLE 2-3 FATIGUE DATA FOR MEDIUM-CARBON STEEL

Number of cycles	Tensile strength, kg_f/cm^2
20 000	3760
40 000	3370
80 000	3090
200 000	2845
1 000 000	2650
2 500 000	2600
8 000 000	2580

Samples were subjected to a load that was applied and removed a large number of times. After the number of cycles listed in the left column, a sample's tensile strength was measured, giving the value in the right column.

problem with the left-hand column. The largest data value here is 400 times as big as the smallest. Suppose that we represent the smallest value by 1 cm on a linear scale. The last data point must then be 400 cm [13 ft] away. This would be a very big graph! Suppose, instead, that we begin by making the largest value fit on a graph 20 cm wide. The smallest value is then represented in 1/400 of this distance, or 0.05 cm [0.02 in]. Since this is smaller than the width of a pencil line, it is much too small to show up on the graph.

When data span a very large range of values like this, they are usually graphed on a scale that is not linear. The most common nonlinear scale, and the only one we use in this book, is called the logarithmic scale.

> Definition: A *logarithmic scale* relates a fixed distance on a graph to a factor-of-10 increase in the quantity being represented.

logarithmic scale

If 1 cm represents 1 s on a logarithmic scale, then the next centimeter represents 10 s, and the next is 100 s. A logarithmic scale has the effect of squeezing together the high values while spreading out the low ones. Figure 2-4 shows a comparison of a linear scale and a logarithmic scale.

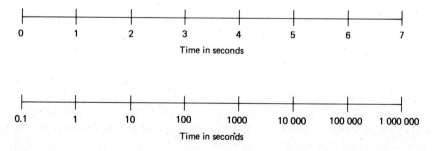

Fig. 2-4 Linear and logarithmic scales. Both begin by representing 1 s by a distance of 1 cm. In the linear scale, each additional centimeter represents an additional second. In the logarithmic scale, each additional centimeter represents a factor-of-10 increase.

Both scales are the same length, and both have their divisions in the same places. But the linear scale represents only 0 to 7 s, while the logarithmic scale represents 0.1 to 1 000 000 s. Notice that it is not possible to represent 0 on a logarithmic scale; extending the divisions to the left would give us 0.01, 0.001, 0.000 1 s, and so on.

Drawing a logarithmic scale is easy if the only numbers we want to represent are simple powers of 10. For numbers in between, we need some help. Figure 2-5 shows a logarithmic scale with some of these numbers indicated. The spacing varies because the scale is nonlinear.

Fig. 2-5 A logarithmic scale, showing the spacing of numbers between the powers of 10.

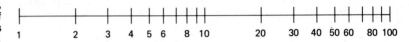

Fortunately, graph paper can be bought where the grid lines are spaced according to a logarithmic scale. If the logarithmic scale is on only one axis, the paper is called *semilog* paper. If the logarithmic scale is on both axes, it is *log-log* paper.

Figure 2-6 shows the data of Table 2-3 graphed on semilog paper. Notice that neither axis starts at zero. For the horizontal logarithmic axis, this would not be possible anyway. For the vertical axis, it is simply a matter of choice. You should take the time to verify that the data points do represent the data of Table 2-3.

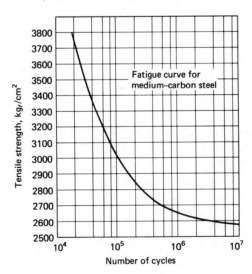

Fig. 2-6 A graph drawn on a semilog grid. The data were taken from Table 2-3.

When the data on both axes span a very large range, both scales must be logarithmic and so log-log paper is used. Figure 2-7 shows such a case. Here we see the amplifier power needed to overcome back-

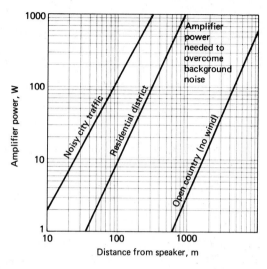

Fig. 2-7 Graphs drawn on a log-log grid.

ground noise under a variety of conditions. Three lines are shown for comparison. In all three cases, the power needed increases as we get farther from the speakers. A technician setting up equipment for an outdoor concert would find this particular information very useful.

Exercises

4. Use Fig. 2-6 to answer the following questions. (*a*) What is the sample's tensile strength after $1\bar{0}0\ 000$ stress cycles? (*b*) What is the tensile strength after 1 million stress cycles? (*c*) What is the tensile strength after $6\bar{0}\ 000$ stress cycles? (*d*) How many stress cycles are needed to reduce the tensile strength of $31\bar{0}0$ kg$_f$/cm^2?
    Answers: (*a*) 3020 kg$_f$/cm^2, (*b*) 2650 kg$_f$/cm^2, (*c*) 3200 kg$_f$/cm^2, (*d*) 78 000 cycles

5. Use Fig. 2-7 to answer the following questions. (*a*) What power is needed to overcome residential noise at a distance 100 m from the speaker? (*b*) A speaker is powered with 50 W. How far will the sound travel in open country? (*c*) A speaker is powered with 80 W. How far will the sound carry in city traffic? (*d*) A speaker can be heard 60 m away in a residential district. By what factor must the power be increased to double the distance?
    Answers: (*a*) 9.0 W, (*b*) 3500 m, (*c*) 80 m, (*d*) about 4 times the power is needed—3 to 13 W

Quantitative information often accompanies design drawings and blueprints (Fig. 2-8). Such drawings (*scale drawings*) are generally scaled-

2-6 Scale Drawings

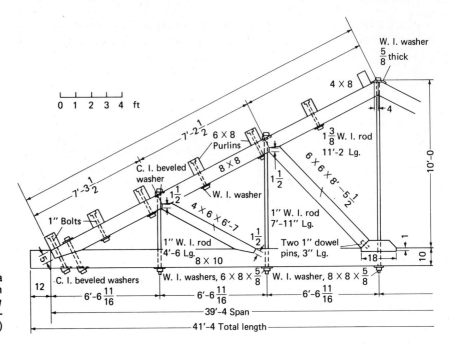

Fig. 2-8 A scale drawing of a timber truss. (Adapted from French & Vierck, *Engineering Drawing*, 10th ed., McGraw-Hill, 1966, p. 584)

down views of the actual object or building. If a blueprint is drawn to a scale of 5 cm = 1 m, for instance, this means that each 5 cm on the drawing represents an actual measurement of 1 m. Even dimensions that aren't listed can then be estimated (approximately) from the drawing. If two walls are shown to be 23.2 cm apart, the actual distance is

$$23.2 \text{ cm} \left(\frac{1 \text{ m}}{5 \text{ cm}} \right) = 4.64 \text{ m}$$

Notice that the problem is set up as a unit conversion.

Maps fall into this category. A map is just a scale drawing of a certain section of the earth's surface. The scale is always indicated somewhere on the map, along with any special symbols that might be used.

2-7 Formulas   Quantitative information may also be represented through formulas. A *formula* is a recipe for calculating one quantity if we know some other ones. Formulas are written with symbols or with words. If we write

$$\text{Speed} = \frac{\text{distance}}{\text{time}} \qquad\qquad (2\text{-}1)$$

we have a formula for finding the speed of an object if we know how long it takes for the object to travel a certain distance. We may also write this as

$$v = \frac{x}{t} \tag{2-2}$$

but now we have to explain what v, x, and t represent. The important formulas in this book are numbered, just as these two are. Most of the time we write our formulas with words. In cases where this becomes too cumbersome, we use symbols and separately list what these symbols represent.

The quantities we use are combined according to standard mathematical rules. There are only four basic *mathematical operations:* addition, subtraction, multiplication, and division. We can also raise a number to a power, which amounts to a repeated multiplication. Thus,

$$(4.0)^2 = (4.0)(4.0) = 16$$

and

$$(3.0)^4 = (3.0)(3.0)(3.0)(3.0) = 81$$

Extracting a root is the reverse of this process:

$$\sqrt{16} = 4.0 \quad \text{(square root)}$$

and

$$\sqrt[4]{81} = 3.0 \quad \text{(fourth root)}$$

But we are really very limited in the number of mathematical operations we can perform. The symbols for the operations are summarized in Table 2-4. Notice that we have followed the significant-digit rule in the examples in this table.

Now we mentioned in Chap. 1 that every physical quantity must have a unit associated with the number. We can't ignore these units in our calculations, because without a unit a number is meaningless. Let's look at an example.

Example 2-1 Finding Gross Vehicle Weight.

A certain truck has a curb weight of 5230 lb. It is loaded with 1.5 tons of gravel. What is the gross weight?

Application: Measuring Distance from a Map

A map is a drawing of a section of the earth's surface. Most maps are drawn to a linear scale; in other words, 1 in or 1 cm on the map represents a certain fixed number of miles or kilometers. Sometimes, however, a map's scale is not linear. You may have seen maps of the world that make Greenland look bigger than Australia (when, in fact, it is only about one-fourth as large). When a map portrays a large section of the spherical earth on a flat sheet, it must have a scale that varies from point to point.

But a map that represents only a small geographical area probably has a linear scale. This scale is written somewhere on the map itself: 1 in = 20 mi, or 1 cm = 5 km, for instance. This scale allows the distance between two points to be estimated by making a measurement on the map.

Suppose that two towns are 12.3 cm apart on a map whose scale is 1 cm = 5 km. The actual distance between the towns is then

$$12.3 \text{ cm} \left(\frac{5 \text{ km}}{1 \text{ cm}} \right) = 61.5 \text{ km}$$

This, of course, is the straight-line distance—sometimes described "as the crow flies."

TABLE 2-4   SYMBOLS FOR THE MATHEMATICAL OPERATIONS

Operation	Symbol	Example
Addition	$a + b$	$2.2 + 1.9 = 4.1$
Subtraction	$a - b$	$56.1 - 24.3 = 31.8$
Multiplication	ab or $a \cdot b$ or $(a)(b)$	$(26)(0.54) = 14$
Division	$\dfrac{a}{b}$	$\dfrac{121}{0.963} = 126$
Raising to second power (squaring)	a^2	$(2.1)^2 = 4.4$
Taking square root	$\sqrt{a}$	$\sqrt{64} = 8.0$

While straight-line distances are useful for planning airplane flights, they are not much help to someone driving on a road or boating a stream. Distance along curved paths can be found by using a specially designed map measurer. This device, shown in the photograph, has a tiny wheel that is rolled along the route. The dial indicates the distance in inches or centimeters, and the geographical distance is then calculated from the map scale. The device is reasonably accurate for streams and for roads in relatively flat regions. For roads in hilly country, however, numerous switchbacks may not show up on the map. The map measurer is not very accurate in such cases.

Our formula is quite simple:

Gross weight = curb weight + load

Inserting the measurements gives

Gross weight = 5230 lb + 1.5 tons

If we hadn't bothered to write in the units, we might be inclined to just add these numbers now. But then the answer wouldn't make a bit of sense. We have to decide on using either pounds or tons for our unit. Using pounds, we have

$$\text{Gross weight} = 5230 \text{ lb} + 1.5 \text{ tons} \left(\frac{2000 \text{ lb}}{1 \text{ ton}} \right)$$

$$= 5230 \text{ lb} + 3\overline{0}00 \text{ lb}$$

$$= 8200 \text{ lb}$$

Notice that we followed the standard procedure for converting units, and we used the significant-digit rule in writing the answer to two places.

Could we have used tons instead of pounds? Certainly. Then the solution is

$$\text{Gross weight} = 5230 \text{ lb} \left(\frac{1 \text{ ton}}{2000 \text{ lb}} \right) + 1.5 \text{ tons}$$

$$= 2.6 \text{ tons} + 1.5 \text{ tons}$$

$$= 4.1 \text{ tons}$$

This is the same answer as before, although it has been written in terms of a different unit. A weight of 4.1 tons is the same as 8200 lb. ◄

Any time a formula tells us to add or subtract two quantities, we have to make sure that the units are the same. We can't add rubber bands and mushrooms and expect the result to mean anything. By the same token, we can't add feet to meters, seconds to kilograms, or gallons to amperes.

But what about multiplication and division? In these cases the units don't have to be the same at all, and often they aren't. Even so, we still carry the units through in the calculation, as Example 2-2 shows.

Example 2-2 Calculating Torque.

Torque may be calculated from the formula

$$\text{Torque} = (\text{force})(\text{torque arm})$$

In the International System, force is measured in newtons (N), although the kilogram-force (kg_f) is another common metric unit. The torque arm is the distance between the point of application of the force and the center of rotation, and in SI units this is expressed in meters (m).

For example, let's suppose that a force of 42 N [9.4 lb] is applied to the end of a torque arm 0.52 m [20 in] long. What is the torque? Using the formula, we have

$$\text{Torque} = (42 \text{ N}) (0.52 \text{m})$$
$$= 22 \text{ N} \cdot \text{m}$$

Notice that this result has units of *newtons times meters*, which we read as *newton-meters*. Since the newton is the SI unit of force and the meter is the SI unit of length, the newton-meter is the official SI unit of torque. ◄
　　　The same procedure applies to formulas involving division. Let's look at an example.

Example 2-3 Calculating Speed.

A certain riverboat travels 98.2 km [61.0 mi] in 5.71 h. What is its average speed?
　　　We saw that speed may be calculated from Eq. (2-1):

$$\text{Speed} = \frac{\text{distance}}{\text{time}}$$

Putting in the measured values gives

$$\text{Speed} = \frac{98.2 \text{ km}}{5.71 \text{ h}}$$

$$= 17.2 \, \frac{\text{km}}{\text{h}}$$

This result has units of kilometers divided by hours, which we read as *kilometers per hour.* ◄
　　　Anytime we calculate from a formula, then, we get the units of the result from the same formula.

Exercises

6.　A certain car weighs 1.51 tons when empty. What is its gross weight if it carries 276 lb of luggage, a driver weighing 189 lb, and a passenger weighing 108 lb?
　　Answer: 3590 lb, or 1.80 tons
7.　What average speed must be maintained to make a 212-mi car trip in 3 h 47 minutes (min)?
　　Answer: 56.0 mph

2-8 Compound Units

In Chap. 1, we said that only seven fundamentally different quantities can be measured. We also listed the SI unit for each of these quantities. When we measure anything else (force, pressure, speed, torque, etc.), we are really measuring some combination of the seven base quantities. Thus *speed* amounts to a distance divided by a time measurement, as we saw in Example 2-3. These separate measurements are combined according to a formula.

What if we use a speedometer to measure speed? We certainly don't use a formula here. But the formula did go into designing the instrument. The result is still the same: Speed is distance divided by time, and the unit of speed is a unit of distance divided by a unit of time.

Units such as kilometers per hour and newton-meters are called compound units, or derived units.

compound, or derived, units

Definition: *Compound units*, or *derived units*, are unit combinations that come from the formula defining a physical quantity.

Units for all quantities other than the seven listed in Table 1-1 are derived. Some derived units are given special names, such as the newton (N) for force and the pascal (Pa) for pressure. We introduce these and other derived units when we discuss their formulas later in this book.

For now, we need to recognize that these compound units exist, and that they may be converted through the procedures we have already developed.

Example 2-4 Converting Torque Units.

In Example 2-2, we calculated a torque and got a result of

$$\text{Torque} = 22 \text{ N·m}$$

Suppose that we want to express this in other units. If we express force in pounds (lb) and torque arm in inches (in), the torque unit becomes pound-inches (lb·in). Let's convert the original answer to this unit.

First we need the conversion relations

$$1 \text{ m} = 39.37 \text{ in}$$

$$1 \text{ N} = 0.224\ 8 \text{ lb}$$

The first came from Table 1-2, and we encounter the second in Table 4-2.

TABLE 2-5 CONVERSION FACTORS FOR COMMON UNITS OF SPEED OR VELOCITY

	$\frac{cm}{s}$	$\frac{m}{s}$	$\frac{km}{h}$	$\frac{ft}{s}$	$\frac{mi}{h}$
1 centimeter per second =	1	0.01	0.036	0.032 808	0.022 369
1 meter per second =	100	1	3.6	3.280 8	2.236 9
1 kilometer per hour =	27.778	0.277 78	1	0.911 34	0.621 37
1 foot per second =	30.48	0.304 8	1.097 3	1	0.681 82
1 mile per hour =	44.704	0.447 04	1.609 3	1.466 7	1

1 ft/min = 0.005 08 m/s
1 in/s = 1 ips = 0.025 4 m/s
1 knot (kn) = 1 **nautical mile per hour** (nmi/h) = 0.514 44 m/s
1 mi/s = 1609.3 m/s

The conversion is then set up as follows:

$$\text{Torque} = 22 \text{ N·m} \left(\frac{39.37 \text{ in}}{1 \text{ m}} \right) \left(\frac{0.224 8 \text{ lb}}{1 \text{ N}} \right)$$

$$= (22)(39.37)(0.224 8) \text{ lb·in}$$

$$= 190 \text{ lb·in}$$

Notice again that we have followed the significant-digit rule in writing this result. A calculation like this can never improve the accuracy of the original measurements, and it would be quite misleading if we quoted the answer as 194.7 lb·in. ◄

Compound units can always be converted by the procedure shown in Example 2-4. But some conversions are needed so often that it may be inconvenient to keep referring to more than one table. To keep things simple, in this book we list separate tables of conversion factors for many of the compound units.

Table 2-5 lists the relationships between the common units for measuring speed (or velocity). The proper SI unit for speed is meter per second (m/s). Another common metric unit, as we have seen, is kilometers per hour (km/h or sometimes kmph). there are also a large number of USCS units in use. The table tells us at a glance, for instance, that 1 mph = 1.609 3 kmph, or that 1 ft/s is the same as 0.304 8 m/s.

Exercises

8. Convert a speed of 60 mph to feet per second (ft/s).
   Answer: 88 ft/s

9. Aluminum sheet from a certain rolling mill is coiled at a rate of 1500 ft/min. Express this (a) miles per hour, (b) kilometers per hour, (c) meters per second.
   Answers: (a) 17 mph, (b) 27 kmph, (c) 7.6 m/s

10. A certain airplane can cruise at 540 knots (kn). Express this in (a) meters per second, (b) kilometers per hour, (c) miles per hour.
    Answers: (a) 280 m/s, (b) 1000 kmph, (c) 620 mph

11. Recording tape is advanced through a certain recorder at a rate of 1-3/4 inches per second (ips, or in/s). Express this in (a) feet per minute, (b) centimeters per second.
    Answers: (a) 8.75 ft/min, (b) 4.44 cm/s

12. Convert a speed of 7.50 mph to yards per minute.
    Answer: 220 yd/min

13. Convert a speed of 5.20 meters per day (m/d) to millimeters per hour.
    Answer: 217 mm/h

Summary We can describe things qualitatively (with words) or quantitatively (with numbers). When accuracy is important, our descriptions are always quantitative: single measurements, data tables, graphs, scale drawings, or formulas. Such quantitative descriptions must always include the measurement units.

   If a set of data spans a very large range of values, it may be impractical to draw a conventional graph. In such cases, a logarithmic graph is often used. Since this has the effect of squeezing together the large values, special care must be taken in reading such a graph. Logarithmic graphs may be semilog, if just one axis has a logarithmic scale, or log-log, if a logarithmic scale is used on both axes.

   A formula is a recipe for finding a quantity if some related quantities are known. The units should be carried through in any calculation based on a formula, because the formula gives the unit of the result as well as its numerical value. All compound units are defined through formulas that relate them back to seven SI base units.

Terms You Should Know

qualitative	circle graph
quantitative	line graph
data table	linear scale

logarithmic scale   formula
scale drawing   speed
mathematical operation   compound unit

Problems

1. The melting point of a lead-tin alloy depends on the relative proportions of lead and tin. The table shown lists the melting point for different percentages of tin, the rest being lead. Graph these data, with the melting temperature on the vertical axis and the percent tin on the horizontal axis.

Percent tin	Melting point, °C
0	326
10	303
20	278
30	255
40	230
50	205
60	187
66	180
70	185
80	198
90	215
100	232

2. A fast-moving car needs a greater distance to stop than a slow-moving car. The total stopping distance may be divided into two parts: the thinking distance, during which the driver moves a foot to the brake pedal, and the braking distance, during which the car is slowing down. The bar graph gives values of thinking distance and braking distance for cars traveling at different speeds on a level road under ideal conditions. On a single grid, draw three line graphs that show the thinking distance, the braking distance, and the total stopping distance at different speeds. Plot the distance on the vertical axis and the speed on the horizontal axis.

3. Do Prob. 2 using USCS units. Express the speed in miles per hour and the distances in feet.

4. Conduct a survey of the makes of cars parked in the student parking lot (or some other parking lot). Count the number of cars of each com-

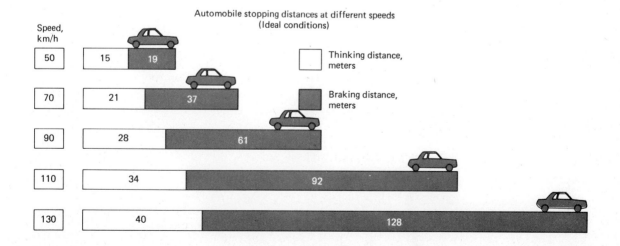

Automobile stopping distances at different speeds
(Ideal conditions)

mon make (Chevrolet, Plymouth, Volkswagen, etc.). Count all the motorcycles together. Make a separate category for unusual cars, in case someone drives a Maserati or a Porsche. Then use your survey data to: (a) Make up a table listing the percentage of cars of each common make; (b) group your data according to General Motors, Ford, Chrysler, American Motors, Japanese manufacturers, German manufacturers, motorcycles, and a separate category for all others. Then show these data on a circle graph.

5. Electrical utility boxes are stamped out by a certain press at a rate of 100 boxes every 7.6 min. There is an average of 5.0-min setup time before each run of 100 boxes. The machine is typically idle 1.00 h each working day (8.00 h) because of lunch and coffee breaks. A certain order requires 3600 boxes. Assuming no waste, how much overtime must the press operator work to complete the order in one working day?

3

Applied Geometry

In this chapter we discuss the quantities called length, area, volume, and plane angle. We look at how these quantities are measured and how they may be calculated. Such measurements and calculations are extremely important in the industrial trades. For instance, electric wire is priced according to its *length*, and its electric resistance also depends on its length. Roofing shingles and floor tile are sold according to the *area* they cover. The strength of a cable and the flow capacity of a pipe each depend on cross-sectional area, and the area of the exterior walls of a building is an important factor in determining the building's heating or cooling requirement. Concrete is mixed by *volume,* and the displacement of a car's engine is a volume measurement. Liquid fuels are bought and sold by volume. The caster and camber in automotive wheel alignment are *angle* measurements. The efficiency of a solar collector depends on its angle of inclination to the ground, its compass direction, and the geographical latitude—which is also an angle. We see, then, that these geometrical quantities are going to be very important to us.

3-1 Geometrical Shapes

Although an infinite variety of geometrical shapes exist, there are a few that we encounter very often in technical applications. If a figure is flat and has straight edges, it is called a *polygon*. Polygons may be regular or irregular. A *regular polygon* is one whose sides are all equal

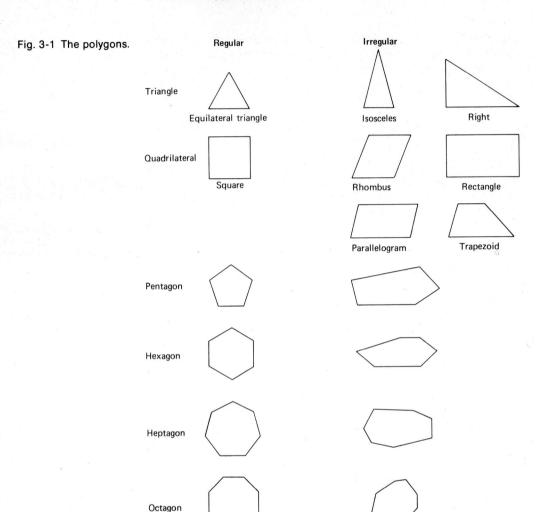

Fig. 3-1 The polygons.

and whose angles are also all equal. Figure 3-1 shows the simplest polygons and the names they have been given. Notice that as the number of sides increases, a regular polygon comes to look more and more like a circle.

3-2 Circles and Circumference

The distance around a polygon is called its *perimeter*, while the distance around a circle is called its *circumference*. The perimeter of a polygon is found simply enough just by adding the lengths of the sides. For the circumference of a circle, we can write the following formula:

$$\text{Circumference of circle} = \pi(\text{diameter}) \tag{3-1}$$

where π (pi) has the value 3.141 592 654. Since wheels, gears, pulleys, axles, and many other machine parts are circular, this formula is very useful. The circumference of a wheel, for instance, is the distance it rolls when it makes one complete revolution (Fig. 3-2).

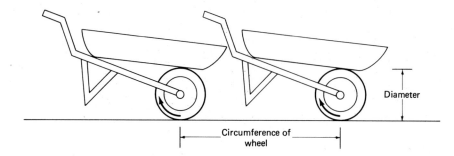

Fig. 3-2 The distance a wheel rolls in one revolution is equal to the wheel's circumference.

Example 3-1 The Belt Drive.

The pulley system shown below has a V-belt that is tight enough not to slip. Let's see how far the belt travels when the large pulley rotates once.

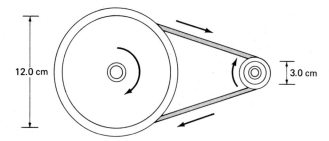

In one rotation, pulley A advances a section of belt whose length is equal to the pulley's circumference. Using Eq. (3-1), we get

$$\text{Circumference A} = \pi(\text{diameter})$$

$$= \pi(12.0 \text{ cm})$$

$$= 37.7 \text{ cm}$$

Now this result is quite a bit bigger than the circumference of pulley B:

$$\text{Circumference B} = \pi(3.0 \text{ cm})$$

$$= 9.4 \text{ cm}$$

So when pulley A rotates once, 37.7 cm of belt is moved along. Since this is more than B's circumference, it causes pulley B to rotate *more* than once. How many times more? Simply the number of times that 37.7 cm of belt would wrap around pulley B. This is

$$\frac{37.7 \text{ cm}}{9.4 \text{ cm}} = 4.0$$

Our conclusion, then, is that B rotates 4.0 times as fast as A. If A is rotating at 312 revolutions per minute (rpm), then B rotates at 1250 rpm. The factor of 4.0 may also be found directly by calculating the ratio of the pulley diameters: (12.0 cm)/(3.0 cm) = 4.0. ◀

Exercises

1. A certain tire has a diameter of 2.1 ft. (*a*) What is its circumference? (*b*) What is the circumference of a tire whose diameter is twice as large?
   Answers: (*a*) 6.6 ft, (*b*) 13 ft
2. What is the circumference of a circular bushing whose radius is 4.213 cm?
   Answer: 26.47 cm
3. What is the perimeter of a square field measuring 98 m on a side? Assume that the field is flat.
   Answer: 392 m

3-3 Pythagorean Theorem

Triangles are commonly encountered shapes because of their natural rigidity. Right triangles are particularly important since most structures have square corners. The Pythagorean theorem tells us how the diagonal side, or *hypotenuse*, of a right triangle is related to the other two sides.

Pythagorean theorem

> *Pythagorean Theorem*: In a right triangle, the square of the hypotenuse equals the sum of the squares of the other two sides.

We may also write this as a formula:

$$(\text{Hypotenuse})^2 = (\text{one side})^2 + (\text{other side})^2 \qquad (3\text{-}2)$$

Example 3-2 shows how this formula is used.

Example 3-2 A Bridge Truss.

A small bridge spans 8.00 m [26.2 ft] between piers. It is to be trussed to

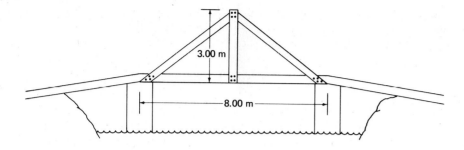

keep it from sagging under load. If the vertical support measures 3.00 m, how long should the diagonal struts be?

We see two right triangles in the picture. (Of course, there are two others on the opposite side of the bridge.) We need to know the length of the hypotenuse. Using Eq. (3-2), we have

$$(\text{Hypotenuse})^2 = (\text{one side})^2 + (\text{other side})^2$$
$$= (3.00 \text{ m})^2 + (4.00 \text{ m})^2$$
$$= 9.00 \text{ m}^2 + 16.0 \text{ m}^2$$
$$= 25.0 \text{ m}^2$$

Notice that this gives us the *square* of the hypotenuse. To get the hypotenuse itself, we have to take the square root:

$$\text{hypotenuse} = \sqrt{25.0 \text{ m}^2}$$
$$= 5.00 \text{ m}$$

We can conclude something else from this result: If we begin with three beams measuring 3.00, 4.00, and 5.00 m and rivet their ends together to form a triangle, we are guaranteed that the triangle has one square corner. ◄

If we know the hypotenuse of a right triangle and we also know the length of one of the sides, we can again use the Pythagorean theorem to find the length of the other side. In this case, our formula must be rewritten:

$$(\text{one side})^2 = (\text{hypotenuse})^2 - (\text{other side})^2 \qquad (3\text{-}3)$$

It's important to carefully distinguish between the hypotenuse and the two sides that form the square corner. This relationship occurs in some surprising places, as Example 3-3 shows.

Application: Structural Rigidity and Triangles

There are two things that affect the overall strength, or load-bearing capacity, of a structure. The first is obviously the strength of the materials being used. The second is the structure's geometry. A bridge may be built of the strongest steels, but if the shape is wrong, it will quickly collapse under load.

Two common ways of gaining structural rigidity are by using arches and triangles. We discuss arches in Chap. 5; here we look at triangles, which are more common.

The triangle is the only rigid polygon. By this, we mean that it is impossible to change its shape without changing the length of at least one side. All other polygons can be distorted without breaking or bending any sides. The diagram shows a rectangular support, which is not rigid. By adding a diagonal brace we split the rectangle into two triangles and give the support rigidity.

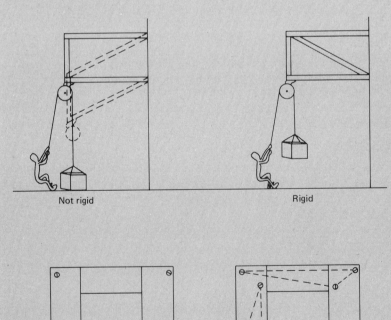

Not rigid   Rigid

Not rigid   Rigid

If we try to improve a polygon's rigidity by putting extra fasteners at the corners, we are still making use of triangles. The shape at the lower left has just one screw in each corner, and it is definitely not rigid. Adding extra screws improves the rigidity *because triangles have been constructed.* For purposes of clarity, only two of the many triangles formed are shown in the drawing.

The California construction codes, designed to help buildings withstand earthquakes, make liberal use of triangles. Geodesic domes are built of regular polygons that are subdivided into triangles. Vehicle chassis, bridges, roof trusses, crane booms, guy wires on antennas, airframes, and a wide variety of other applications also use this principle.

Example 3-3 Motion of a Piston and Crankshaft.

In Fig. 3-3 we see a cutaway drawing of one cylinder of a conventional internal-combustion engine. As the piston moves up and down, the connecting rod causes the crankshaft to rotate. The problem is to find how far the piston moves as it turns the crackshaft 90° from top dead center. This engine has a stroke of 8.58 cm [3.38 in], and the connecting rod is 13.60 cm [5.35 in] in length.

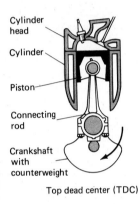

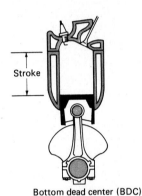

Cylinder head

Cylinder

Piston

Connecting rod

Crankshaft with counterweight

Top dead center (TDC)   90° after TDC   Stroke   Bottom dead center (BDC)

Fig. 3-3 Motion of the piston and crankshaft in a conventional internal-combustion engine. Most of the piston's downward motion takes place in the first quarter-revolution of the crankshaft.

The second diagram shows the situation when the crankshaft has moved 90° from the dead center. There is a right triangle formed by the center of the wrist pin, the center of the crank arm, and the center of rotation of the crankshaft. The bottom side of this triangle equals the radius of the crankshaft circle. From Fig. 3-3 we can see that this is just half of the stroke, or 4.29 cm.

We can now use Eq. (3-3) to find the vertical side of this triangle:

$$(\text{One side})^2 = (\text{hypotenuse})^2 - (\text{other side})^2$$
$$= (13.60 \text{ cm})^2 - (4.29 \text{ cm})^2$$
$$= 185.0 \text{ cm}^2 - 18.4 \text{ cm}^2$$
$$= 166.6 \text{ cm}^2$$

Then taking the square root, we get

$$\text{Vertical side} = 12.91 \text{ cm}$$

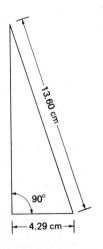

This is the distance between the center of the wrist pin and the crankshaft's center of rotation. At top dead center, this distance was 13.60 cm + 4.29 cm, or 17.89 cm. So in the first 90° of crankshaft rotation, the piston has moved 17.89 cm − 12.91 cm, or 4.98 cm.

Application: Constructing Square Corners

For layout work on small pieces, a square corner can be drawn with the help of a tool appropriately known as a *square*. The photograph shows the squares used by carpenters, machinists, and drafters. For larger work, a rafter square may be used. This is basically a large version of the carpenter's square.

None of these tools gives good accuracy on very large work. If a very large rectangle is to be laid out (say a garage foundation), the best method is to check that the diagonals are equal. These diagonals don't actually have to be measured; an easy procedure is to stretch a string between two opposite corners and then check the string's length against the distance between the other two opposite corners.

Another way to check a square corner on large work is to use the Pythagorean theorem. If a triangle has sides measuring 3.00 and 4.00 m and a hypotenuse of 5.00 m, we are guaranteed that we have a right triangle. In fact, the units could be feet, yards, or anything else instead of meters. As long as the three sides are in the ratio of 3:4:5, the triangle must have a 90° angle. We can use this principle to build a square if none is available. Or we can check a corner by laying off 3 units along one side, 4 units along the other side, then measuring the hypotenuse. When the corner is square, the hypotenuse measures 5 units.

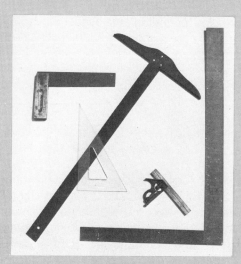

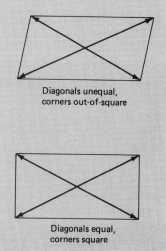

Diagonals unequal,
corners out-of-square

Diagonals equal,
corners square

Most people would guess that the piston moves half of its stroke in the first 90° of crankshaft rotation. We have just proved otherwise. Half of the stroke is 4.29 cm, but the piston actually moves 4.98 cm in turning the crankshaft 90°. In the next 90° the piston travels the rest of its stroke, or just 3.60 cm.

One result of all this is that the piston needs to travel faster in the first quarter-turn than in the second. This tends to make the crankshaft vibrate rather than rotate smoothly, which is one reason that engines need flywheels. ◀

Exercises

4. Find the hypotenuse of a right triangle whose sides measure (a) 5.0 and 12.0 m, (b) $6\bar{0}$ and $8\bar{0}$ in, (c) 1.00 and 1.00 km, (d) 72 cm and 1.2 m, (e) 49.31 and 76.82 m.
   Answers: (a) 13 m, (b) $10\bar{0}$ m, (c) 1.41 km, (d) 140 cm, or 1.4 m, (e) 91.28 m

5. A certain radio tower is 28.5 m high. To keep it from swaying in the wind, three guy wires are fastened at a point 26.2 m above the ground. The wires are anchored to the ground 18.7 m from the tower's base. How long are the wires?
   Answer: 32.2 m each

6. A certain ladder measures 10.0 m, fully extended. It is leaned against a wall with its base 2.0 m from the wall. How high will it reach?
   Answer: 9.8 m

3-4 Areas There are as many different formulas for finding area as there are different geometrical shapes. In almost all cases, though, we measure area in *square units*. A square unit is the area of a perfect square measuring one unit on a side. Thus 1 square centimeter (cm²) is the area of a square that measures 1 cm on each side, and 1 square foot (ft²) is the area of a square measuring 1 ft on each side. When we say we have an *area* of 12.1 cm², we mean that 12.1 one-centimeter squares would cover the area. We don't necessarily mean that the area is a square, however. It could be a triangle or a circle and still be as big as the area covered by the 12.1 unit squares.

In Fig. 3-4, we see a comparison of the sizes of three unit squares. According to the International System, the unit used to measure length is the meter (m). The SI unit of area is therefore the square meter, which we abbreviate as m². In fact, we abbreviate all square units as the square of the length unit: square feet is ft² and square centimeters is cm², for instance.

The area of a square or rectangle may be found from the

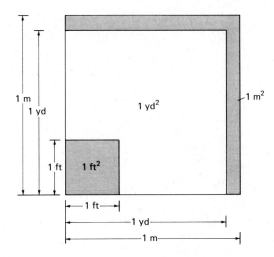

Fig. 3-4 Relative sizes of the square foot (ft²), the square yard (yd²), and the square meter (m²).

well-known formula

$$\text{Area of rectangle} = (\text{one side})(\text{other side}) \qquad (3\text{-}4)$$

Thus a square measuring 3.0 ft on each side has an area of (3.0 ft)(3.0 ft), or 9.0 ft². But since 3.0 ft is the same as 1.0 yd, this same area is (1.0 yd)(1.0 yd), or 1.0 yd². We conclude, then, that there is 9 ft² in a 1 yd². We may verify this by looking at Fig. 3-5. Each side of the square yard is divided into 3 ft, and this division gives a total of 9 ft².

This means that we have to be extra careful with area units. There is 39.37 in in 1 m for instance, but there is *not* 39.37 in² in 1 m². The correct number is (39.37)², or 1550. With this in mind, we can work out

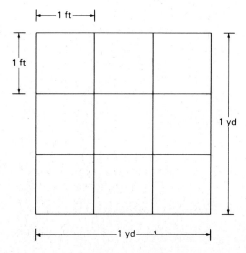

Fig. 3-5 There is 9 ft² in 1 yd².

TABLE 3-1 CONVERSION FACTORS FOR UNITS OF AREA

	cm²	m²	in²	ft²
1 square centimeter =	1	0.000 1	0.155 00	0.001 076 4
1 square meter =	10 000	1	1550.0	10.764
1 square inch =	6.451 6	6.451 6 × 10⁻⁴	1	0.006 944 4
1 square foot =	929.03	0.092 903	144	1

1 acre = 43 560 ft² = 4046.9 m²   1 mi² = 640 acres = 2.589 99 × 10⁶ m²
1 are = 100 m²   1 circular mil (cmil) = 7.854 0 × 10⁻⁵ in²
1 hectare = 2.471 1 acres = 10 000 m²   = 5.067 1 × 10⁻¹⁰ m²

the relationship among any area units. For purposes of convenience, some of the more common ones are listed in Table 3-1.

We have a formula for the area of a square or rectangle, but for other shapes we need to use different formulas. In Table 3-2 we see formulas for the areas of some common shapes. The symbols in these formulas are defined in the sketches. Notice that each formula requires us to multiply two lengths, or a length by itself. Each of these formulas therefore gives a result in square units.

TABLE 3-2 AREA FORMULAS

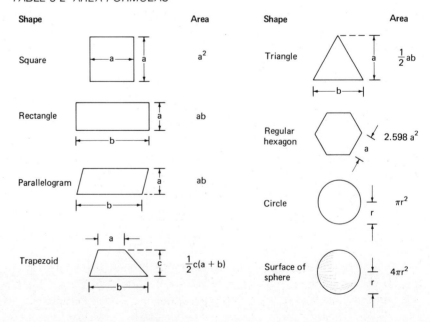

Shape	Area	Shape	Area
Square	a^2	Triangle	$\frac{1}{2}ab$
Rectangle	ab	Regular hexagon	$2.598\,a^2$
Parallelogram	ab	Circle	πr^2
Trapezoid	$\frac{1}{2}c(a+b)$	Surface of sphere	$4\pi r^2$

Example 3-4 Cross Sectional Area of a Pipe.

The rate at which a fluid flows through a pipe is governed by a number of factors. One is the cross-sectional area of the pipe. All other factors being equal, the flow rate will double if this area is doubled. Suppose that we have two pipes with inside diameters of 4.50 and 9.00 cm. What are their cross-sectional areas?

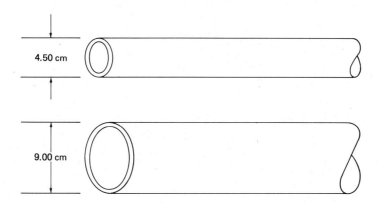

We'll assume that the pipes have circular cross sections. From Table 3-2, the formula for the area of a circle is

$$\text{Area of circle} = \pi r^2$$

where r is the radius. If the diameter of the small pipe is 4.50 cm, its radius is half this, or 2.25 cm. The area is then

$$\text{Area of small pipe} = \pi(2.25 \text{ cm})^2$$
$$= \pi(5.06 \text{ cm}^2)$$
$$= 15.9 \text{ cm}^2$$

For the large pipe, a diameter of 9.00 cm means a radius of 4.50 cm. Then

$$\text{Area of large pipe} = \pi(4.50 \text{ cm})^2$$
$$= \pi(20.25 \text{ cm}^2)$$
$$= 63.6 \text{ cm}^2$$

Notice that the large pipe has twice the diameter of the small one, but its *area* is 4 times as great. In this case, the large pipe could handle 4 times the fluid flow of the small pipe under the same conditions. ◄

Exercises

7. Aluminum flashing comes in rolls 18.0 in wide and 25.0 ft long. Find the surface area one roll will cover in (a) square feet, (b) square meters.
    Answers: (a) 37.5 ft², (b) 3.48 m²
8. A circular gear blank has a diameter of 9.31 cm. Find the area of this circle in (a) square centimeters, (b) square inches.
    Answers: (a) 14.6 cm², (b) 2.27 in²
9. A certain stage is in the shape of a regular hexagon, with each side measuring 2.92 m. How many square meters of carpet are needed to cover the stage?
    Answer: 22.2 m²
10. A style of square floor tile measures 25.0 cm on a side. How many tiles are needed to cover the floor of a rectangular room if the room's dimensions are 3.80 m by 4.92 m?
    Answer: 300 tiles
11. A certain house has an outside perimeter of 122 ft and a height to the roofline of 16 ft. One "square" of aluminum siding covers 100 ft² of surface. (a) What is the outside surface area of the house? (b) Approximately how many squares of siding are needed to cover the house?
    Answers: (a) 2000 ft², (b) less than 20 squares, allowing for windows and doors

3-5 Volumes   Just as areas are measured in terms of the area of a unit square, *volumes* are measured in terms of the volume of a unit cube. A cube that measures 1 m on each side has a volume of 1 m³. Thus the cubic meter (m³) is the SI unit of volume. Some non-SI units for volume are the cubic foot (ft³), the cubic yard (yd³), and the cubic inch (in³).

In Fig. 3-6, we see a unit cube measuring 1 yd on a side. Each side is divided into 3 ft. If we count the total number of small cubes (including the ones that are hidden in the picture), we conclude that there is 27 ft³ in 1 yd³. Again, this means that we have to be careful in relating cubic units. There is 2.54 cm in 1 in, for instance, but there is certainly not 2.54 cm³ in 1 in³. The correct number is (2.54)³, or 16.39.

In addition to the cubic units, there are a number of other units for volume. These are sometimes referred to as *capacity* units: gallons, liters, bushels, quarts, and so on. Although the liter (L) is not a proper SI unit, it is commonly used in metricated countries because the cubic meter is too big for many purposes. For instance, it would take nearly 3 years to drink 1 m³ of milk at the rate of 1 L per day. Table 3-3 lists conversion factors for some of the common units of volume.

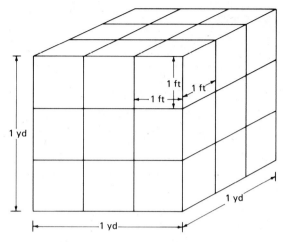

Fig. 3-6 There is 27 ft³ in 1 yd³.

TABLE 3-3   CONVERSION FACTORS FOR UNITS OF VOLUME AND CAPACITY

	L	**m³**	**gal**	**ft³**
1 liter =	1	0.001	0.264 17	0.035 315
1 cubic meter =	1000	1	264.17	35.315
1 liquid gallon =	3.785 4	0.003 785 4	1	0.133 68
1 cubic foot =	28.317	0.028 317	7.480 5	1

1 barrel, liquid = 42 gal = 0.158 99 m³
1 board foot (fbm) = 144 in³ = 0.002 359 7 m³
1 cm³ = 10⁻⁶ m³
1 in³ = 16.387 cm³ = 1.638 7 × 10⁻⁵ m³
1 yd³ = 27 ft³ = 0.764 55 m³
1 ounce (oz), fluid = 29.574 cm³ = 2.957 4 × 10⁻⁵ m³
1 quart (qt), fluid = 2 liquid pints (pt) = 32 oz = 0.946 35 L
1 register ton = 100 ft³ = 2.831 7 m³

A solid shape whose faces are all rectangles is called a *rectangular prism.* A cube is a special rectangular prism whose faces are all squares. The volume of these solids may be calculated from the formula

Volume of rectangular prism = (length)(width)(height)   (3-5)

For solids having other shapes, different formulas must be used. Table 3-4 lists formulas for the volumes of some of the more common shapes. As usual, the units should be included in any calculation based on these formulas.

TABLE 3-4 FORMULAS FOR VOLUMES OF COMMON SOLIDS

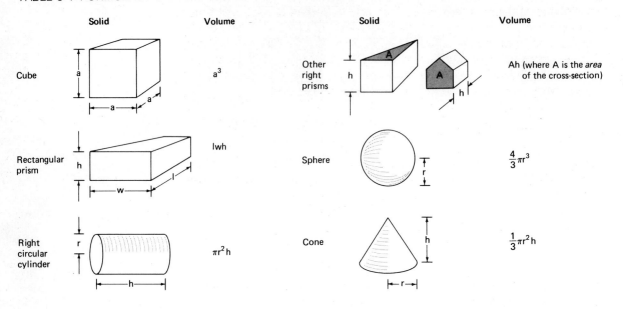

Solid	Volume	Solid	Volume
Cube	a^3	Other right prisms	Ah (where A is the *area* of the cross-section)
Rectangular prism	lwh	Sphere	$\frac{4}{3}\pi r^3$
Right circular cylinder	$\pi r^2 h$	Cone	$\frac{1}{3}\pi r^2 h$

Example 3-5 Volume of a Concrete Pad.

A concrete pad is to be poured for a shop floor. The pad has the shape shown. How much concrete should be ordered?

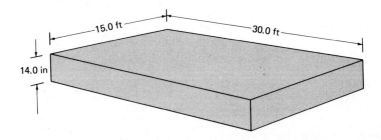

Although a pad this large must contain steel reinforcing rods, the volume of these rods is small compared to the total volume of the pad. Therefore a volume of concrete equal to the pad's total volume should be ordered.

Since the shape is a rectangular prism, we use Eq. (3-5):

$$\text{Volume} = (\text{length})(\text{width})(\text{height})$$
$$= (15.0 \text{ ft})(30.0 \text{ ft})(14.0 \text{ in})$$

Before we do the arithmetic, we need to look at the units. We are multiplying feet times feet times inches. This gives square feet times inches (ft²·in), which is *not* a commonly used volume unit. If we convert the last unit from inches to feet, the volume will work out in cubic feet. This would be all right, except that concrete is ordered by the cubic yard. This means that we need a second unit conversion from cubic feet to cubic yards. Putting in these conversion factors gives

$$\text{Volume} = (15.0\,\text{ft})(30.0\,\text{ft})(14.0\,\text{in}) \left(\frac{1\,\text{ft}}{12\,\text{in}} \right) \left(\frac{1\,\text{yd}^3}{27\,\text{ft}^3} \right)$$

Notice that all the unwaranted units cancel: "in" cancels "in," and "ft·ft·ft" in the numerator cancels "ft³" in the denominator. Now we have

$$\text{Volume} = \frac{(15.0)(30.0)(14.0)}{(12)(27)}\,\text{yd}^3$$

$$= 19.4\,\text{yd}^3$$

We left all the arithmetic until this last step, because this way the calculator has to be picked up only once. ◄

Let's now look at a similar problem that uses metric units.

Example 3-6 Capacity of a Storage Tank.

The tank shown is in the shape of a cylinder (technically, a *right circular cylinder*). Let's calculate its capacity in liters.
From Table 3-4, the formula for the volume of a cylinder is

$$\text{Volume of cylinder} = \pi r^2 h$$

where h is the height and r is the radius. The radius of the tank is half its diameter, or 0.541 m. Putting in the numbers, we get

$$\text{Volume} = \pi (0.541\,\text{m})^2 (2.153\,\text{m})$$

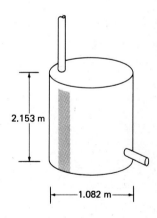

2.153 m

1.082 m

This calculation gives us the volume in cubic meters. To get liters, we need to introduce a conversion factor from Table 3-3. This gives

$$\text{Volume} = \pi (0.541\,\text{m})^2 (2.153\,\text{m}) \left(\frac{1000\,\text{L}}{1\,\text{m}^3} \right)$$

$$= \pi (0.541)^2 (2.153)(1000)\,\text{L}$$

$$= 1980\,\text{L} \quad ◄$$

Application: Bubble Levels and Plumb Lines

If buildings didn't have horizontal floors, everything would tend to slide toward the downhill side. If walls weren't vertical, there would be extra expense in bracing them so they wouldn't fall over. For these simple reasons, most of our structures are built on horizontal and vertical lines.

What we call *vertical* is the direction in which gravity pulls. This obviously varies from point to point on the earth's surface. The easiest way to find the local vertical is with a plumb line—a string with a weight called a "plumb bob" at the end. The string is hung from a point, and when the plumb bob comes to rest, its center is vertically beneath this point of suspension. The string then defines a vertical line. Although it takes some patience to use a plumb line, and wind can sometimes be a problem, this is the truest indication of the vertical.

The horizontal is at a right angle to the vertical. The surface of a calm lake or pond is horizontal because water always flows downhill and "seeks its own level." For this reason "level" and "horizontal" have come to mean the same thing.

The horizontal may be found with a tool known as a "bubble level." A sealed curved glass tube (called a vial) contains water or oil and an air bubble. The vial is mounted in a long, straight board or metal bar. When the bar is level, the bubble comes to rest between two lines marked on the glass. The accuracy of the bubble level is limited by its length: short ones are not very accurate. Many bubble levels contain an extra vial that gives an approximate indication of the vertical as well.

Exercises

12. A cylindrical tank is 30.0 cm in diameter and 80.0 cm tall. Find its volume in (*a*) cubic centimeters, (*b*) liters.
    Answers: (*a*) 56 500 cm³, (*b*) 56.5 L
13. Find the volume, in liters, contained by a perfect cube 1.00 ft on a side.
    Answer: 28.3 L
14. A holding tank is in the shape of a rectangular prism 12.5 m long, 6.22 m wide, and 2.32 m deep. Find the volume held by the tank in

For very large jobs, such as leveling a foundation, a variation of this principle can be used. A piece of clear plastic hose is stretched between opposite corners of the foundation and then filled with water from one end. Since water seeks its own level, the water levels in both ends of the hose form a perfect horizontal. This guide can then be used to level the footer and the first courses of block or brick.

      (a) cubic meters, (b) liters, (c) gallons, (d) cubic feet.
      Answers: (a) 180̄ m³, (b) 180̄ 000 L, (c) 47 700 gal, (d) 6370 ft³

15. A certain car engine has a displacement of 302 in³. Calculate the displacement in (a) cubic meters, (b) liters.
      Answers: (a)4950 cm³, (b)4.95 L

16. One gallon of a certain paint covers 450 ft² of area. Calculate the area in square meters covered by 1 L of the same paint.
      Answer: 11 m²

3-6 Angles

Angle measurements occur in a wide variety of technical applications. Figure 3-7 shows the proper angles for grinding a drill point. Figure 3-8 shows the caster angle for pivoting a wheel. *Caster angle* is a part of the front-end alignment specification on motor vehicles.

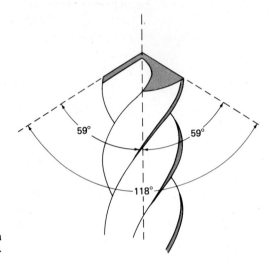

Fig. 3-7 Angles for grinding a drill point.

The units for angle measurement are summarized in Table 3-5. Although the SI unit for angles is the *radian*, the *degree* is commonly used throughout the world. Most of the angles in this book are specified in degrees.

One important angle measurement is compass direction. If a bar magnet is suspended from a string or pivoted on a bearing, it aligns itself

(A) Positive caster angle on a bicycle.   (B) Negative caster angle on a supermarket buggy.

Fig. 3-8 A caster angle is used on most steerable wheels. With positive caster, the wheel has a natural tendency to straighten itself out after a turn. With negative caster, the wheel must be steered out of any turn.

TABLE 3-5   CONVERSION FACTORS FOR UNITS OF PLANE ANGLE

	rad	°	′	″	r
1 radian =	1	57.296	3437.8	$2.062\,6 \times 10^5$	0.159 15
1 degree =	0.017 453	1	60	3600	0.002 777 7
1 minute =	$2.908\,9 \times 10^{-4}$	0.016 667	1	60	$4.629\,6 \times 10^{-5}$
1 second =	$4.848\,1 \times 10^{-6}$	$2.777\,8 \times 10^{-4}$	0.016 667	1	$7.716\,0 \times 10^{-7}$
1 revolution (r) =	6.283 2	360	2.160×10^4	$1.296\,0 \times 10^6$	1

with the earth's magnetic field. This magnetic field is quite different from the earth's gravitational field, whose direction is found from a plumb line. Since the earth's magnetic field runs nearly north and south, magnetic compasses are very useful for navigation. Figure 3-9 shows how the magnetic bearings, in angular degrees, compare with the compass directions.

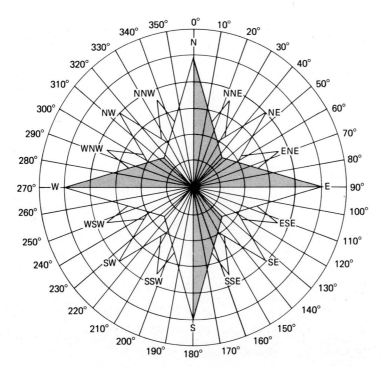

Fig. 3-9 Magnetic bearings and compass direction.

Points on the earth's surface are located on maps by their longitude and latitude. These are angular measurements based on the earth's nearly spherical shape.

latitude

Definition: *Latitude* is the angular distance north or south of the equator, as measured from the earth's center.

longitude

Definition: *Longitude* is the angular distance east or west of a certain north-south line (the prime meridian) that runs through Greenwich, England.

The meaning of these definitions can be seen more clearly in Fig. 3-10. In the top diagram, a slice has been drawn through the earth at the equator. This is used as the reference for expressing latitude measurements. Points on the equator itself are at 0° latitude, while the north pole is 90° north latitude and the south pole is 90° south latitude. The point labeled *P* has a latitude of 30°N, which is its angle from the plane of the equator.

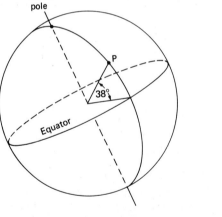
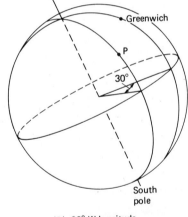

Fig. 3-10 Points are located on the earth's surface by specifying two angles called the *latitude* and *longitude*. Point *P* has a latitude of 38°N and a longitude of 30°W, which places it near the Azores Islands in the Atlantic Ocean.

(A) 38° N latitude

(B) 30° W longitude

The arc drawn from the north pole to the south pole through point *P* is called *P*'s *meridian*. The longitude of point *P* is the angle between its meridian and a reference meridian that passes through Greenwich, England. This reference meridian is called the *prime meridian*. Points west of Greenwich are designated as so many degrees west longitude, while points east of Greenwich have east longitude. Longitude measurement goes from 0 to 180°, east and west. At 180° the meridian is exactly on the other side of the earth from Greenwich; this is called the *international date line*.

By using longitude and latitute, any point on the earth's surface can be found. The island of Hawaii, for instance, is at about 19°N, 155°W.

3-7 Time

Until very recently, all time measurements were based on the motion of the earth. The time for the earth to circle the sun is called a year, while the time for the earth to rotate once on its axis is a day. A year corresponds to 365.242 2 days.

Recent measurements have shown that tidal friction is slowing the earth's rotation rate so that each year is 5.3 milliseconds (ms) shorter than the previous year. This is when the second is defined according to an exact 24-h day. While this 5.3 ms won't matter much to most of us, for very accurate measurements it can lead to problems. That is why the second has been redefined in terms of the natural frequency of the cesium atom (Table 1-1). For our purposes, however, we can consider a day to consist of 24 h, where 1 h is exactly 3600 s long.

Summary

Geometrical measurements are very important in the trades and commerce. But for these measurements to be useful, their units must be clearly understood. In Chap. 1, we talked about the common units for measuring distance or length. In this chapter, we have added area, volume, and plane angle. In the chapters to follow, we have many occasions to refer to all these quantities and their units.

Terms You Should Know

polygon	area
quadrilateral	volume
rhombus	rectangular prism
parallelogram	right circular cylinder
trapezoid	sphere
pentagon	cone
hexagon	caster angle
perimeter	latitude
circumference	longitude
Pythagorean theorem	meridian

Application: The Ackermann Correction (Toe-Out on Turns)

The alignment of a vehicle's front wheels is a geometrical problem. The steering linkage must be built and adjusted so the wheels run on parallel tracks when the car is going straight. If this were not done, the wheels would scuff and slip along the pavement. Handling would be poor and tread wear uneven.

When the vehicle is turning, the situation is quite different. If the wheels were still parallel in a turn, they would be steering around different centers, as shown in the top diagram. These centers are along the axis of the rear axle, so there is no problem with the rear wheels. But the front wheels will scuff and slip throughout the turn.

This problem was first remedied by Rudolph Ackermann in 1818 (on carriages, of course). He introduced the principle of toe-out on turns. The lower diagram shows how the outside wheel must execute a larger turning circle than the inside wheel; the difference in radius is the width of the vehicle—about 1.5 m [5 ft] for cars. If the inside wheel is angled 23° from forward, the outside one will typically be angled only 20°. The 3° difference is the toe-out for the turn, and it makes the two turning centers coincide.

Ordinarily, the toe-out itself is not adjusted. Instead, the related camber and caster angles are set during the wheel alignment. If this does not bring the toe-out within specifications, it usually indicates a bent steering arm.

When an alignment is done on a stationary vehicle with the wheels pointing straight, the wheels are actually set slightly closer at the front than at the back. This "toe-in" amounts to 1.5 to 5 mm [1/16 to 3/16 in] as measured at hub height. The idea is this: It is not possible to set the wheels *exactly* parallel, because no measurement is perfectly accurate. If the wheels are toed out even slightly, the vehicle will tend to turn on its own. The slight toe-in prevents the possibility of an accidental toe-out when the vehicle is going straight. Thus the toe-out occurs *only* when the wheels are steered into a turn.

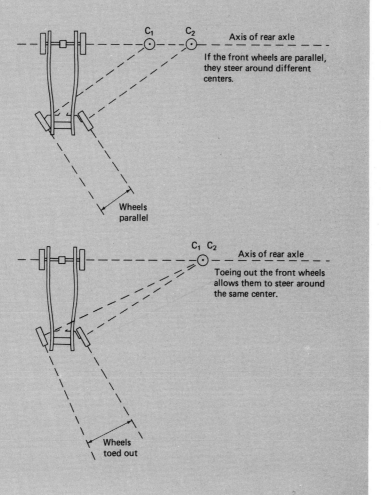

C_1 C_2 Axis of rear axle

If the front wheels are parallel, they steer around different centers.

Wheels parallel

C_1 C_2 Axis of rear axle

Toeing out the front wheels allows them to steer around the same center.

Wheels toed out

Problems

1. Find the number of square kilometers in a square nautical mile.
2. A roller 8.51 cm in diameter drives a conveyor belt as shown. The roller rotates 60.0 times each minute. Find the speed of the belt in meters per second.

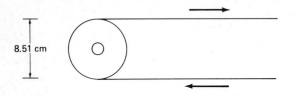

3. A certain large truck has wheels 1.3 m in diameter. How many revolutions do the wheels make each minute when the truck is traveling at a speed of 60 kmph? (*Hint:* Change the 60 kmph to meters per minute.)
4. A roof truss is to be built for a small building. Which is the better of the two designs shown?

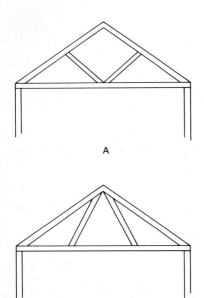

A

B

5. One "square" of roofing shingles covers 100 ft² of roof area. Estimate the number of squares of shingles needed to cover the roof of the house shown.

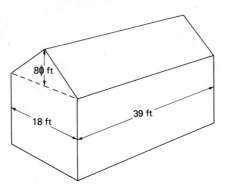

6. A large cylindrical smokestack is 68 m [220 ft] tall. Its diameter is 7.5 m [25 ft]. (*a*) What is the outside surface area of the stack? (*b*) If 1 L of paint covers 9.5 m² of surface, how much paint is needed to paint the stack? (*c*) How many gallons of paint are needed?
7. A flat piece of property is in the shape of a trapezoid. The two parallel sides measure 421.3 m and 676.9 m. The perpendicular distance between these sides is 398.5 m. Find the area of this property in (*a*) square meters, (*b*) hectares, (*c*) acres.
8. A rectangular swimming pool measures 32.5 ft by 80.0 ft. The depth slopes continuously from 3.0 ft at one end to 9.0 ft at the other. (*a*) How many gallons of water are needed to fill the pool? (*b*) How many additional gallons are needed to raise the water level an additional 2.0 in?

4 Matter

Matter is anything that occupies space. This includes a large variety of things: boxcars, cows, air, oil, and dust, to name a few. Matter can be changed from one form to another: a boxcar can be cut into scrap, remelted, and formed into refrigerator casings. Or the cow can be butchered into steak and hamburger. The matter itself, however, is never destroyed in such processes. It is always totally accounted for in the end products. In this chapter we discuss how matter is measured, what it is made of, and how it is transformed.

Introduction

The *mass* of an object is a measure of the amount of matter in it. A river barge has a larger mass than a canoe, and a car's suspension spring has more mass than a clock's hairspring. But bigger objects don't always have more mass than small objects. A ball bearing, for example, can have more mass than a much larger piece of styrofoam. Many materials, from steel to dry cement to salami, are priced according to their mass.

4-1 Mass and Weight

We measure an object's mass by comparing it with standard masses on a *balance*. Figure 4-1 shows a type of balance commonly found in laboratories. You are already familiar with the balances used in doctor's offices and delicatessens. Such balances usually have the standard masses built right into the instrument.

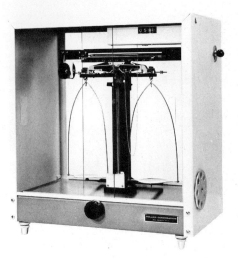

Fig. 4-1 A laboratory balance
for measuring small masses.
(Courtesy Voland Corporation)

The SI unit of mass is the *kilogram* (kg). Figure 4-2 shows the U.S. prototypes of the standard kilogram. All mass measurements made in the United States are indirect comparisons with the mass of one of these platinum-alloy cylinders kept at the National Bureau of Standards. Other units of mass have been defined in terms of the kilogram. Table 4-1 lists some of the more commonly used mass units and their relationships.

Now it may seem that this quantity we call *mass* is really the same thing as weight. Not so. As we said, mass is a measure of the amount of matter in an object. But *weight* is something else: it is a measure of the gravitational attraction the earth has for an object. This gravitational attraction varies slightly from point to point on the earth's surface: it is greater near the poles than at the equator, for instance, and it is greater at sea level than on a mountain or in an airplane. In fact, if we take an object far enough from the earth, its weight vanishes completely and it becomes "weightless." Its mass, on the other hand, remains the same wherever it is.

How much does an object's weight vary as it is moved from one place to another? If it stays on the earth's surface, the weight change is less than 1 percent. This may not sound like much, but it is enough to complicate commercial transactions based on weight. If coffee is shipped from Colombia to New York, for instance, it weighs more when it is unloaded than when it was loaded. If steel is shipped the other way—New York to Colombia—it loses weight. Because of this, goods are almost always bought and sold according to their mass rather than their weight.

An object's weight is measures on a *spring scale*, examples of which are shown in Fig. 4-3. This device balances the tension in a spring against the attraction of gravity. At places where gravity is stronger, the

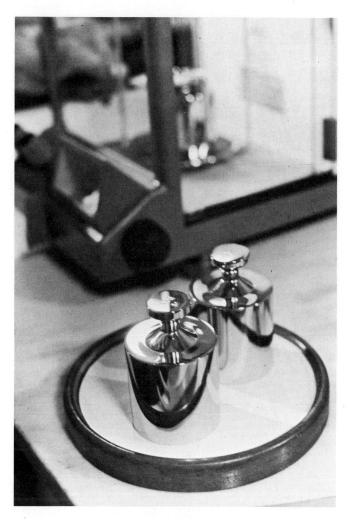

Fig. 4-2 The U.S. prototypes of the international kilogram. (Courtesy National Bureau of Standards)

spring stretches farther. Spring scales always measure weight rather than mass. But if a device that looks like a scale has the label "no springs," then it is really a balance and it measures mass rather than weight.

The following definitions summarize this difference between mass and weight.

Definition: *Mass* is the measure of the amount of matter in an object. An object's mass cannot be changed by moving it from one place to another.

mass

TABLE 4-1   CONVERSION FACTORS FOR COMMON UNITS OF MASS

	g	kg	t	lb$_m$	slug
1 gram =	1	0.001	10^{-6}	$2.204\ 6 \times 10^{-3}$	$6.852\ 2 \times 10^{-5}$
1 kilogram =	1000	1	0.001	2.204 6	0.068.522
1 metric					
ton (t) =	10^6	1000	1	2204.6	68.522
1 pound-mass					
(lb$_m$), avoir-					
dupois =	453.59	0.453 59	$4.535\ 9 \times 10^{-4}$	1	0.031 081
1 slug =	14 594	14.594	0.014 594	32.174	1

1 atomic mass unit (u) = $1.660\ 5 \times 10^{-27}$ kg
1 carat, metric = 200 mg
1 dram, troy = 60 grains (gr) = 3.887 9 g
1 gr = 0.064 799 g
1 ounce-mass, avoirdupois = 28.350 g
1 ounce-mass, troy = 31.103 g
1 pound-mass, troy = 12 troy ounce-masses = 373.24 g
1 ton, long = 2240 lb$_m$ = 1016.0 kg
1 ton, short = 2000 lb$_m$ = 907.18 kg

Fig. 4-3 Spring scales for measuring weight and other forces.

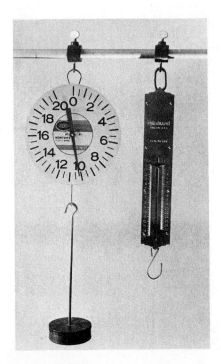

Definition: *Weight* is the measure of the earth's gravitational attraction for an object. An object's weight may change when it is moved from one place to another.

Unfortunately, the words "mass" and "weight" are often used interchangeably when, in fact, they shouldn't be. A cereal package may be labeled "Net Weight 454 g," when in fact the cereal is being sold by mass, and it is the mass that is 454 g. If it is to be shipped across state lines, it is illegal to price the cereal according to its weight.

Exercises

1. Find the number of grams in 14.2 kg.
    Answer: 14 200 g
2. Convert a mass of 98.3 slugs to kilograms.
    Answer: 1430 kg
3. Calculate the number of metric carats in 1 g.
    Answer: 5.00 metric carats
4. Find the number of grains in 14.72 g.
    Answer: 227.2 gr
5. Calculate the number of milligrams in 1 troy ounce.
    Answer: 31 103 mg

Weight is just one example of what we call force. Some other forces are friction, spring tension, magnetic attraction and repulsion, aerodynamic lift, and fluid surface tension. Although these forces arise in very different ways, they have one thing in common: they are all a push or a pull.

4-2 Weight as a Force

Definition: A *force* is a push or a pull in a particular direction.

The directional aspect of a force is something we return to later. For now, we need to appreciate that there are many kinds of force and that weight is just one of them. *Any* force can be measured with a spring scale, and the SI unit for *any* force is the *newton* (N). All other units for measuring force have been defined in terms of the newton. Table 4-2 lists the conversion relations between some of the more common force units. Notice that the English unit of force is the *pound* (lb) and that a newton is roughly a quarter-pound (about the weight of an apple).

Example 4-1 Thrust of a Jet Engine.

Jet and rocket engines are rated according to their maximum *thrust*, which is the force they develop when operating at full throttle. Each of a

TABLE 4-2   *CONVERSION FACTORS FOR UNITS OF FORCE*
            (INCLUDING WEIGHT)

	N	**kg$_f$**	**lb**
1 newton =	1	0.101 97	0.224 81
1 kilogram-force (kg$_f$) =	9.806 65	1	2.204 6
1 pound =	4.448 2	0.453 59	1

1 dyne (dyn) = 10^{-5} N
1 gram-force (g$_f$) = 0.001 kg$_f$ = 9.806 65 × 10^{-3} N
1 ounce-force = 0.278 01 N
1 ton-force = 907.18 kg$_f$ = 8896.4 N
1 ton-force, metric = 1000 kg$_f$ = 9806.7 N

Boeing 707's four engines can produce a thrust of 18 000 lb. Let's express this thrust in the SI unit.

According to Table 4-2, the conversion relation is

$$1 \text{ lb} = 4.448 \text{ 2 N}$$

Then

$$\text{Thrust} = 18\ 000 \text{ lb} \left(\frac{4.448 \text{ 2 N}}{1 \text{ lb}} \right)$$

$$= 8\bar{0}\ 000 \text{ N}$$

This number has nothing to do with the mass or the weight of the engine; rather, it is a measure of the push the engine can supply. If, however, the engine were mounted vertically, this $8\bar{0}\ 000$-N thrust could lift a weight of $8\bar{0}\ 000$ N, or 18 000 lb. ◄

We said that an object's weight is different from its mass. Yet it should be obvious that the two are related: things that have a large mass are also heavy, and it takes a large force to lift them. We can write the relationship this way:

$$\text{Weight} = (\text{mass}) (\text{acceleration due to gravity}) \qquad (4\text{-}1)$$

or, if we prefer symbols,

$$w = mg \qquad (4\text{-}2)$$

where m is the mass and w is the weight. The quantity g, which we call the acceleration due to gravity, is a measure of the strength of the earth's

Application: Explosives

Explosives are chemical compounds that decompose very rapidly when they have been heated. This rapid chemical decomposition is accompanied by a change of phase and the liberation of a large amount of additional heat that causes the released gases to expand further. For instance, the detonation of 1 kg of blasting gelatin quickly liberates nearly 1 m³ of water vapor, carbon dioxide, and nitrogen, and the initial pressure may be as much as 13 000 times the pressure of the atmosphere.

Blasting gelatin, dynamite, and TNT are examples of "brisant" or "shattering" types of explosives. They are generally detonated with blasting caps—small metal tubes filled with mercury fulminate or lead ozide. The blasting cap is detonated electrically or with a safety fuse. The explosion of the cap decomposes part of the main explosive, which almost instantaneously raises the temperature as high as 4700°C. This extremely high temperature immediately causes the decomposition and phase change of the entire explosive. Brisant-type explosives get their shattering power from the extreme speed of this process. They are well suited to mining, road construction, and demolition applications.

Gunpowder (black powder) and cordite (smokeless powder) are propellant-type explosives. Although they also release large quantities of gas, they do this too slowly to be effective in blasting applications. Propellant-type explosives are used in ammunition, where the gases propel a projectile down a gun barrel, and also in solid-fuel rockets.

gravitational attraction.* This quantity varies from point to point on the earth. Table 4-3 lists values of g for various latitudes and elevations above sea level. By referring to this table, we can calculate an object's weight from its mass.

Example 4-2 Finding Weight from Mass.

A certain jetliner carries 72 120 kg of fuel. Let's find the weight of this fuel when the plane is cruising at an altitude of 10 000 m at 20° north latitude.

* This book uses italic g to represent the acceleration due to gravity. The symbol "g" is the abbreviation for the *gram*, which is a unit of mass.

TABLE 4-3   VALUES OF THE ACCELERATION DUE TO GRAVITY
(g) AT VARIOUS LATITUDES AND ALTITUDES

Latitude	Altitude, m					
	0	1000	2000	3000	4000	10 000
0°	9.780	9.777	9.774	9.771	9.768	9.750
10°	9.782	9.779	9.776	9.773	9.770	9.751
20°	9.786	9.783	9.780	9.777	9.774	9.756
30°	9.793	9.790	9.787	9.784	9.781	9.762
40°	9.802	9.799	9.796	9.792	9.789	9.771
50°	9.811	9.808	9.805	9.801	9.798	9.780
60°	9.819	9.816	9.813	9.810	9.807	9.788
70°	9.826	9.823	9.820	9.817	9.814	9.795
80°	9.831	9.828	9.824	9.821	9.818	9.800
90°	9.832	9.829	9.826	9.823	9.820	9.801

The units are newtons per kilogram (N/kg).
The "standard" value g_0 is 9.806 65 N/kg.

From Table 4-3, the value of g under these conditions is

$$g = 9.756 \ \frac{N}{kg}$$

Using Eq. (4-1) or (4-2), we get

$$\text{Weight} = 72\ 120 \text{ kg} \left(9.756 \ \frac{N}{kg} \right)$$

$$= 703\ 600 \text{ N}$$

If the plane carried the same amount of fuel at sea level and 50° latitude, the weight would be

$$\text{Weight} = 72\ 120 \text{ kg} \left(9.811 \ \frac{N}{kg} \right)$$

$$= 707\ 600 \text{ N}$$

Notice that the same fuel weighs $4\bar{0}00$ N more in the second instance! Expressed in pounds, this weight difference is

$$4\bar{0}00 \text{ N} = 4\bar{0}00 \text{ N} \left(\frac{1 \text{ lb}}{4.448\ 2 \text{ N}} \right)$$

$$= 9\bar{0}0 \text{ lb}$$

This corresponds to the weight of about five passengers. ◀

Occasionally we know an object's weight, but we need to find the mass. To handle such problems, we can rewrite Eq. (4-1) as

$$\text{Mass} = \frac{\text{weight}}{\text{acceleration due to gravity}} \qquad (4\text{-}3)$$

Again, if we are interested in accuracy, we need to know the latitude and altitude. Example 4-3 shows the use of this formula.

Example 4-3　Mass on a Chain.

A certain chain has a breaking strength of 1250 N. In other words, this is the *force* (pull) necessary to break it. How much mass can the chain support if it is used at 30° north latitude and sea level?
　　We need to find the mass that weighs 1250 N at this point on the earth. From Table 4-3, the value of g is

$$g = 9.793 \ \frac{\text{N}}{\text{kg}}$$

Using Eq. (4-3), we have

$$\text{Mass} = \frac{1250 \ \text{N}}{9.793 \ \text{N/kg}}$$
$$= 128 \ \text{kg}$$

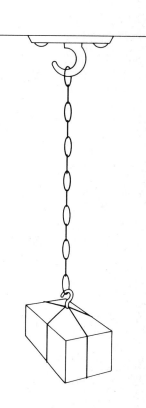

　　Notice that we retained three-digit accuracy in this result, since the breaking strength was given with three significant digits. But, in fact, it makes little sense to worry about extreme accuracy in a case like this. For one thing, a chain should never be used with a load near its breaking strength, because then a sudden jerk could cause it to fail. Furthermore, we didn't include the weight of the chain itself in the calculation.
　　To do an approximate calculation, which will seldom be wrong by more than a fraction of 1 percent, we can use the so-called standard acceleration due to gravity. This is the internationally accepted average of g, symbolized by g_0. As indicated in Table 4-3, its value is 9.806 65 N/kg. Using this figure, the calculation becomes

$$\text{Mass} \simeq \frac{1250 \ \text{N}}{9.806 \ 65 \ \text{N/kg}}$$

Application: Combustion of Gasoline

Gasoline is a complex mixture of around 100 different hydrocarbons that boil in the temperature range 40 to 200°C [100 to 400°F]. It may also contain small amounts of additives to prevent knocking, to absorb moisture in the fuel tank, and to lubricate the valve stems and the upper part of the cylinder walls in engines. For the most part, however, gasoline can be considered a mixture of compounds of hydrogen and carbon with the average formula C_8H_{18}.

Air is a mixture of about 23 parts oxygen and 77 parts nitrogen, by mass. (There are also small amounts of other gases present.) When gasoline is burned slowly in the open air, the air's nitrogen is left unaffected. The oxygen combines with the gasoline's carbon and hydrogen and forms carbon dioxide and water vapor. Both of these products are colorless, odorless, and harmless.

In an engine, however, the combustion takes place very rapidly. For an engine rotating at 3600 rpm, the fuel-air mixture must burn in less than 1/100s. If it burns slower, the flame will envelop the exhaust valves when they open, and may overheat and even melt them.

For complete combustion, 15.5 g of air should be mixed with each gram of fuel, to give 16.5 g of exhaust gases. If there is

where the symbol $\simeq$ is read "approximately equals." The result is 127 kg, which is very close to our original answer. ◄

Exercises

6. A certain car in Albuquerque weighs 3350 lb. This weight is the same as the total supporting force supplied by the tires. Express this force in newtons.
   Answer: 14 900 N
7. A two-story frame house in Pittsburgh weighs 120 000 lb. This weight is also the total force bearing on the foundation footer. Express this force in newtons.
   Answer: 530 000 N

less than this amount of air, the mixture is said to be rich; it will burn faster but incompletely. If there is more than this amount of air, the mixture will burn completely, but part of this combustion will take place in the exhaust system. Thus there will be a loss of power.

When combustion is incomplete, the exhaust gases contain unburned hydrocarbons, particulates (soot), and carbon monoxide (CO). Carbon monoxide is dangerous because of its affinity for the iron atoms in the blood's hemoglobin, and breathing this odorless gas can cause asphyxiation. These three pollutants are easily avoided by burning a lean mixture.

Unfortunately, a lean mixture does more than reduce an engine's power. It raises the temperature high enough (greater than 1000°C) to burn some of the nitrogen in the air. Nitrogen forms a number of oxides: N_2O, NO, N_2O_3, NO_2, N_2O_5, and so on. The first of these is called "laughing gas." Commonly used by dentists, it is dangerous only in high concentrations. But the others are no laughing matter—they are poisonous to humans, contribute to smog formation, and in the presence of water form highly corrosive nitric acid.

Designing an engine for pollution control is thus a complicated problem. Rich mixtures produce some pollutants, while lean mixtures reduce fuel economy and often produce other pollutants.

8. A certain airplane has a weight of 26 700 N when it is cruising just above the beach at Assateague Island. This weight is also the total lift force the wings must supply to sustain level flight. Calculate this lift force in pounds.
   Answer: 60$\overline{0}$0 lb

9. A cubic meter of aluminum has a mass of 2699 kg. Find its weight in newtons (*a*) at sea level at the equator, (*b*) at 30° north latitude, elevation 2$\overline{0}$00 m, (*c*) at 50° north latitude and sea level.
   Answers: (*a*) 26 4$\overline{0}$0 N, (*b*) 26 420 N, (*c*) 26 480 N

10. A truck has a weight of 22 300 N when it is at 40° north latitude and an elevation of 1$\overline{0}$00 m. Find its mass in kilograms.
    Answer: 2280 kg

4-3 Metrication and the Newton-Kilogram Controversy

We have seen that the SI unit of force or weight is the *newton* (N), while the SI unit of mass is the kilogram (kg). The weight of 1 kg in newtons is just the numerical value of g: at sea level at the equator, 1 kg weighs 9.780 N, while at 50° north latitude and an altitude of 2000 m, 1 kg weighs 9.805 N. On an average, 1 kg weighs 9.806 65 N. This value of 9.806 65 N/kg is symbolized by g_0 and is referred to as the "standard acceleration due to gravity."

Now although the official unit of force is the newton, it has become the custom in many parts of the world to use the *kilogram-force* (kg_f) as the metric unit of force. This allows a great simplification, since a mass of 1 kg has a weight of 1 kg_f to within an accuracy of about ±0.5 percent. In fact, at places where g is very close to g_0, a mass of 1 kg may weigh almost exactly 1 kg_f. The true definition of the kilogram-force, however, is based on the newton:

$$1 \text{ kg}_f = 9.806\ 65 \text{ N} \qquad \text{exactly} \qquad (4\text{-}4)$$

The U.S. National Bureau of Standards (NBS) is encouraging the use of the newton to measure forces. Even so, the kilogram-force has begun to appear on some instruments and in some shop manuals in the United States. If we pretended that the kilogram-force doesn't exist, sooner or later you would encounter this unit anyway and become totally confused. So in this book we use two metric force units: the newton, which is the official SI unit, and the kilogram-force, which at this time is sometimes used in industry and the trades. Having these two metric force units is really not that complicated. It is much like measuring volume in either cubic meters (m^3) or liters (L). In this book, we frequently use the kilogram-force as the force unit when it simplifies our calculations. Notice that there are almost 10 N in 1 kg_f; using a 10-to-1 conversion will result in an error of only 2.0 percent.

A similar situation exists with the USCS units. In the United States, the standard unit for force and weight is the pound (lb), while the engineering unit for mass is called the *slug*. We mentioned earlier that goods are commonly bought and sold by mass, but what person in the street has ever heard of the slug? Does anyone ever buy a slug of potatoes? Of course not. Instead, the pound has also been made into a mass unit called the *pound-mass* (lb_m). A mass of 1 lb_m weighs 1 lb at points where g equals g_0. At other places, it weighs slightly more or less than 1 lb depending on whether locally the earth's gravity is stronger or weaker than the standard value. The pound-mass can be converted to kilograms according to the exact conversion

$$1 \text{ lb}_m = 0.453\ 592\ 37 \text{ kg} \qquad \text{exactly}$$

and the pound, which is a force unit, can be converted to the kilogram-force by

$$1 \text{ lb} = 0.453\ 592\ 37 \text{ kg}_f, \quad \text{exactly}$$

Unfortunately, the pound-mass is not always abbreviated as lb_m on commercial products. Instead, we often see just "lb," and are thus led to believe we are dealing with weight rather than mass. But remember that practically anything which is being *priced* by the pound is really referred to the pound-mass, whether it is abbreviated this way or not.

Similarly, the kilogram-force is not always abbreviated as kg_f. Instead, it is often written simply as "kg," which leads to a confusion between mass and weight. But remember that if a push or a pull is being measured in kilograms, the unit is really the kilogram-force (kg_f).

Table 4-4 summarizes these units of mass and force. It is very important that you, the student, make these distinctions between force and mass units very carefully. They are used throughout the rest of this book.

Exercises

11. In the United States, dry Portland cement is sold in sacks having a mass of 94 lb_m. Express this mass in kilograms.
    Answer: 43 kg
12. A common building brick has a mass of 2.50 lb_m. (*a*) Express this mass in kilograms. (*b*) Find the approximate weight in newtons, based on g_0. (*c*) Find the approximate weight in kilogram-force, based on g_0.
    Answers: (*a*) 1.13 kg, (*b*) 11.1 N, (*c*) 1.13 kg_f
13. A certain steel cable has a breaking strength of 28 $\bar{0}00$ lb. (*a*) Express this in newtons. (*b*) Express this in kilogram-force. (*c*) Approximately what mass, in kilograms, can the cable support without breaking?

TABLE 4-4   *THE USCS AND METRIC UNITS FOR FORCE AND MASS*

Unit system	Mass unit	Force unit	Mass-weight relationships (on earth's surface)
SI	kilogram (kg)	newton (N)	1 kg weighs about 9.81 N
Customary metric	kilogram (kg)	kilogram-force (kg_f)	1 kg weighs about 1 kg_f
U.S. engineering	slug	pound (lb)	1 slug weighs about 32.2 lb
USCS	pound-mass (lb_m)	pound (lb)	1 lb_m weighs about 1 lb

Answers: (*a*) 125 000 N, (*b*) 12 700 kg$_f$, (*c*) about 12 700 kg

14. A steel ingot weighs 11 tons. (*a*) What is its weight in pounds? (*b*) What is its weight in newtons? (*c*) What is its weight in kilogram-force? (*d*) What is its approximate mass in pound-mass? (*e*) What is its approximate mass in slugs? (*f*) What is its approximate mass in kilograms?

Answers: (*a*) 22 000 lb, (*b*) 98 000 N, (*c*) $\overline{10}$ 000 kg$_f$, (*d*) about 22 000 lb$_m$, (*e*) about 680 slugs, (*f*) about $\overline{10}$ 000 kg

4-4 The Structure of Matter

All matter is made up of very large numbers of very small particles. Although scientists are still searching to find the smallest of these building blocks, we begin with three of the most important. These have been named the *proton*, the *electron*, and the *neutron*. These particles are very, very tiny—certainly much smaller than can be seen with the most powerful microscopes. If protons could be packed together with their sides touching, it would take 10^{22} (10 000 000 000 000 000 000 000) of them to cover the period at the end of this sentence. The neutron is about the same size as the proton. The electron is unusual in that its size can't actually be measured; it behaves something like a cotton ball.

The masses of these particles are listed in Table 4-5. Notice that protons and neutrons have the same mass, while the mass of the electron is much smaller. Also listed is a property called *electric charge*. Now we can't say exactly what electric charge is, any more than we can say what time or length is. But we can measure electric charge; its SI unit is the coulomb (C). And we also know that electric charge can be positive or negative: protons have a positive charge, while electrons have a negative charge. Neutrons have no charge at all.

Electric charge is important because it gives rise to a force called the *electric force*. Let's list some of the properties of this force:

Properties of the Electric Force

1. Charges with opposite signs attract one another.
2. Charges with like signs repel one another.

TABLE 4-5 PROPERTIES OF THE ELEMENTARY PARTICLES

Particle	Mass	Diameter	Electric charge
Proton	1.67×10^{-27} kg	2×10^{-15} m	1.602×10^{-19} C
Electron	9.11×10^{-31} kg	Indefinite	-1.602×10^{-19} C
Neutron	1.67×10^{-27} kg	2×10^{-15} m	0

3. The force is strong when the charges are close and weak when they are far away.
4. The force is stronger when the charges are larger.
5. Charges add algebraically. When an object has equal numbers of protons and electrons, it acts as though it has no charge.
6. There are normally equal numbers of protons and electrons in any sizable chunk of matter. For every proton, somewhere there is an electron.

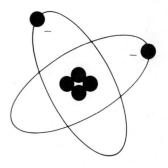

Fig. 4-4 Diagram of a helium atom. Two protons and two neutrons form the nucleus, which is orbited by two electrons.

The electric force is very strong force—stronger than gravity, for instance. This force is responsible for holding electrons and protons together in miniature solar systems called *atoms*. Figure 4-4 shows a diagram of an atom of helium. The center, composed of protons and neutrons, is called the *nucleus*. The electrons orbit very rapidly and form a fuzzy spherical shell around the nucleus. In any neutral atom, there are just as many orbiting electrons as there are protons in the nucleus.

The simplest possible atom has a nucleus of a single proton and a single orbiting electron. The substance composed of such atoms is called *hydrogen*. At ordinary temperatures, it is a colorless and explosive gas. It is also the lightest element and was once commonly used to fill dirigibles.

Table 4-6 lists the 10 simplest kinds of atoms. Substances made up of a single kind of atom are called elements.

TABLE 4-6   THE TEN LIGHTEST ELEMENTS

Number of protons	Chemical symbol	Element name	Properties at ordinary temperature and pressure
1	H	Hydrogen	Gas, lightest element, highly flammable, burns to yield water vapor
2	He	Helium	Very light gas, difficult to liquefy, does not burn or react chemically
3	Li	Lithium	Soft solid, the lightest metal, reacts explosively with water
4	Be	Beryllium	Solid, very light hard metal, toxic
5	B	Boron	Solid, hard and brittle nonmetal
6	C	Carbon	Solid, burns to form carbon dioxide and carbon monoxide, found in all living things
7	N	Nitrogen	Gas, the main component of air (accounting for 78% by volume), very difficult to burn
8	O	Oxygen	Gas, necessary for combustion, accounts for 21% of the atmosphere by volume
9	F	Fluorine	Gas, poisonous, extremely corrosive and causes severe burns
10	Ne	Neon	Gas, chemically inert

Application: Inert Gas Welding

When heated, metals react chemically with oxygen in the air and form oxides. A surface coating of oxide makes it difficult or even impossible to solder, braze, or weld a metal. One way of preventing this oxidation on heating is to use a flux. Different fluxes are used for different metals and different temperatures; some actually dissolve oxides that are already present.

Another way to prevent oxidation is to surround the work with an *inert gas*. This, in fact, is the only practical way to weld easily oxidized metals like aluminum. Even with steel it gives a cleaner, stronger weld in a shorter time than other techniques.

Of the 92 naturally occurring elements 6 are *inert*; that is, they do not ordinarily form chemical compounds. They are all very difficult to liquefy, and they are found in small amounts in the atmosphere. These six inert gases are helium, neon, argon, krypton, xenon, and radon. Radon is radioactive, which makes it infeasible for welding applications. Krypton and xenon are very rare and therefore expensive. Argon comprises nearly 1 percent of the atmosphere. Together with helium and neon, it is produced as a by-product of air liquefaction. All inert-gas welding is done with helium or argon. Argon is usually preferred because when ionized, it conducts electricity better than helium.

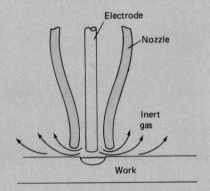

The function of the inert gas is to form a shield around the welding electrode and the hot part of the work. To do this, the gas must flow continuously. The diagram shows a typical gas tungsten-arc outfit.

Definition: An *element*, or a *chemical element*, is a substance that contains only a single kind of atom.

Altogether 92 different elements are found in nature. They have atoms containing up to 92 protons. In addition, another 12 elements have been created under laboratory conditions. Everything we see around us is made up of combinations of the 92 naturally occurring kinds of atoms. In fact, 98.5 percent of the earth's solid crust is composed of just eight elements. Their proportions are shown in a circle graph in Fig. 4-5.

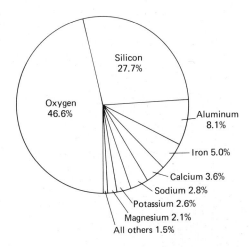

Fig. 4-5 Proportions of the most common elements in the earth's crust.

If there are so few elements, and even fewer very common ones, how do we account for the vast variety of materials we encounter? The answer is that most kinds of atoms have a natural tendency to attach to other kinds of atoms. This attachment is called *chemical combination*. We won't discuss the details of how this happens; the interested student can find the explanation in any introductory chemistry book. The point is that it *does* happen, and such chemical combination is the result of the electric force just described.

For instance, if hydrogen is mixed with oxygen and then ignited, every two atoms of hydrogen combine with one atom of oxygen to form water. Water is an example of a *chemical compound*, symbolized by the formula H_2O. Similarly, carbon can combine with hydrogen in a variety of ways: C_2H_2 is called acetylene, CH_4 is methane, C_3H_8 is propane, C_8H_{18} is octane, and there are many others. The compounds of carbon and hydrogen are called *hydrocarbons*. Hydrocarbons are the main ingredients in most of our fuels.

Chemical compounds have properties that are much different from the properties of the elements that went into them. The smallest combination of atoms that exhibits the properties of the compound is called a *molecule.*

molecule

Definition: A *molecule* is the smallest piece of a substance that exhibits the properties of the substance. A molecule of a chemical compound is a combination of atoms. A molecule of an element is generally the same as a single atom.

Molecules are very small, which means that it takes a very large number of them to make up a sizable piece of matter. A cup of water, for instance, contains about 10^{25} water molecules.

4-5 The Phases of Matter

Matter can exist in four different states or *phases*: solid, liquid, gas, and plasma. If the temperature is low enough, all substances but helium will go into the solid state. Some substances—carbon, water, and iron, for instance—can exist in more than one solid state. Carbon can exist as graphite, which is the slippery black material used in pencils, or as diamond, which finds use as a gemstone and as an abrasive. The difference in properties results from the different ways the carbon atoms line up in the solid. In fact, it is possible to change graphite into diamond, and this has actually been done on a small scale.

In a solid, the molecules hold on to one another through very strong electric forces. This is what gives the solid its definite shape and relatively hard surface. But the molecules themselves don't sit still. Instead, they vibrate as they hold on to one another. As the solid's temperature is increased, this molecular vibration grows increasingly violent.

If a solid is heated to a high enough temperature, it usually does one of two things. Either it decomposes chemically, or it melts into a liquid. Wood, for instance, consists mostly of cellulose, which is a complicated chemical compound containing carbon, hydrogen, and oxygen. When wood is heated enough, the molecular vibrations become so violent that the cellulose decomposes into water and carbon. The water is driven off as steam, and the remaining carbon burns to yield carbon dioxide. The part of the wood that is not cellulose is left behind as ash.

If a solid doesn't decompose on heating, it will eventually undergo a phase change to the liquid state. This melting takes place at different temperatures for different substances. It happens when the molecules vibrate so rapidly that they can no longer hold onto one another strongly. In a liquid, the molecules have broken into small groups that bump and slide over other groups. Like a solid, a liquid has a definite volume.

Unlike a solid, a liquid assumes the shape of its container. Most liquids pour very easily. Some, like molasses, pour very slowly. A few, like glass, pour so slowly that they seem to be solid. But even glass will pour if left alone long enough. This has been a problem with restoring the stained-glass windows on some medieval cathedrals: after many centuries, each piece of glass has grown very thick on the bottom and very thin on top.

In a typical liquid, the molecules themselves are moving very fast. Some liquids like warm sugar ($C_{12}H_{22}O_{11}$) decompose chemically on further heating. Most liquids simply begin to pour more easily as the molecules move faster and faster. But eventually a point is reached where the molecules separate from one another completely and go off in their own directions. At this point, the liquid has become a gas.

Like a liquid, a gas assumes the shape of its container. Unlike a liquid, a gas has no definite volume. Instead, it expands to completely fill its container, top to bottom. Notice that any substance will become a gas if it is heated to a high enough temperature. Thus we can have gaseous water, mercury, or even iron.

As a gas is heated further, the molecules continue to move faster. Eventually the molecules break apart into atoms. At very high temperatures, the atoms themselves begin to come apart and electrons move off on their own. Atoms that have lost electrons are called ions.

> Definition: An *ion* is an atom whose number of electrons does not equal the number of protons in the nucleus.   *ion*

When the temperature is high enough to form ions, the gas glows a characteristic color. Neon glows a bright red, argon glows green, and air glows pale blue. The substance has now changed into the *plasma* state. Some familiar plasmas are shown in Fig. 4-6. Every flame or fire is a plasma. So are the surface of the sun and lightning bolts in the atmosphere. Matter will exist as a plasma only when subjected to a continuous source of heat, such as the large electric current in lightning or the thermonuclear reaction in the sun. Without such a source of heat, a plasma will quickly cool and the electrons will recombine with the ions to form an ordinary gas.

Many substances are mixtures of different kinds of molecules. Gasoline, **Mixtures** for instance, contains about 100 different kinds of hydrocarbon molecules and may also contain antiknock compounds and other chemical additives. Paint is a mixture of water or oil (called the *vehicle*) and various chemicals (called the *pigment*) that give it its body and color. And air, as we mentioned earlier, is a mixture of many different gases.

Fig. 4-6 Some familiar plasmas. (Lightning courtesy National Oceanic and Atmospheric Administration)

We can mix any number of gases in any proportions we like, and they will stay mixed unless we take great pains to separate them.* This is not true of liquids. For instance, water will mix with alcohol, but it will not mix with oil or gasoline. If two liquids will mix, we say they are *miscible*. Otherwise, they are *immiscible*.

When immiscible liquids are poured together (Fig. 4-7), one floats on the other, and each keeps its original volume. But when miscible

* This is not intended to encourage the reader to mix two chemically reactive gases like hydrogen and chlorine.

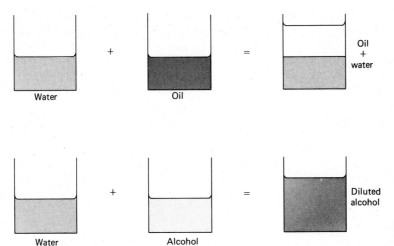

Fig. 4-7 Two liquids may be miscible or immiscible.

liquids are poured together, the molecules themselves mingle with one another, and even the smallest drop of the mixture has the same properties as any other part of the mixture. In this case, the total volume of the mixture is usually *less* than the sum of the volumes of the separate liquids. Thus 1 L of water mixed with 1 L of alcohol gives only around 1.8 L of the mixture. The *mass* of the mixture, however, is always equal to the sum of the masses of the liquids mixed together.

The reason that plastic is so difficult to recycle is that the plastic in a comb is different from the plastic in a milk bottle, which is different from the plastic in a phonograph record. If these different plastics are melted together, they are found to be immiscible. To recycle plastics, they first must be sorted according to type. Solids cannot mix with one another in the same sense that liquids or gases do. Gunpowder, for instance, is a mixture of powdered sulfur, charcoal, and saltpeter (potassium nitrate). Yet if some gunpowder is placed under a magnifying glass, the small particles of charcoal and sulfur can be seen separately from the tiny saltpeter crystals. The molecules themselves are not mixed. This is why gunpowder is such a slow-burning explosive.

Metal *alloys* like brass or solder are solid mixtures, but they are mixed while the metals are liquid. The mixing of the molecules is fairly complete, which causes the alloy to have properties different from the metals that go into it. Table 4-7 lists some common metal alloys and the elements which compose them.

When a solid is mixed with a liquid, as salt in water, the mixture is called a *solution*. Solutions tend to be transparent, even when they are colored. In a true solution, the solid never settles out of its own accord. Thus we don't expect to find vast salt deposits at the bottom of the ocean. If, however, a very strong solution is chilled, we may see some of the

TABLE 4-7   *SOME COMMON ALLOYS*

Alloy	Composition	Properties
Red brass	Copper, with smaller amounts of tin, lead, and zinc	A free-cutting brass with good casting and finishing properties
Yellow brass	Copper and zinc, with small amounts of lead and tin	Easily machined, cheaper but not as strong as red brass
Hard bronze	Most copper, with small amounts of tin and zinc	Very strong, used for bushings, bearings, valve guides, etc.
Monel metal	Nickel and copper, with small amounts of iron, silicon, manganese, and carbon	Very strong, excellent corrosion resistance
Nichrome	Nickel and chromium	High melting point, used in electric resistance heating elements
Carbon steel	Iron with up to 1% carbon	Easily stamped and formed, poor machining properties
Stainless steel	Iron, chromium, nickel, up to 1% carbon	Hard, high corrosion resistance
Brazing solder	Copper and zinc with a trace of lead	Melts at 849 to 870°C (1560 to 1600°F)
Silver solder	Silver, copper, and zinc	Hard and strong, higher melting point than soft solder
Soft solder	Lead and tin	Soft, melts between 180 and 320°C (356 to 600°F), depending on proportions of tin and lead

solid dropping out of solution. This is because the amount of solid a liquid can hold in solution depends on the temperature. Hot water dissolves much more sugar or salt than cold water does.

Some mixtures of solids and liquids are not transparent and tend to separate when left alone. These are not solutions, but *suspensions*. Muddy water is a suspension of dirt in water, and paint is a suspension of pigment in a vehicle. If particles of dirt or other foreign matter are suspended in diesel fuel, they can accelerate the wear of the closely fitting parts in the injection system. For this reason, diesel engines usually have a two-stage fuel filtering system. The final filter is capable of removing suspended particles with diameters down to 3 micrometers (μm) or 0.000 12 m.

It is also possible to have a suspension of a solid in a gas. This is what smoke is. If the particles in the suspension are small enough, it may take a very long time for them to settle out. Figure 4-8 shows a particulate analyzer used to measure the number of soot particles suspended in an automobile's exhaust gases.

We may also have solutions of liquids in gases. Air will dissolve a certain amount of water; and the warmer the air, the more water it will hold. Now water is not truly a gas until it has been heated past 100°C [212°F], its normal boiling point. Thus the moisture in the air is a liquid dissolved in a gas. When air is cooled, its capacity for holding water is reduced, and water may drop out of the solution as dew, rain, or fog. The amount of water dissolved in the air is described by the *relative humidity*. When the relative humidity is 60 percent, for instance, this means that the air is holding 60 percent as much water as it is capable of holding at that temperature.

Finally, we can have a solution where a gas is dissolved in a liquid. Carbon dioxide, for instance, dissolves very well in water, and we find it in soda pop, beer, and sparkling wines. Air dissolved in water supports fish and other aquatic life. But unlike the other solutions we

Fig. 4-8 A particulate analyzer is used to measure the solid particles suspended in an engine's exhaust gases. (Courtesy General Motors Corp.)

Application: Vapor Barrier in Walls

With the high cost of energy, it makes economic sense in most climates to insulate the walls as well as the attic in homes and other buildings. Few people, however, worry about whether the air in the insulation is inside air or outside air. In fact, it makes a difference, because warm air generally holds more moisture than cold air. If moisture-laden air comes into contact with a cold surface, condensation develops. This can lead to peeling paint or, over a period of years, to rotten studs and sheathing.

The diagram shows the cross section of an insulated wall. We'll assume that the outside air is colder than the inside air, which is the case that leads to the problems. If the moist inside air passes through the walls, it will form condensation on the inside of the sheathing. But if the walls are filled with dry outside air, no condensation can form. This is why insulation bats have a foil or tarred paper covering on the side that is stapled to the studs. This is referred to as a "vapor barrier," or sometimes a "moisture barrier." Any tears in this foil should be repaired with tape before the wallboard is hung. Knowledgeable builders also place a few small vent holes in the outside wall, so outside air will pass in before any inside air leaks out.

What if existing walls are to be insulated with blown-in insulation? Some such insulations form their own moisture barriers. If this is not the case, the wallboard or plaster should be sealed with shellac or varnish and then wallpapered or painted in the usual way.

Occasionally, people will try to remedy the problem of peeling exterior paint by covering the building with aluminum or vinyl siding. This just covers up the problem without solving it.

have mentioned, a hot liquid will hold *less* gas than a cold liquid. This is why air bubbles form on the sides of a pan of hot water long before the water reaches a boil. This principle can be used to operate a refrigerator from a flame. The refrigerant in the absorption-type refrigerator is a gas dissolved in a liquid. The liquid is heated by a flame or by solar energy, which drives the refrigerant gas out of solution and forces it through the system.

Paint peels only because moisture has seeped in behind it, and eventually this moisture will lead to deterioration of the wood. The only ways to remedy such a problem are to install vapor barriers or to live in an unheated house.

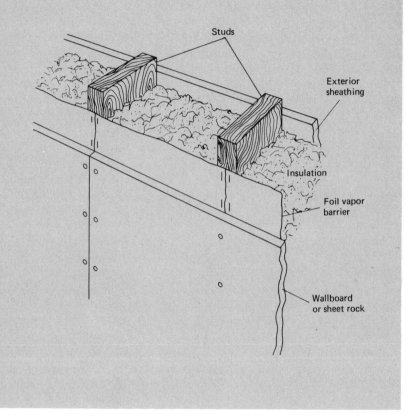

Studs

Exterior sheathing

Insulation

Foil vapor barrier

Wallboard or sheet rock

Molecules of the same type always have a certain amount of electrical attraction for one another. In a gas, this attraction is of little importance, since gas molecules move so fast that they are constantly bouncing off one another and off the walls of their container. But when a gas is cooled enough, the molecules slow down to the point where the intermolecular attraction pulls them together into groups. At this point, the gas has condensed into a liquid. If the liquid is cooled further, the molecules

4-7 Adhesion and Cohesion

slow down even more and a point is reached where the intermolecular forces bind them together in the rigid mass we call a solid. At this point, the liquid has frozen.

The force of attraction between molecules of the same kind is called *cohesion*. If you happen to break a thermometer, you may find many small drops of mercury scattered about the floor. But if you begin to push these drops together, you find that they combine very easily to form bigger and bigger drops. This is because of the high cohesion of mercury. Particles of dry sand in a pile have no tendency to stick together; but if the sand is wet, the natural cohesion of the water allows the sand to hold its shape. An impression in damp sand can be used as a mold for molten metal in a process known as sand casting.

Molecules of different kinds sometimes attract one another, and sometimes they don't. If they do, the attraction is called *adhesion*. Substances that "stick" to a variety of other materials are called *adhesives*. Table 4-8 lists some of the common types.

Summary   Matter is made of elementary particles that group under the influence of forces which are mainly electrical in origin. Protons, neutrons, and electrons group to form atoms, and atoms may group to form molecules. At low enough temperatures, molecules hold onto one another very tightly to form solids. At higher temperatures, motion of the molecules overcomes their attraction for one another, and the matter goes into the liquid, gas, or even plasma state.

All the properties of matter, and all the ways it behaves, are determined by the atoms and molecules it is made of. The mass of a brick, for instance, is the sum of the masses of all the elementary particles in it. And the strength of a glue is determined by the forces of attraction between the glue molecules and the molecules of the materials being joined. There are many other examples that we encounter in this book. We frequently return to this description of the atomic and molecular structure of matter to explain things that would otherwise remain very mysterious.

Terms You Should Know

mass	newton
weight	slug
force	pound
spring scale	pound-mass
balance	kilogram-force
kilogram	proton

TABLE 4-8   COMMON TYPES OF ADHESIVES

1. *Contact cements*
   Applied and allowed to set before work is pressed together, then forms bond immediately on contact. Used for plastic laminates, foam, hardboard, metal, and wood. Good weather resistance but flammable.
2. *Epoxy adhesives*
   Can be used on virtually any materials to produce a strong and waterproof bond. Requires mixing a resin with a hardener. Usually too expensive and awkward for large projects.
3. *Formaldehyde adhesives*
   Exceptionally strong bonds with wood, but poor water resistance. Require mixing with water and clamping of work. Commonly used with indoor furniture.
4. *Gums and pastes*
   Suitable only for paper and cardboard, but very inexpensive. Poor water resistance.
5. *Latex adhesives*
   Usually used on fabrics or carpet to form a strong and waterproof yet flexible bond.
6. *Mastic adhesives*
   Sometimes called "construction adhesives." Used for bonding ceiling and floor tile, plywood wall panels, and similar materials to wood, concrete, or asphalt. Available in cartridges that fit into a caulking gun. Waterproof, but flammable and have long curing time.
7. *Plastic cements*
   For small glass, wood, and plastic parts. Waterproof, but flammable. Work must be clamped.
8. *Polyvinyl resin adhesives*
   Usually called "white glue." Good for wood and other porous materials but poor resistance to dampness. Work must be clamped for best results.
9. *Resorcinol adhesives*
   Very strong bonds with wood and similar materials, and common in boatbuilding because of high water resistance. Require mixing with resin, and clamping of work.
10. *Rubber-base cements*
   Good for nonstructural bonding between wood and concrete, paper and wallboard, rubber and rubber, and ceramic on walls. Good weather resistance but flammable.
11. *Silicone sealants*
   Low strength, but useful for waterproof mating of poorly fitting parts.

electron
neutron
electric force
atom
nucleus
chemical element
chemical compound
hydrocarbon
molecule
phases of matter

ion
plasma
miscible
immiscible
solution
alloy
suspension
adhesion
cohesion

Problems

1. A standard .30-caliber bullet has a mass of 110 grains (gr). How many of these bullets can be cast from 1 kg of lead, assuming no waste?

2. A freighter carries a cargo of 186 lathes, each having a mass of 1712 kg. It leaves a port at 40° north latitude and docks at the equator. How much weight, in newtons, has the cargo lost? How much mass has it lost?

3. A large truck has a mass of 6582 kg. How many metric tons of steel can it carry if it is not to exceed a total mass of 15 t?

4. A bridge span 500 ft long can support a total vehicle load of 350 tons. (This unit is the ton-force.) As a temporary repair, the entire road surface is covered with a layer of asphalt 2 in thick. This adds a weight of 580 lb to each foot of the span. What total vehicle load can the resurfaced bridge support?

5. Gear blanks are to be made of hard bronze by sand casting. Each blank will have a mass of 0.87 kg, and 42 are needed. The bronze is to have a composition of 87.1 percent copper, 9.9 percent tin, and the rest zinc. Find the mass of copper and tin needed, assuming no waste.

5
Mechanical Properties of Solids

Introduction

The buildings we live in, the cars we drive, and the tools we use are all solids. They are composed of molecules that hold one another very tightly. As a result, solids tend to keep their shape and size under a variety of conditions, which is what makes them so useful. In this chapter, we discuss some of the mechanical properties of solids. In particular, we see how to predict the mass and weight of a solid object and how different solids behave under the influence of forces.

5-1 Density

You may have heard this riddle: "Which weighs more, a pound of feathers or a pound of steel?" Of course, the answer is that they weigh the same; a pound of anything weighs a pound. Yet 1 lb of feathers would certainly occupy a larger *volume* than 1 lb of steel. Or if we had equal *volumes* of feathers and steel, the steel would certainly weigh quite a bit more than the feathers.

　　To describe the property that a unit volume of some materials is heavier than a unit volume of others, we use a quantity called density. This quantity is actually based on mass rather than weight. It is defined as follows:

density

Definition: *Density* is the ratio of an object's mass to its volume.

$$\text{Density} = \frac{\text{mass}}{\text{volume}} \qquad (5\text{-}1)$$

For a substance like cork, which has a low density, a small mass occupies a large volume. For a substance like lead, which has a high density, a large mass fits into a small volume. Figure 5-1 shows two trucks carrying the same mass. Yet the load has a much smaller volume on the truck carrying a high-density material.

Fig. 5-1 Mass and density. Both trucks are carrying about the same mass. The moving van is loaded with furniture, bedding, clothing, and other low-density solids. The flatbed trailer carries just two small coils of steel, which have a high density.

In SI units, mass is measured in kilograms and volume in cubic meters, so the SI unit of density is kilograms per cubic meter (kg/m^3). It is, however, more common to list densities in terms of another metric unit: grams per cubic centimeter (g/cm^3). Table 5-1 lists the densities of common solids in the customary USCS and metric units.

Knowing a solid's density enables us to estimate the mass if we know the volume:

$$\text{Mass} = (\text{density})(\text{volume}) \qquad (5\text{-}2)$$

Or the same relationship can be written to allow us to calculate the volume when we know the mass:

$$\text{Volume} = \frac{\text{mass}}{\text{density}} \qquad (5\text{-}3)$$

TABLE 5-1 DENSITIES OF SOME COMMON SOLIDS*

Solid	g/cm³	lb$_m$/ft³
Aluminum	2.699	168.5
Aluminum, Al-Clad 17 ST alloy	2.96	185
Brass, yellow, cast	8.44	527
Brick (average)	1.9	120
Bronze, gun metal	8.78	548
Cement, dry	1.5	94
Cement, set	2.7–3.0	170–190
Concrete (average)	2.4	150
Copper, cast	8.30–8.95	518–559
Copper, hard-drawn	8.89	555
Douglas fir, dry	0.446	27.8
Glass, common	2.4–2.8	150–175
Gold, cast	19.3	1200
Ice (0°C)	0.917	57.2
Lead	11.3	705
Oak, white, dry	0.710	44.3
Paper	0.70–1.15	44–72
Pine, white, dry	0.373	23.3
Platinum	21.37	1334
Sand	1.5	94
Silver, cast	10.5	655
Solder, plumbing	9.4	590
Steel, 1 percent carbon	7.83	489
Steel, stainless	7.75	484
Tungsten	18.6–19.1	1160–1190
Tungsten carbide	14.0	874
Uranium	18.7	1170

* The temperature is 20°C [68°F] unless stated otherwise.
Note: 1 g/cm³ = 1 kg/L = 1000 kg/m³

Let's look at some examples of how these equations can be used to solve everyday problems which involve calculating the density mass or volume of a solid.

Example 5-1 Mass of Sand in a Bin.

A bin made of concrete blocks has the inside dimensions shown in the diagram. It is filled level to the top with sand. What is the total mass of sand?

From Table 5-1 we see that sand has a density of 94 lb$_m$/ft³. To use Eq. (5-2), we also need the volume. Since this is a rectangular prism, the

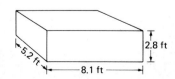

2.8 ft
5.2 ft
8.1 ft

volume is

$$\text{Volume} = (\text{length})(\text{width})(\text{height})$$
$$= (8.1 \text{ ft})(5.2 \text{ ft})(2.8 \text{ ft})$$
$$= 120 \text{ ft}^3$$

The mass of sand is then

$$\text{Mass} = (\text{density})(\text{volume}) \qquad\qquad (5\text{-}2)$$
$$= 94 \, \frac{\text{lb}_m}{\text{ft}^3} \, (120 \text{ ft}^3)$$
$$= 11 \, 000 \text{ lb}_m$$

What is this sand's weight? Strictly speaking, it depends on the latitude and altitude. But there is no need for extreme accuracy in a case like this; it would be close enough for any practical purpose to say that the sand weighs 11 000 lb. Errors in dimensioning and leveling the top and bottom are probably much greater anyway than the 0.5 percent maximum error in this weight approximation. ◀

Example 5-2 Length of a Steel Coil.

A steel sheet is rolled to a thickness of 1.14 mm. After shearing and coiling, it measures 2.32 m in width and is found to have a mass of 2730 kg. A customer needs to know the length of the coil in order to estimate how many electric stove shells can be stamped from it. Obviously, no one wants to unroll the coil just to measure its length. It is much easier in cases like this to calculate the answer.

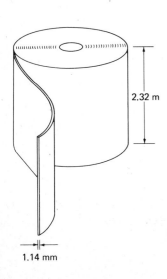

According to Eq. (5-3), the actual volume of the steel is

$$\text{volume} = \frac{\text{mass}}{\text{density}}$$

We are working in metric units here. From Table 5-1 the density of steel is 7.83 g/cm³. Then

$$\text{Volume} = \frac{2730 \text{ kg}}{7.83 \text{g/cm}^3}$$

2.32 m

1.14 mm

The numerator of this fraction has the unit kilograms while the denominator has the unit grams. This calls for the conversion factor 1000 g = 1 kg. Inserting this conversion factor gives

$$\text{Volume} = \frac{2730 \text{ kg}}{7.83 \text{ g/cm}^3}\left(\frac{1000 \text{ g}}{1 \text{ kg}}\right)$$

$$= 349\,000 \text{ cm}^3$$

We already know that the sheet is 2.32 m in width and 1.14 mm in thickness. Expressing both of these in centimeters gives us a 232-cm width and a 0.114-cm thickness. The length must then be

$$\text{Length} = \frac{349\,000 \text{ cm}^3}{232 \text{ cm (0.114 cm)}}$$

$$= 13\,000 \text{ cm}$$

This is the same as 130 m, or 430 ft. It would certainly have been a big job to completely unroll this coil! Yet the coil itself would have a diameter of only around 50 cm [20 in], even with the hole in the center. ◀

Exercises

1. Find the pound-mass of 1.23 ft³ of bronze.
    Answer: 674 lb$_m$
2. Calculate the mass in grams of 358 cm³ of AlClad 17 ST alloy.
    Answer: 1060 g
3. What volume, in cubic feet, is occupied by 120 lb$_m$ of ice?
    Answer: 2.10 ft³
4. Find the volume in cubic centimeters of 1.68 kg of uranium.
    Answer: 89.8 cm³
5. A cylindrical concrete column is 86 cm in diameter and 2.32 m high. (a) Find its mass in kilograms. (b) What is its approximate weight in pounds?
    Answers: (a) 3200 kg, (b) 7100 lb
6. What volume of aluminum will have the same mass as 1.000 cm³ of platinum?
    Answer: 7.918 cm³

Springs are usually made from various steel alloys. Valve springs and scale springs are made from high-carbon steel, while large coil and leaf springs are fashioned from silicon-manganese steel. For applications with high operating temperatures, chromium-vanadium steel may

5-2 Springs and Elasticity

Application: Archimedes and the Gold Crown

Around the year 250 B.C., Hiero, king of Syracuse, commissioned a golden crown to be made for him. He weighed out an exact amount of gold from the royal treasury and gave it to a goldsmith for the project. When the smith delivered the finished product, Hiero checked its weight to make sure that all the gold had been used. The weight checked perfectly.

But shortly after, someone suggested that the goldsmith might have substituted a cheaper metal, perhaps silver, for the interior of the crown. The king had to know for sure. So he called on Archimedes to determine whether the crown was solid gold, but without ruining the crown by cutting it apart. Today, this type of problem falls under the title "nondestructive testing." For Archimedes, it must have been quite a challenge.

While mulling over the problem, Archimedes decided to go to the public baths for an afternoon of relaxation. As he lowered his body into the bath, the water overflowed. It was then that the solution struck him. He jumped up and ran naked through the streets to his home, shouting *"Heureka,"* or "I have found it."

The king, of course, demanded an explanation. Archimedes set up a demonstration in the royal chambers. He lowered the crown into a jar of water and caught the overflow. The volume of water that spilled over had to be the same as the volume of the crown. He then refilled the jar and lowered an equal mass of pure gold in it, again catching the overflow. If the crown was pure gold, the two overflow volumes should have been the same. But if the crown had silver in it, which is less dense, the crown's volume would have to be larger than the volume of an equal mass of pure gold. This, in fact, is exactly what happened. We can only speculate on the fate of the dishonest goldsmith.

be used, and for corrosive atmospheres springs may be made of stainless steel.

In electrical applications such as relays, springs are often made of nonferrous alloys like phosphor bronze, Monel metal, and beryllium-copper wire. Occasionally, small springs may also be made of plastics such as nylon.

To be used in a spring, a material must have *elasticity,* that is, the ability to remember its original shape when it is deformed. Within limits, a spring may be stretched, compressed, twisted, or bent, and yet still return to its original shape when released. And the process may be repeated any number of times.

When a spring (or any other material, for that matter) is being stretched, we say that it is in *tension.* When it is being squashed, we say it is in *compression.*

Figure 5-2 shows the behavior of a large coil spring in tension. The original length of the spring isn't given, but it is probably around 20 cm. The graph shows the force necessary to stretch the spring by different amounts. For instance, it takes a force of 30 kg$_f$ to stretch this spring 2.0 cm. If the 30-kg$_f$ force is then removed, the spring goes back to its original unstretched length. This happens because the spring is *elastic.*

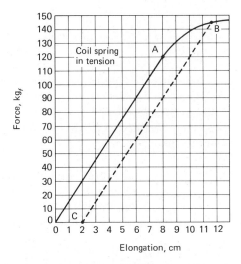

Fig. 5-2 Behavior of a large coil spring under tension. Point *A* is the spring's elastic limit. If the spring is stretched to point *B,* a permanent deformation (*C*) remains when the force is removed.

Now from your own experiences with door springs and the like, you know that a spring can be stretched only so far before it becomes permanently deformed. For the spring we have been talking about, this happens for forces greater than 120 kg$_f$, which corresponds to elongations beyond 8.0 cm. This point has been labeled *A* in Fig. 5-2. It is called the elastic limit.

Definition: The *elastic limit* is the point beyond which a spring will suffer a permanent deformation. It may be described by the maximum allowable *force,* or by the maximum *deformation* that is not permanent.

elastic limit

The elastic limit in Fig. 5-2 can be stated as either 120 kg$_f$ or 8.0 cm.

Suppose that our spring is stretched by a 145-kg$_f$ force. The graph shows that it will experience a total stretch of 11.6 cm (point *B*). When this force is removed, the spring contracts according to the dotted line. At point *C*, there is no longer any force on it; yet it is still 2.0 cm longer than its original length. This 2.0 cm is the permanent stretch that resulted from exceeding the elastic limit.

Let's now talk about the behavior of the spring below the elastic limit. We see that a 15-kg$_f$ force stretches the spring by 1.0 cm. Doubling the force to 30 kg$_f$ doubles the stretch to 2.0 cm. And for each additional 15 kg$_f$, there is an additional stretch of 1.0 cm. This 15-kg$_f$/cm stretch is called the spring's *elastic constant*, or its *spring constant*, or simply its *stiffness*.

stiffness　　　Definition: A spring's *stiffness* is the ratio of the deforming force to the elastic deformation it produces:

$$\text{Stiffness} = \frac{\text{force}}{\text{deformation}} \qquad (5\text{-}4)$$

Figure 5-3 shows a simple way to measure the stiffness and elastic limit of a door spring. If a spring has a high stiffness, it takes a large force to produce a small stretch. If the stiffness is low, a small force produces a large deformation. As usual, Eq. (5-4) may be rewritten in two other forms:

$$\text{Force} = (\text{stiffness})(\text{deformation}) \qquad (5\text{-}5)$$

$$\text{Deformation} = \frac{\text{force}}{\text{stiffness}} \qquad (5\text{-}6)$$

The first formula allows us to calculate the force needed to produce a certain deformation, while the second allows us to calculate the deformation resulting from a given force. These formulas apply as long as the elastic limit has not been exceeded. Although we have been talking about deformation as a stretch, it may also be a compression in these formulas. Let's look at an example.

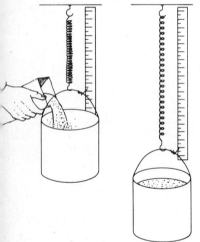

Fig. 5-3 Measuring the stiffness and elastic limit of a door spring. A lightweight plastic bucket is hung from the spring, and a piece of wire is twisted (or better, soldered) onto the handle as a pointer. This pointer moves in front of a meter stick (or yardstick) that has been clamped in place alongside the spring. Sand may be used.

Example 5-3 Stiffness of an Automobile Spring Suspension.

A person with a mass of 84.5 kg climbs into the center of a car with a conventional spring suspension. This causes the frame to drop 0.92 cm closer to the ground. What is the stiffness of the suspension?

Although the car is supported by four springs, they work together and it is the combined stiffness that is of interest. The force causing the additional spring compression is the person's weight—approximately 84.5 kg$_f$. Then using Eq. (5-4), we have

$$\text{Stiffness} = \frac{\text{force}}{\text{deformation}}$$

$$= \frac{84.5 \text{ kg}_f}{0.92 \text{ cm}}$$

$$= 92 \text{ kg}_f/\text{cm}$$

Suppose that this car weighs 1450 kg$_f$. How much will the springs expand if they are removed? The answer is that they will expand just as much as they've been compressed. We can use Eq. (5-6):

$$\text{Deformation} = \frac{\text{force}}{\text{stiffness}}$$

$$= \frac{1450 \text{ kg}_f}{92 \text{ kg}_f/\text{cm}}$$

$$= 16 \text{ cm}$$

Note that this result doesn't give us the actual size of the springs; it simply tells us how much their length changes when they are compressed by the weight of the car. This 16 cm (or 6.3 in) is approximately the same as the distance the bumper must be lifted with a bumper jack before the wheel leaves the ground. ◄

When a spring is designed to be compressed rather than stretched, very often it is made so that it "bottoms out" before the elastic limit is exceeded. For springs used in tension, such as expansion connectors on power lines, care must be taken that the elastic limit is never exceeded.

Exercises

7. Under a force of 1.6 kg$_f$, a certain spring is compressed by 2.3 cm. What is the spring's stiffness?
   Answer: 0.70 kg$_f$/cm

8. A certain spring has a stiffness of 56 kg$_f$/cm. How much force is needed to stretch it 4.5 cm?
   Answer: 250 kg$_f$

9 A coil spring is 21.2 cm in length. A force of 5.63 kg$_f$ stretches it to 24.7 cm. (*a*) Find the stiffness. (*b*) Find the force needed to stretch the spring to a 30.0-cm length. (*c*) Find the spring's length

Application: Forming Metals

When a metal is poured into a mold in its molten state, it is said to be *cast*. This is the first step in forming all metals. Some castings, such as engine blocks and large work rolls, are then machined on their working surfaces. But direct casting is generally feasible only for very large items; in most manufacturing the metal is put through additional forming steps.

If metal is going to be formed further, it is usually cast into *ingots*. (Another procedure is continuous casting, which is discussed in Chap. 15.) The ingots are then reheated to make sure that the temperature, and therefore the elastic limit, is the same throughout. The hot ingots are then *rolled* into slabs, which may be further rolled into billets, bars, plate, or sheet. To be formed by rolling, a metal must have a large plastic range when in compression. Such metals are said to be *malleable*. Malleable metals may also be formed by *extrusion*, which is a pressing operation that forces a metal to flow plastically between two dies. Another pressing operation where the flow is not so extreme is called *forging*. Hot forging is used for thick items like crankshafts, while cold forging is done on small pieces made from sheet.

Small rod and wire are formed by drawing through a die. To be formed by drawing, a metal must have a large plastic range when in tension. Such metals are said to be *ductile*. Metals like aluminum and copper are more malleable than ductile. Steel is both malleable and ductile, particularly when heated above 750°C [1400°F].

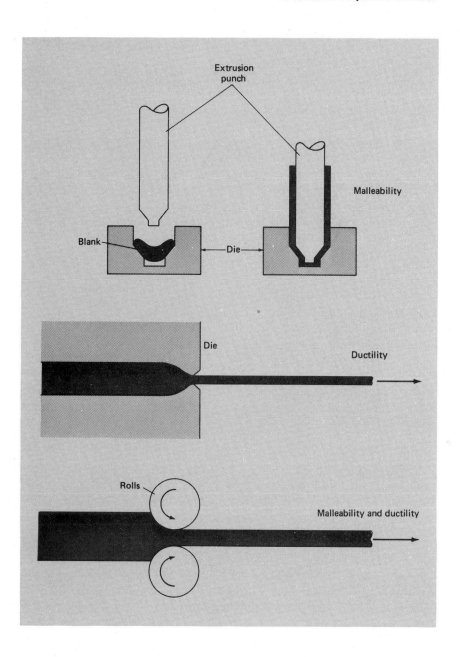

under a stretching force of 22.0 kg$_f$. In all cases, assume that the elastic limit is not exceeded.

Answers: (a) 1.61 kg$_f$/cm, (b) 14.2 kg$_f$, (c) 34.9 cm

10. A certain spring has an elastic limit of 42.5 kg$_f$ of tension. Its stiffness is 8.60 kg$_f$/cm. How much can it be stretched without permanently deforming it?

   Answer: 4.94 cm

11. Under a compression of 860 lb, a coil spring is 11 in in length. Its stiffness is 120 lb/in. What is the spring's length when the force is removed?

   Answer: 18 in

12. A certain spring reaches its elastic limit when it has been stretched 22 cm. Its stiffness is 0.64 kg$_f$/cm. What kilogram-force is needed to exceed the elastic limit?

   Answer: 14 kg$_f$

5-3 Stress and Strain

Every solid, regardless of its composition, behaves to some extent like a spring. Even materials like brick and wood can be deformed slightly; then they spring back when the force is removed.

When a weight is hung from a cable or a bridge is supported by a pier, the deforming force does not act at just one point. Instead, it is spread over the entire cross-sectional area of the cable or pier. The ratio of the force to the cross-sectional area on which it acts is called the *stress*.

stress

Definition: *Stress* is the ratio of a force to the cross-sectional area on which it acts.

$$\text{Stress} = \frac{\text{force}}{\text{area}} \tag{5-7}$$

Suppose that we hang a 10-kg$_f$ weight (22 lb, or 98 N) by a piece of steel wire of the type used in bridge cables. If the wire has a diameter of 1 mm, this force will break it. But if the wire is 2 mm in diameter, it will have strength to spare. In both cases the force is the same, but in the second case this force is spread over a larger area. With a larger area, the *stress* is smaller.

Example 5-4 Stress in a Jack Post.

A jack post is used to hold up a sagging floor joist. The weight supported by the post is 650 kg$_f$ [1400 lb]. The post is a hollow steel cylinder with

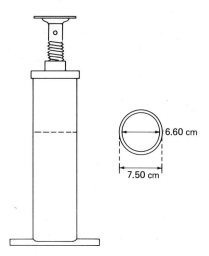

an outside diameter of 7.50 cm [2.95 in] and an inside diameter of 6.60 cm [2.60 in]. What is the stress in the post?

The cross section of the post is shown in the diagram. To find its area, we can calculate the area of the outside circle and then subtract the area of the inside circle. This gives

$$\text{Area} = \pi(\text{outside radius})^2 - \pi(\text{inside radius})^2$$

$$= \pi(3.75 \text{ cm})^2 - \pi(3.30 \text{ cm})^2$$

$$= 44.2 \text{ cm}^2 - 34.2 \text{ cm}^2$$

$$= 10.0 \text{ cm}^2$$

To find the stress, we use Eq. (5-7):

$$\text{Stress} = \frac{\text{force}}{\text{area}}$$

$$= \frac{650 \text{ kg}_f}{10.0 \text{ cm}^2}$$

$$= 65 \text{ kg}_f/\text{cm}^2$$

In other words, each square centimeter of the cross section of the post supports 65 kg_f of weight. Altogether, 10.0 cm² supports the total 6500 kg_f. ◀

Since stress is force divided by area, the unit of stress is a unit of force divided by a unit of area. In Example 5-4 this came out as kilogram-force per square centimeter. However, any force unit divided by any area unit gives a stress unit: pounds per square inch, newtons per square centimeter, kilogram-force per square meter, pounds per square foot, and so on. The SI unit of stress is newtons per square meter. This has been given a special name, the *pascal* (Pa):

$$1 \text{ Pa} = 1 \frac{\text{N}}{\text{m}^2} \tag{5-8}$$

Now if we spread 1 N of force over an area of 1 m², the stress developed is very small indeed. This is approximately the stress we get by spreading a bedsheet over a mattress. Thus the pascal (Pa) is much too small a unit to use in most applications. Instead, it becomes necessary to create multiples of this unit by using the SI prefixed in Table 1-3: 1 kPa = 1000 Pa, and 1 MPa = 1 000 000 Pa. The largest steady stress ever produced under test conditions was around 1.7×10^5 MPa, which is equivalent to 2.5×10^7 lb/in², or 1.8×10^6 kg$_f$/cm². (This stress is some 700 times greater than the stress the weight of Mount Everest places on the crust of the earth.)

Laboratory scientists measure stresses in pascals, kilopascals, and megapascals. Remember, this *is* the official stress unit. But in industry and the trades, the pascal and its multiples are often inconvenient. The pascal is based on an area of 1 m², while most machine and structural parts have areas measured in square centimeters. By using a metric stress unit based on the square centimeter instead of the square meter, many unit conversions (and mistakes) can be avoided. Thus the kilogram-force per square centimeter, which we encountered in Example 5-4, is a commonly used unit. Although it is a metric unit, it is not considered an SI unit.

Table 5-2 lists the numerical relations between the most common units of stress. The following exercises are based on this table and Eq. (5-7).

Exercises

13. A 65$\bar{0}$-lb weight is supported by a bar with a cross-sectional area of 9.70 in². (a) Calculate the stress in the bar. (b) Calculate the stress in the same bar if the weight is doubled. (c) Calculate the stress produced by the 65$\bar{0}$-lb weight if the cross-sectional area of the bar is doubled.
    Answers: (a) 67.0 psi, (b) 134 psi, (c) 33.5 psi
14. A weight of 1250 N is supported by a column with a cross-sectional area of 48.7 cm². Calculate the stress in the column in (a) newtons per square centimeters, (b) kilopascals.
    Answers: (a) 25.7 N/cm², (b) 257 kPa

TABLE 5-2   *CONVERSION FACTORS FOR COMMON UNITS OF STRESS*

	$\dfrac{kg_f}{cm^2}$	$\dfrac{kg_f}{m^2}$	Pa	$\dfrac{lb}{ft^2}$	$\dfrac{lb}{in^2}$
1 kilogram-force per square centimeter =	1	10 000	98 066.5	2048.2	14.223
1 kilogram-force per square meter =	0.000 1	1	9.806 65	0.204 82	$1.422\ 3 \times 10^{-3}$
1 pascal =	$1.019\ 7 \times 10^{-5}$	0.101 97	1	0.020 885	$1.450\ 4 \times 10^{-4}$
1 pound per square foot =	$4.882\ 4 \times 10^{-4}$	4.882 4	47.880	1	$6.944\ 4 \times 10^{-3}$
1 pound per square inch =	0.070 307	703.07	6894.8	144	1

$$1\ Pa = 1\ \frac{N}{m^2}$$

$$1\ kPa = 1000\ \frac{N}{m^2} = 0.1\ \frac{N}{cm^2} = 0.145\ 04\ \frac{lb}{in^2}$$

$$1\ MPa = 10^6\ \frac{N}{m^2} = 100\ \frac{N}{cm^2} = 145.04\ \frac{lb}{in^2}$$

15. A 130-kg_f weight is supported by a post with a 23-cm² cross-sectional area. Find the stress in the post in (a) kilogram-force per square centimeter, (b) kilopascals.
    Answers: (a) 5.7 kg_f/cm², (b) 550 kPa
16. A steel cable 1.50 cm in diameter has a circular cross section. Calculate the stress in the cable if it is used to lift a weight of 2230 kg_f.
    Answer: 1260 kg_f/cm²
17. A small airplane has a mass of 1940 kg. Its wings have a total area of 84 000 cm². Calculate the average stress on the wings when the plane is in level flight in (a) kilogram-force per square centimeter, (b) pounds per square inch.
    Answers: (a) 0.023 kg_f/cm², (b) 0.33 psi
18. A concrete column has a square cross section measuring 38 cm on a side. The column supports a mass of 12.6 metric tons (t). What is the approximate stress in the column?
    Answer: 8.7 kg_f/cm²
19. The pier of a bridge supports a total weight of 5200 t. The contact surface between the bridge and the pier has a total area of 11 m².

Calculate the stress at the top of the pier in (a) kilogram-force per square meter, (b) kilogram-force per square centimeter, (c) kilopascals, (d) pounds per square inch.
Answers: (a) 470 000 kg_f/m^2, (b) 47 kg_f/cm^2, (c) 4600 kPa, (d) 670 psi

5-4 Strain

Stress is important because objects tend to deform in proportion to the stress on them. The jack post in Example 5-3 will have its length shortened slightly by the 65-kg_f/cm^2 compressive stress on it. If this stress is doubled, the amount of compression is doubled. Notice that the stress can be doubled in two ways: either by doubling the weight on the post or by using a post with only half the cross-sectional area.

Strain is the measure of an object's deformation under stress. It may be defined this way:

strain

Definition: *Strain* is the fractional deformation experienced by an object under stress. For objects in tension or compression (rather than twisting or bending), it is given by the formula

$$\text{Strain} = \frac{\text{change in length}}{\text{original length}} \qquad (5\text{-}9)$$

A rubber band experiences a large strain even when the stress is small. For most construction materials, however, only very small strains can be tolerated—even when the stress is large. Figure 5-4 shows the actual amount that different metals stretch when subjected to the same stress. This elongation is given in units of micrometers (μm) (1 μm = 10^{-6} m). Notice that the elasticities of these metals are quite different; some are much "stiffer" than others.

5-5 Strengths of Materials

We said that all solids behave like springs. They deform under stress, and when the stress is released, they "remember" their original shape. If the stress is too large, however, the solid's elastic limit may be exceeded, just as with a spring. In that case, the material will retain a permanent "set," or deformation.

Figure 5-5 shows a stress-strain curve for structural steel in tension. This is considered to be a "soft" or a "mild" steel. The strain is in units of meters of deformation per meter of original length. At stresses up to about 2500 kg_f/cm^2, the steel is elastic; removing the stress allows the steel to spring back to its original shape. At higher stresses, the steel becomes *plastic*. Now the stress has overcome the forces holding the molecules together, and these molecules begin to slide over one another. When the stress is removed, the steel still springs back a little, but not all the way to its original shape. If the steel is used in a structure, its elastic limit should not be exceeded. On the other hand, if the steel is to be formed (bent, rolled, extruded, etc.), the elastic limit must be exceeded to get the steel to change its shape. But even here there is a limit. When the steel is stressed to its *ultimate strength* (point *B* on the graph), it begins to flow very rapidly and the stress must actually be reduced to get it to stop. With only a slight additional deformation, the steel breaks. The breaking strength is usually not much different from the ultimate strength.

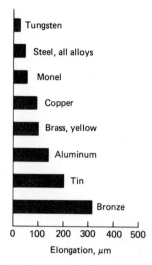

Fig. 5–4 Elastic elongations of different metals subjected to the same tensile stress of 100 kg$_f$/ cm². All samples are 1.00 m in length. The stress does not exceed any of the elastic limits.

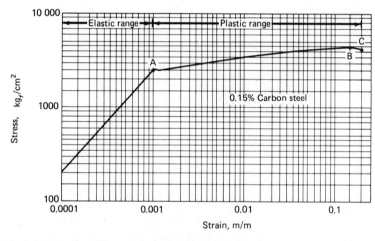

Fig. 5–5 Stress-strain curve for low-carbon structural steel in tension. The steel is elastic up to point *A*, which is the elastic limit. At higher stresses it becomes plastic. Point *B* is the ultimate strength, and *C* is the breaking strength, also called the fracture or rupture strength.

Not all solids behave as low-carbon steel does. Some materials, like brass and copper, have an ultimate strength more than 3 times as high as their elastic limit. These make poor structural materials, but the large *plastic range* makes it easy to form them into small parts. Other materials, like Monel metal, masonry, high-carbon steel, and timber, exhibit practically no plastic behavior. Their breaking strength is nearly the same as their ultimate strength and just slightly higher than their elastic limit. Under stress, these materials either behave elastically or fail.

So far, we have been talking about solids in tension. How do they behave when they are compressed? For low-carbon steel, there is no

Application: Arches in Architecture

The earliest large buildings were made of stone, often mortarless but sometimes with lime-mortar joints. Such masonry is very strong in compression, but very weak in tension. The buildings were therefore made massive enough that despite wind and ground settling, no part of the structure would ever go into tension. Still, there was a problem in providing openings. The ancient Greeks used large numbers of massive columns spaced fairly close together and capped with stone slabs. Over the centuries, most of these slabs failed under the bending stresses, leaving only the columns standing upright today.

The Romans made a big improvement with the invention of the circular arch. The stones forming the arch were wider at the top than at the bottom, so any load on the arch pressed the stones together more tightly and actually made the span stronger. It was virtually impossible for any part of this arch to go into tension. Many Roman arched structures are still in use today.

Still, the circular arch did not have the same strength at all points along the span; it was still weakest near the center. The Gothic arch, invented in medieval Europe, solved this problem. By making the arch higher than wide, all points across the span could bear the same load. But trimming the stone into such complicated shapes was anything but an easy task! The cruck construction of old English homes was based on the Gothic arch, but using wood rather than stone for the arch.

Most masonry homes have long been built on the Greek design, although they may not appear so at first glance. The Georgian house can be looked at as a series of six columns across the front that bear the load. The windows are set in the bays between these columns, which do virtually nothing to support the structure. In fact, modern frame home construction uses 8-ft-high columns (2×4 studs) on 16-in centers, spanned by a 2×4 top plate, which is basically the Greek design adapted to modern cut lumber.

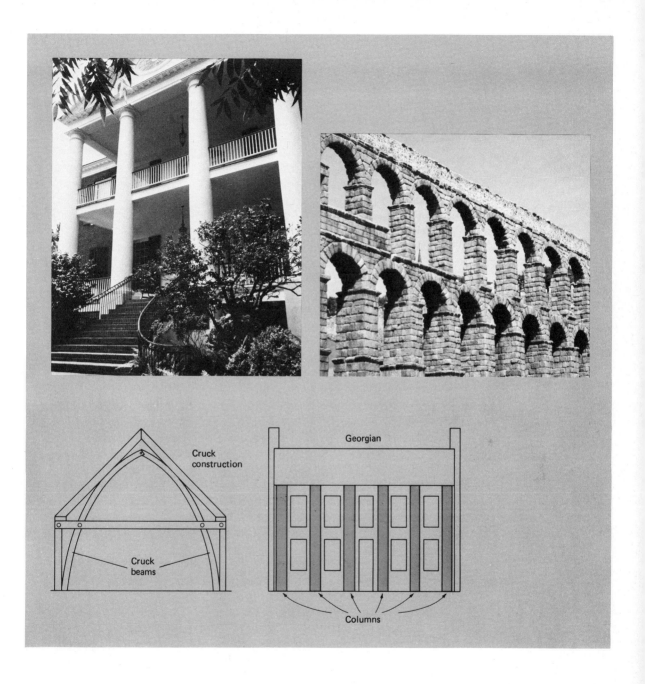

Cruck
construction

Cruck
beams

Georgian

Columns

difference at all. Stresses beyond 2500 kg$_f$/cm², it begins to flow plastically, and beyond 4600 kg$_f$/cm² it fractures. Figure 5-6a shows an automobile that has been crushed for scrap. To do this, a stress higher than the elastic limit had to be applied. In the process, the ultimate strength was also exceeded in many places, causing the metal to break apart. It is unlikely that anyone will be upset about the metal failure in this case.

In Fig. 5-6b a van door has been cut and pressed from a piece of steel (1000 such doors can be made from the coil in the background). Care must be taken in designing the pressing dies so the steel is never stressed past its ultimate strength as it is formed.

For other materials, like cast iron and masonry, the ultimate strength is much greater in compression than in tension. Still, these materials fracture if the stress is high enough. There are also solids like aluminum and brass that have *no* ultimate strength in compression. As they deform plastically, their cross section increases and a greater stress must be supplied to compress them further, which causes them to spread out even more. They never do rupture. Figure 5-7 shows the tremendous work roll used to form aluminum into thin sheet. If the roll weren't so large, its own elastic limit could be exceeded in applying the stress to flatten the aluminum.

Fig. 5-6 Stressing steel past its elastic limit.

Fig. 5-7 Work roll used to form aluminum sheet 5.5 m wide. (Courtesy ALCOA)

Table 5-3 lists the *elastic* ranges and values of ultimate strength for different materials. Notice that the *stress* determines whether a material's elastic limit is exceeded, not the force. Figure 5-8 shows a familiar example. Here a C-clamp is being used to glue two pieces of wood. As the clamp is tightened, its anvils stress the wood past its elastic limit, and the resulting depressions will later be visible in the work. This can be avoided by inserting another wood block between each anvil and the work. Since the same force is now spread over a larger area between the block and the work, the stress on the work is reduced and the elastic limit is not exceeded.

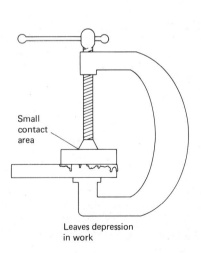

Small contact area

Leaves depression in work

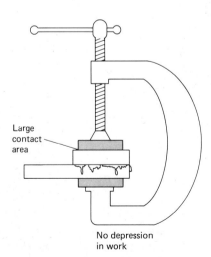

Large contact area

No depression in work

Fig. 5-8 Use of a C-clamp to glue two pieces of wood.

*TABLE 5-3   ELASTIC LIMIT AND ULTIMATE STRENGTH OF SOME COMMON
SOLIDS* [Units are kilogram-force per square centimeter (kg$_f$/cm²).]

Solid	Elastic limit		Ultimate strength	
	Tension	Compression	Tension	Compression
Aluminum	840	840	1 800	∞
Brass	630	630	2 000	∞
Brick, best hard	30	840	30	840
Brick, common	4	70	4	70
Bronze	2 800	2 800	5 300	∞
Cement, Portland, 1 mo old	28	140	28	140
Cement, Portland, 1 yr old	35	210	35	210
Concrete, Portland, 1 mo old	14	70	14	70
Concrete, Portland, 1 yr old	28	140	28	140
Copper	700	700	2 500	∞
Douglas fir	330	330	500	430
Granite	49	1 300	49	1300
Iron, cast	420	1 800	1 400	5600
Lead	10	10	200	∞
Limestone and sandstone	21	630	21	630
Monel metal	6 300	6 300	7 000	∞
Oak, white	310	310	600	520
Pine, white, eastern	270	270	400	345
Slate	35	980	35	980
Steel, bridge cable	6 700	6 700	15 000	∞
Steel, 1 percent C, tempered	5 000	5 000	8 400	8 400
Steel, chrome, tempered	9 100	9 100	11 000	11 000
Steel, stainless	2 100	2 100	5 300	5 300
Steel, structural	2 500	2 500	4 600	4 600

Note: 1 $\frac{kg_f}{cm^2}$ = 98.1 kPa = 14.22 $\frac{lb}{in^2}$

Example 5-5 Stress in a Brick Chimney.

A brick chimney is to be built 12.5 m [41.0 ft] high. It will have the same
cross section for its entire height. We want to know if the bottom course of
bricks will be able to support the weight of all the bricks above.

Now if we know the actual cross-sectional dimensions of the chimney, we can calculate its volume and from that get the chimney's total weight. But it isn't really necessary to go to this much trouble. Table 5-3 tells us that common brick can support a compressive stress of up to 70 kg_f/cm^2. All we need to do, then, is calculate the weight supported by a 1-cm^2 area on the bottom course. If this weight is greater than 70 kg_f, the chimney will crumble. If it is less than 70 kg_f, the chimney will stand.

The volume of a prism 1 cm^2 in cross section and 12.5 m high is

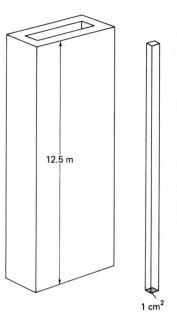

12.5 m

1 cm^2

$$\text{Volume} = (1 \text{ cm}^2)(12.5 \text{ m}) \left(\frac{100 \text{ cm}}{1 \text{ m}} \right)$$

$$= 1250 \text{ cm}^3$$

The mass of this volume of brick is given by Eq. (5-2):

$$\text{Mass} = (\text{density})(\text{volume})$$

From Table 5-1, we find that the density of brick is 1.9 g/cm^3.

Then

$$\text{Mass} = 1.9 \frac{g}{cm^3} (1250 \text{ cm}^3)$$

$$= 2400 \text{ g}$$

which is the same as 2.4 kg. The weight pressing down on our square centimeter is therefore very close to 2.5 kg_f. This gives us an actual stress, from Eq. (5-7), of

$$\text{Stress} = \frac{\text{force}}{\text{area}}$$

$$= \frac{2.4 \text{ kg}_f}{1 \text{ cm}^2}$$

$$= 2.4 \text{ kg}_f/cm^2$$

Since this is nowhere near the 70-kg_f/cm^2 elastic limit of brick, the chimney will stand like the rock of Gibraltar. In fact, the chimney could be over 300 m [1000 ft] high, and the stress would not exceed the elastic limit on the bottom courses of brick. On the other hand, a chimney this high would have to be set on pilings into bedrock because an earth base would not support the stress.

Example 5-6 Strength of a 2 × 4.

The 2 × 4 has long been used as the standard timber size for studs in conventional construction. Unfortunately, its size is anything but standard; through the years it has shrunk from 2.0 in by 4.0 in in cross section to the present 1½ in by 3½ in. Let's calculate the amount of weight such a 2 × 4 can support when standing on end.

If the 2 × 4 is cut from pine, its elastic limit (Table 5-3) is $270 kg_f/cm^2$. Although the ultimate strength is slightly higher, anything in excess of the elastic limit causes a permanent deformation, which is highly undesirable in construction. Since the elastic limit is a stress, we may begin with Eq (5-7):

$$\text{Stress} = \frac{\text{force}}{\text{area}}$$

This may be rewritten as

$$\text{Force} = (\text{stress})(\text{area})$$

The area of the 2 × 4 is

$$\text{Area} = (1.5 \text{ in})(3.5 \text{ in}) \left(\frac{2.54 \text{ cm}}{1 \text{ in}}\right)\left(\frac{2.54 \text{ cm}}{1 \text{ in}}\right)$$

where we have introduced two conversion factors to change the inches to centimeters. This gives an area of 33.9 cm^2, so

$$\text{Force} = 270 \frac{kg_f}{cm^2} (33.9 \text{ cm}^2)$$

$$= 9100 \text{ } kg_f$$

So the total weight the 2 × 4 can support is approximately 10 tons! Of course, this assumes that there are no knots and that there are fire-stops and/or sheathing that keep it from bending. Even so, we see that lumber is a very strong building material. ◀

An increase in temperature reduces the elastic limit of most solids. This is why metals are often heated before being put through a forming operation. For machine parts operating at high temperatures, this reduction in elastic limit has to be kept in mind; parts are often known to fail because of excessive temperatures.

If a metal gets hot enough, it may flow plastically under its own weight even though it hasn't melted yet. This is exactly what happened when the roof of Washington Cathedral was covered with pure lead. High temperatures generated by the absorption of summer sunshine caused the roof to slip and droop in only a few years. Eventually the roof was replaced with an alloy of 94 percent lead and 6 percent antimony, which has a higher elastic limit. There has been no further problem.

Exercises

20. The following solids are subjected to the compressive forces listed. Determine in each case whether the solid behaves elastically, whether it behaves plastically, or whether it breaks.

Solid	Cross-sectional area, cm^2	Compressive force
(a) Monel metal	1.25	7 200 kg_f
(b) Brass	0.511	660 kg_f
(c) Brass	1.20	660 kg_f
(d) White oak	24.0	14 000 kg_f
(e) Structural steel	24.0	14 000 kg_f
(f) Aged concrete	1200	170 t

Answers: (a) elastic, (b) plastic, (c) elastic, (d) breaks, (e) elastic, (f) breaks

21. The following solids are subjected to the tensile forces listed. Determine in each case whether the solid behaves elastically, whether it behaves plastically, or whether it breaks.

Solid	Cross-sectional area	Tensile force
(a) White pine	16.2 cm^2	4100 kg_f
(b) Copper	0.006 1 cm^2	5.8 kg_f
(c) Bronze	1.05 cm^2	5 600 kg_f
(d) Aluminum	1.2 in^2	25 000 lb
(e) Aluminum	0.62 in^2	25 000 lb
(f) Cast iron	56.3 in^2	52 tons

Answers: (a) elastic, (b) plastic, (c) breaks, (d) plastic, (e)breaks, (f) elastic

Application: Reinforced Concrete and Masonry

Concrete is strong in compression but weak in tension. When a concrete slab is bent even slightly, the outside of the bend goes into tension and begins to fail. Since this effectively reduces the thickness of the slab, the cracks progress through until the entire section crumbles.

In applications like tall buildings, dams, retaining walls, and roads, the concrete is subject to some bending. In these cases, reinforced concrete must be used. This is made by pouring the freshly mixed concrete over a grid of steel reinforcing rods or bars. Since the steel is strong in tension, and since it becomes one solid mass with the concrete, the resulting structure can withstand bending stresses.

Concrete is a mixture of dry Portland cement, fine aggregate (sand), coarse aggregate (gravel), and water. Proportions vary with the application, but a typical mix is 1 part cement to 2 parts sand, 3.5 parts gravel, and 2 parts water, by volume. The water combines chemically with the cement and becomes part of the final hardened mass. An excess of water weakens the concrete; however, it is often needed to make the mixture easy to pour. After pouring, the concrete should be troweled to work the excess water to the surface (the denser cement and aggregates go to the bottom). This also gives a smooth, finished surface. Concrete tends to get continuously stronger with age, but it is quite weak when just freshly hardened. The reinforcing rods help prevent failure under compression while the concrete is aging.

Brick and block walls are also reinforced with steel if there is any chance they might be subject to bending. It is much easier to reinforce such walls to begin with than to try to repair them later. The idea is the same as with reinforced concrete: Reinforcing trusses, bars, or corrugated strips are embedded in the mortar joints to make the final hardened mass strong in tension and bending.

22. A steel bridge cable has a diameter of 1.8 cm. (*a*) What is the largest weight this cable can support without deforming permanently? (*b*) What is the largest weight the cable can support before it breaks?
Answers: (*a*) 17 000 kg$_f$, (*b*) 38 000 kg$_f$

23. The head of a hammer has a diameter of 2.5 cm. With what force must the hammer strike a flat bar of aluminum to leave a circular impression of the entire head?
Answer: More than 4100 kg$_f$

5-6 Bending

So far, we have talked about how solids behave in tension and compression. Some of the most important applications, however, involve bending. A bridge bends when a car drives onto it; if it is well designed, it should bounce back when the car has crossed. You certainly have noticed how trees bend and sway in the wind. To a lesser extent, frame structures made from trees do the same thing in a storm. The frame of a truck bends slightly when it is loaded, and a fiber glass fishing rod or diving board bends considerably under load.

Figure 5-9 shows a timber beam subjected to a bending stress. As the beam sags, the top goes into compression and the bottom goes into tension. The dotted line represents the border between the portion being compressed and the portion being stretched. As we saw earlier, many solids are stronger in compression than they are in tension. These materials, such as masonry, aluminum, and cast iron, will fail from the convex or outside face of the bend. Structural steel, on the other hand, is just as strong in tension as in compression. This makes structural steel ideal for applications where there are large bending loads—bridges, automobile frames, and skyscrapers, for instance. Masonry is never used where there is the possibility of bending.

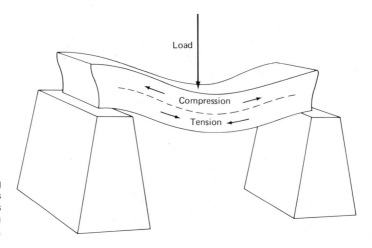

Fig. 5-9 A timber beam bending under a load. The top is compressed while the bottom is in tension. The actual deflection has been exaggerated.

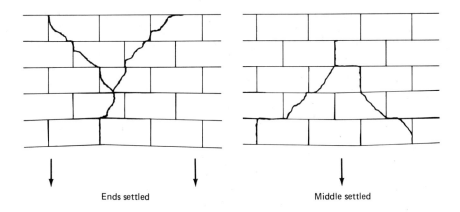

Ends settled   Middle settled

Fig. 5-10 Cracks will develop in a concrete-block wall when the footer settles unevenly and subjects the wall to bending stressed. Although concrete block will withstand up to 140 kg$_f$/cm² in compression, it will fail under a much smaller stress if it is bent.

Figure 5-10 shows what happens to a concrete-block wall when the footer settles unevenly. If the ends of the wall settle and the middle doesn't, cracks develop from the top. If the middle settles and the ends don't, cracks develop from the bottom. This is why backyard mechanics are cautioned not to jack up a car's axle with concrete blocks while working underneath. If the surface under the block is just slightly uneven, the block will fracture and the vehicle will fall. It is much safer to use wooden blocks or, better yet, specially made jack stands (Fig. 5-11).

We said that a tall building bends in a high wind. This wind loading is an important design factor. The medieval Gothic cathedrals were braced with elaborate buttresses to prevent any part of the masonry from going into tension under the action of the wind. Even so, some of these huge stone structures were known to collapse soon after being built. Figure 5-12 shows the stress developed by wind blowing against a solid, flat surface at different speeds. Notice that at a speed of 50 kmph [31 mph] the wind load is 4 times as great as at 25 kmph [15.5 mph].

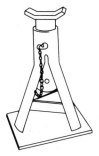

Fig. 5-11 Jack stands for safely supporting a car. A conventional jack can twist or slip, while a concrete block can fracture.

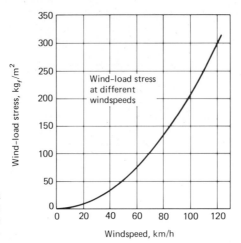

Fig. 5-12 Stress produced by a steady wind blowing perpendicular to a solid surface at different wind speeds.

Because of *wind loading*, skyscrapers must be built with structural-steel frames. Even so, the building bends somewhat in the wind. This bending can cause problems with windows, which have a lower elastic limit than structural steel. If the building bends too much, the seals around the windows may fail, or the windows themselves may shatter as they are forced to deform with the structure. This is exactly what happened in a well-publicized case in Boston, where windows kept shattering and popping out of the John Hancock building. The solution in that case was to mount the building on large shock absorbers, so the entire structure tipped slightly in the wind rather than bending.

Summary A force can never act at a geometrical point; instead it is always spread over some area. The ratio of the force to the area on which it acts is called the *stress*. If the stress acts to stretch a solid, we say the solid is in *tension*. If the stress acts to squash it, we say it is in *compression*. Stresses may also produce bending and/or twisting in solids.

Solids always deform under the action of stresses. This deformation, expressed as a fraction of the original dimension, is called the *strain*. The strain may be either *elastic* or *plastic*. The maximum stress that can be applied without causing plastic behavior is called the *elastic limit*. For some solids, the elastic limit in tension is different from that in compression.

Stresses are often produced by a solid's own weight. If we know its volume, a solid's weight can be calculated from its *density*.

The mechanical properties of solids are very important in the design and frabrication of tools, machines, buildings, roads, bridges, furniture, and so on.

Terms You Should Know

density	pascal
elasticity	strain
tension	ultimate strength
compression	plastic range
elastic limit	elastic range
stiffness	wind loading
stress	

Problems

1. A foundry is to cast 10 000 brass hinges, each having a volume of 28.3 cm³. Scrap brass is bought according to its mass. Allowing 10 percent extra for waste, how many kilograms of brass are needed?

2. A coil of aluminum sheet is 0.250 mm thick, 50.0 cm wide, and 30.0 m in uncoiled length. (*a*) What is the mass of this coil in kilograms? (*b*) What is its mass if production error causes it to be rolled to a thickness of 0.260 mm instead of 0.250 mm?

3. Gold plating is sometimes used on electric contacts because it does not corrode like copper and other electric conductors. The plating is typically 7.60 μm in thickness. Gold is priced by the troy ounce. How many troy ounces of gold are needed to plate both sides of 10 000 gold contacts, each with a face area of 20.0 mm²?

4. A certain spring stretches 16.7 cm under a tension of 12.2 kg$_f$. Its elastic limit is 28 kg$_f$ of tension. (*a*) What is the spring's stiffness? (*b*) What force is needed to stretch it 28.3 cm? (*c*) How far does it stretch under a tension of 8.05 kg$_f$?

5. A coil spring has a stiffness of 56.2 kg$_f$/cm. Its unstretched length is 19.80 cm, and it "bottoms out" at a length of 11.03 cm. What compressive force will cause it to bottom out?

6. A piece of bridge cable wire is 0.63 cm in diameter. What is the maximum load that it can support if its elastic limit is not to be exceeded?

7. A piano has a mass of 415 kg. It is to rest on a pine floor. How much total contact area should there be between the legs and the floor, if an impression is not to be left in the pine?

8. To allow a margin of safety, most building codes specify that brick in cement mortar should not be placed under a compressive stress greater than 325 psi. According to this code, what is the largest load that can be placed on the brick column shown?

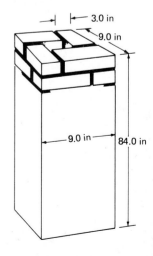

6
Fluids

Liquids and gases are two different states of matter. As we saw in Chap. 4, a liquid can change its shape very easily while keeping the same volume. A gas, on the other hand, expands to fill its container top to bottom. Liquids are always visible, even when they are colorless, but a colorless gas cannot be seen directly. And liquids are always denser than gases. It may seem, then, that liquids and gases have nothing in common.

But if we look at their behavior more closely, we find that there are a large number of similarities. Liquids and gases both tend to flow downhill, and from regions of high pressure to regions of low pressure. Both are transported through pipes and conduits, pressurized by pumps and compressors. They buoy up light objects, and they place a drag on objects moving through them. Liquids and gases both transfer heat through a process known as convection, and both can be used to power turbines or other engines. In fact, under certain conditions in high-pressure boilers, it is difficult to say whether the water is actually liquid or gas (steam) because the transition is not abrupt. For these reasons, it is convenient to group liquids and gases together as fluids.

fluid   Definition: A *fluid* is a substance that flows.

In this chapter, we discuss some of the properties of fluids in general, as well as some of the differences in the behavior of liquids and gases.

It is easy to apply a force to a solid object; we simply grab onto it and push or pull. Applying a force to a fluid is a bit more difficult. If we step on a puddle of water, as in Fig. 6-1, the water simply splashes out of the way and our foot goes through to the bottom.

Fig. 6-1 A force can be applied at a small spot on a solid, but not on a fluid.

6-1 Pressure

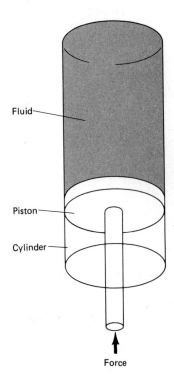

Fig. 6-2 Using a movable piston in a cylinder to apply a force to a fluid.

To apply a force to a fluid, then, we have to confine it or somehow prevent it from flowing very rapidly. The most common way of doing this is with a piston and cylinder (Fig. 6-2). This piston fits tightly enough that the fluid cannot flow out around it. Now the force is spread over the entire cross-sectional area of the piston, and the fluid is being compressed. We can describe this quantitatively by giving the value of the fluid pressure.

Definition: *Fluid pressure* is the ratio of the applied force to the cross-sectional area on which it acts:

fluid pressure

$$\text{Pressure} = \frac{\text{force}}{\text{area}} \qquad (6\text{-}1)$$

If we apply a pressure to a confined fluid, the entire fluid is compressed. We can say that the fluid "transmits" the pressure to all its parts and to the container that holds it as well.

Notice that fluid pressure is defined the same way that we defined the *stress* on a solid. The only difference is that a solid may go into tension, but a fluid cannot (at least not to any great extent). The pressure on a fluid must always be a compression. The units of pressure, however, are the same as the units of stress: pascals (Pa), kilogram-force per

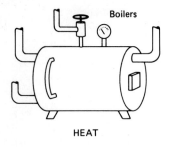

Boilers

HEAT

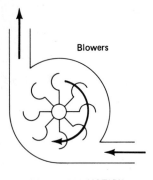

Blowers

MECHANICAL MOTION

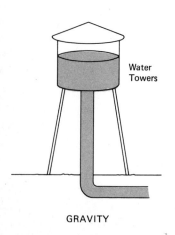

Water
Towers

GRAVITY

Fig. 6-3 Some ways of
generating a fluid pressure.

square centimeter, pounds per square inch, and so on. Conversion factors between these units have been listed in Table 5-2.

Example 6-1 Calculating Fluid Pressure in a Cylinder.

A certain cylinder is 5.62 cm in diameter, is filled with oil, and has a tightly fitting piston. What is the oil pressure when a force of 214 kg_f is applied to the piston?

Since the cylinder is probably circular in cross section, its area is

$$Area = \pi(radius)^2$$

$$= \pi(2.81 \text{ cm})^2$$

$$= 24.8 \text{ cm}^2$$

Using Eq. (5-1), the pressure is

$$Pressure = \frac{214 \text{ kg}_f}{24.8 \text{ cm}^2}$$

$$= 8.63 \text{ kg}_f/\text{cm}^2$$

As we will see later, this result is the *gauge pressure.* Using Table 5-2, we may also express it as 123 psi, 846 000 Pa, or 846 kPa. ◀

A piston in a cylinder is not the only way of generating a fluid pressure. Some others are shown in Fig. 6-3. Various mechanical devices like blowers and compressors are often used. A fluid pressure can also be generated by the force of gravity, as with water impounded behind a dam or stored in a tank. Another way is to use heat, as with a boiler, a gas refrigerator, or a chemical explosion.

Regardless of how the pressure is generated, a fluid flows from where the pressure is high to where the pressure is low. This is why air leaks out of tires, but it never leaks *in.* It is also why exhaust gases flow out of a car's tail pipe, why water flows out of spigots, and why clouds travel toward low-pressure regions.

A pressurized fluid exerts a force perpendicular to any surface in contact with it, the *fluid force.* The amount of force can be found by rewriting Eq. (6-1):

$$Fluid \text{ force} = (pressure)(area) \qquad (6-2)$$

Example 6-2 illustrates the use of this equation.

Example 6-2 Force on a Connecting Rod.

The piston of a certain engine has a cross-sectional area of 178 cm^2. At the beginning of the power stroke, the pressure in the cylinder reaches a peak of 22.4 kg$_f$/cm². What force does the piston transmit to the connecting rod at this point?

The force of the hot gases expanding against the piston is found from Eq. (6-2):

$$\text{Fluid force} = (\text{pressure})(\text{area})$$

$$= 22.4 \ \frac{\text{kg}_f}{\text{cm}^2} \ (178 \ \text{cm}^2)$$

$$= 3990 \ \text{kg}_f$$

Since the piston is near top dead center, this entire force is transmitted to the connecting rod and then to the crankshaft. Notice that in USCS units this amounts to nearly 8800 lb! The entire piston and crankshaft assembly must be capable of withstanding this force without appreciable deformation. ◄

Exercises

1. Find the pressure in kilogram-force per square centimeter when the following forces act on a fluid through pistons of the given areas.
   (a) Force = 6.22 kg$_f$, area = 0.089 2 cm²
   (b) Force = 864 kg$_f$, area = 12.4 cm²
   (c) Force = 4460 N, area = 0.167 m²
   (d) Force = 96.2 lb, area = 2.07 in²
   Answers: (a) 69.7 kg$_f$/cm², (b) 69.7 kg$_f$/cm², (c) 0.272 kg$_f$/cm², (d) 3.27 kg$_f$/cm²

2. Find the pressure in pounds per square inch when the following forces act on a fluid through pistons of the given areas.
   (a) Force = 3260 lb, area = 38.1 in²
   (b) Force = 185 lb, area = 2.16 in²
   (c) Force = 312 kg$_f$, area = 78.7 cm²
   (d) Force = 78.3 N, area = 0.023 1 m²
   Answers: (a) 85.6 psi, (b) 85.6 psi, (c) 56.4 psi, (d) 0.492 psi

3. Express the results in Exercise 2 in kilopascals (kPa).
   Answers: (a) 590 kPa, (b) 591 kPa, (c) 389 kPa, (d) 3.39 kPa

4. Find the force in kg$_f$ when a fluid under the given pressures acts on pistons of the given areas.
   (a) Pressure = 16.7 kg$_f$/cm², area = 39.3 cm²
   (b) Pressure = 198 psi, area = 48.2 in²
   (c) Pressure = 562 kPa, area = 122 cm²
   Answers: (a) 656 kg$_f$, (b) 4330 kg$_f$, (c) 699 kg$_f$

5. Express the results in Exercise 4 in pounds.
   Answers: (a) 1450 lb, (b) 9540 lb, (c) 1540 lb

6. Express the results in Exercise 4 in newtons.
   Answers: (a) 6440 N, (b) 42 500 N, (c) 6860 N

6-2 Atmospheric Pressure

The surface of the earth is at the bottom of an ocean of air. Since air is a gas, the atmosphere does not occupy a well-defined volume. In other words, there is no upper surface to the ocean of air; it simply trails out into space, getting "thinner" at greater distances from the earth. Even so, about 90 percent of the atmosphere is below an altitude of 16 km, which is only about 0.1 percent of the earth's diameter. The earth's gravity tends to hold its atmosphere fairly tightly.

Saying that gravity holds onto the atmosphere is the same as saying that air has weight. The combined effect of this atmospheric weight is to produce an *atmospheric pressure* on any surface in contact with it. At sea level, the average or "standard" value of this atmospheric pressure is 1.033 227 kg_f/cm^2 [14. 695 95 psi, or 101.325 kPa]. At higher altitudes, the atmospheric pressure tends to be lower. Figure 6-4 shows how standard atmospheric pressure decreases with altitude. Variations of up to ±5 percent of these standard values have been recorded with changing weather conditions.

We don't ordinarily notice the effects of atmospheric pressure acting on our bodies. Why? Because our species has evolved in the constant presence of the atmosphere. But lower the atmospheric pressure—by climbing a mountain or flying at high altitude—and we quickly notice the difference. We get out of breath and tire easily, and at extremely low pressures the small blood vessels in the eyes, ears, and nose begin to hemorrhage. This is why the cabins of high-altitude aircraft must be pressurized.

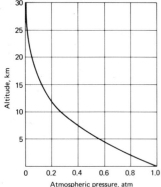

Fig. 6-4 Variation of atmospheric pressure with altitude.

Hold your hand palm-up in front of you. The upper surface area is about 160 cm² [25 in²]. The force of the atmosphere pushing down on this surface area is (160 cm²) (1.0 kg_f/cm^2), which works out to 160 kg_f [350 lb]! This is certainly a tremendous force. Why, then, doesn't atmospheric pressure flatten your hand (and the rest of your body) against the earth? The answer is that your hand has a bottom surface as well as a top. And the atmosphere pushes against *all* surfaces that come in contact with it. So while the top of your hand is being pushed down with a 160-kg_f force, the underside is being pushed *up* with the same force. The two forces balance each other, just as with two arm wrestlers whose muscles strain while neither one moves.

Can we ever get the atmosphere to push only one way, say, down but not up? Yes. This is the idea behind the so-called suction cup and suction clamp shown in Fig. 6-5. By removing the air under the rubber cup, we eliminate the atmospheric pressure on the underside. Atmospheric pressure on the other side then holds the cup firmly against the surface it's been placed on. If the cup has a surface area of 3.5 cm², for instance, the 1.0-kg_f/cm^2 atmospheric pressure generates a holding force of 3.5 kg_f [7.7 lb]. This is the force needed to pull off the cup. Of course, the surface needs to be smooth to prevent air from leaking back

Fig. 6-5 Suction cups and suction lamps depend on atmospheric pressure to hold them in place.

under the cup. It helps to moisten the cup first, preferably with soapy water or glycerine.

The same principle is used when the gauge blocks in Fig. 1-1 are "wrung" together. The surfaces of the blocks are so smooth and flat that this "wringing" forces air out from between them. Atmospheric pressure then helps hold the blocks together (surface cohesion also helps). A number of gauge blocks may be combined in this way to build up any required length standard.

Although we often hear the word "suction," in physical terms there is no such thing. What holds a suction cup is not suction at all, but atmospheric pressure. The action of atmospheric pressure can be seen in the demonstration in Fig. 6-6. A small amount of water is placed in an open 1-gal oil can. The water is heated to boiling, so the can fills up with steam and the air is driven out. The can is then removed from the flame; and when the water stops boiling, the cap is screwed on tightly. At this point, the pressure of the steam inside balances the atmospheric pressure outside. But as the steam begins to condense, the inside pressure drops and the atmosphere slowly crushes the can. The process can be sped up by splashing cold water on the outside of the can. In the early 1700s an engine was successfully built to operate on this principle. It was appropriately named the "atmospheric engine."

There are many other ways to see the effect of atmospheric pressure. Figure 6-7 shows a simple demonstration. Fill a jar or bottle to its brim with water. Lay a piece of cardboard over the opening, and hold it in place while you turn the bottle upside down. Now let go of the cardboard. The water will stay up in the bottle. To flow out, it must push the cardboard against the atmospheric pressure, and it simply isn't

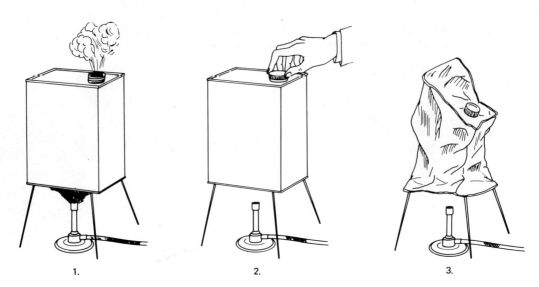

1.   2.   3.

Fig. 6-6 Crushing a can with atmospheric pressure.

heavy enough to do this. The same thing happens if we try to pour oil from a can that has a single hole punched in it. The oil drips out very slowly, and the flow stops completely once in awhile as an air bubble flows inside to equalize the pressure. To speed things up, we can punch additional holes in the can. Putting one in the bottom, as in Fig. 6-8, works best.

6-3 Vacuums A space with absolutely nothing in it is said to be a *vacuum.* In practice, a perfect vacuum cannot be achieved; even in the reaches of outer space there are still a dozen or so gas molecules in each cubic centimeter. But it is easy to achieve a *partial vacuum,* which is a gas with a pressure less than atmospheric pressure. This can be done by condensing steam, by withdrawing a piston from a closed cylinder, by lighting a fire in an enclosed sample of air (thus using up the oxygen), or by a number of other techniques. Once we have a partial vacuum in a container, air tries to get *in* rather than out. This principle is used in the conventional piston engine during its intake stroke. As the piston is withdrawn from the cylinder, a partial vacuum is created and outside atmospheric pressure forces air through the carburetor and intake manifold and into the cylinder. The principle is also used in the containment vessels at nuclear power plants. Pumps keep the entire inside of the structure at a slight vacuum so that the atmosphere can leak in but any radioactive gases generated inside can't leak out. Furnaces and boilers also maintain the firebox at a pressure below atmospheric, so flames won't leap out when the firebox door is opened.

Later, we see other applications that depend on the partial vacuum. For now, let's remember that a vacuum cannot "suck." Instead, it is the atmosphere that pushes.

Fig. 6-7 Atmospheric pressure will hold the water in an inverted bottle if there is a rigid surface for the pressure to act on. Will this work if the bottle contains some air as well as the water?

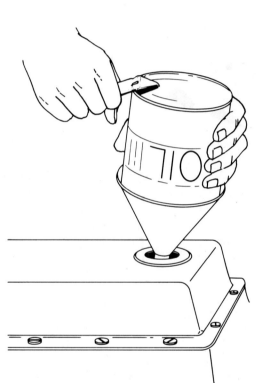

Fig. 6-8 Atmospheric pressure can prevent a liquid from flowing under its own weight. Punching a hole in the can's bottom equalizes the atmospheric pressure on the top and bottom surfaces of the oil.

6-4 Gauge and Absolute Pressure

We have seen that it is possible for gases and even liquids to be at pressures below atmospheric. In most cases, however, it is atmospheric pressure that controls how a fluid moves. If a tire is punctured, air leaks out until the inside pressure drops to atmospheric—then the leaking stops. If the valve on a CO_2 tank is opened, carbon dioxide flows until the tank's pressure drops to atmospheric—then the flow stops. To move the piston in a single-acting hydraulic cylinder (on a hydraulic lift, for instance), the oil pressure must be greater than the atmospheric pressure on the outside of the piston. Because the atmosphere is all around us, a great many processes depend on the *difference* between a fluid's pressure and the atmospheric pressure.

Application: The Lift Pump

This is a very old invention, but one that still sees a fair amount of use. The version in the drawing is hand-operated. They key to the device is a pair of check valves, which allow water to pass in the upward direction but do not permit a reverse flow. When the piston is pushed down, valve 1 is forced open while the weight of the water in the cylinder keeps valve 2 shut. This action puts the piston beneath the water in the cylinder. Now the piston is drawn upward. Check valve 1 immediately closes and allows the water to be lifted above the spout. At the same time, valve 2 is forced open by the atmospheric pressure on the water in the well, and water flows upward to fill the cylinder beneath the piston. The process is repeated to pump the water continuously.

There is a limit to how far water can be lifted by this device. Since atmospheric pressure is 1.03 kg_f/cm^2, the atmosphere cannot support a column of water whose back pressure is greater than this value. The back pressure of a 1-m column of water is easily calculated to be 0.10 kg_f/cm^2. Thus the maximum height of the water column is 1.03/0.10 m, or 10.3 m [33.8 ft]. There will be a slight variation in this figure, as a result of changes of temperature and barometric pressure. Also a slight additional pressure is needed to operate the lower check valve, so in practice it is better to count on no more than about 9 m [30 ft] of lift. Since this limit is based on atmospheric pressure, it makes no difference whether the pump is hand-operated or electric-powered, or whether it uses a piston or vanes or rotors. To lift the water farther (without using multiple stages), it is necessary to use force pumps or jet pumps.

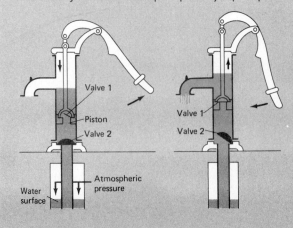

This has led to two different ways of expressing pressures. One way is to start with zero pressure representing a perfect vacuum. Then atmospheric pressure is 1.033 2 kg$_f$/cm², or slightly more or less depending on the altitude and the weather. This system gives us absolute pressure.

Definition: *Absolute pressure* is a pressure scale that uses zero to represent a perfect vacuum.   *absolute pressure*

The other way is to use zero to represent atmospheric pressure. Then a vacuum becomes a negative pressure of up to −1.033 2 kg$_f$/cm², and positive pressures are those greater than atmospheric. This system is called gauge pressure.

Definition: *Gauge pressure* is a pressure scale that uses zero to represent atmospheric pressure.   *gauge pressure*

When we follow the recommended inflation pressures for our car tires, we are using gauge pressure. Zero gauge pressure means no inflation at all, which means that the air inside the tire is at the same pressure as the air outside. Table 6-1 compares the gauge and absolute pressures in some common applications.

TABLE 6-1   *PRESSURES DEVELOPED IN SOME COMMON APPLICATIONS**

	Gauge pressure, kg$_f$/cm²	Absolute pressure, kg$_f$/cm²
Oxygen cylinder	141	142
Diesel injector	25	26
Gas-engine compression test	9.8	10.8
Car tire	2.0	3.0
Pressure cooker	1.0	2.0
Standard atmosphere	0	1.0
Manifold vacuum at idle speed	−0.6	0.4
Perfect vacuum	−1.0	0

*Notice that the distinction between gauge and absolute pressure is very important when the pressure is low.

Note: $1 \dfrac{kg_f}{cm^2} = 98.1 \text{ kPa} = 14.22 \dfrac{lb}{in^2}$

Notice that the term "gauge pressure" does not simply mean pressure that is read from a gauge. In fact, either gauge or absolute pressure may be measured with a suitable pressure gauge. If the instrument reads zero when it is open to the atmosphere, it is indicating gauge pressure. If it reads something near 1 kg$_f$/cm² [15 lb/in²] when open to the atmosphere, then it is indicating absolute pressure. In the old USCS units, where pressure is expressed in pounds per square inch, it is common to see the pressure unit abbreviated as "psi." The abbreviation "psig" then means pounds per square inch gauge, while "psia" means pounds per square inch absolute. There are no corresponding abbreviations for the metric units.

With very high pressures, as in compressed-gas cylinders, the difference between gauge and absolute pressure is of little importance. For pressures closer to atmospheric, however, the difference is very important. A petroleum still designed to operate at an absolute pressure of 0.75 kg$_f$/cm² will not function at all at a gauge pressure of 0.75 kg$_f$/cm².

Example 6-3 The Pressure Relief Valve.

In Fig. 6-9 we see a simple gravity-operated *pressure relief valve*. The valve is attached to a boiler to prevent the steam pressure from rising to a dangerous level. Let's suppose that the valve is to open when the steam pressure exceeds 5.5 kg$_f$/cm², gauge. If the pipe has an inside diameter of 2.6 cm, what should the cap weigh?

We can use Eq. (6-2) to find the upward force against the inside of the cap:

$$\text{Fluid force} = (\text{pressure})(\text{area})$$

Calculating the cross-sectional area of the pipe gives 5.3 cm². Then

$$\text{Fluid force} = 5.5 \ \frac{\text{kg}_f}{\text{cm}^2} \ (5.3 \ \text{cm}^2)$$

$$= 29 \ \text{kg}_f$$

This upward force has to open the valve. The cap must therefore weigh about 29 kg$_f$.

What about the atmospheric pressure on the top of the cap? It doesn't figure in. Certainly this 1-kg$_f$/cm² pressure is there, helping hold down the cap. But then the absolute steam pressure is higher than the gauge pressure by this same 1 kg$_f$/cm², and this is helping push *up* the cap. The atmospheric 1 kg$_f$/cm² balances out on both sides of the cap, just as it does on our outstretched hand. We don't need to worry about it

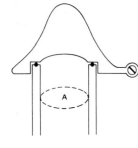

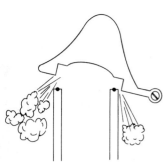

Fig. 6-9 The gravity-operated pressure relief valve.

at all. This is the beauty of using gauge rather than absolute pressure when one side of the system is exposed to the atmosphere. ◀

Notice that the cap in Example 6-3 was rather heavy—some 64 lb, in old USCS units. This makes the gravity-operated pressure relief rather cumbersome. For the low pressures used in the early days of steam technology, the weight needed was much smaller. Today, the gravity-operated relief valve has been replaced in most applications by the spring-operated valve. This is the familiar device found on home hot-water heaters. The device is shown in Fig. 6-10. The principle is the same as for the gravity-operated valve, except that the counterforce is supplied by the compression of a spring rather than the weight of a cap.

Exercises

7. How much force will atmospheric pressure generate on a surface 2.5 m² in area?
   Answer: 26 000 kg$_f$
8. A jar has a lid 5.5 cm in diameter. The jar has been vacuum-sealed to preserve its contents. What is the force of the atmosphere pressing in on the lid?
   Answer: 98 kg$_f$
9. The draft in a certain chimney flue is maintained at a gauge pressure of −0.032 kg$_f$/cm². During a rainstorm, the actual atmospheric pressure drops to 0.981 kg$_f$/cm². What is the absolute pressure in the flue?
   Answer: 0.949 kg$_f$/cm², absolute
10. The sketch shows a simple steam-driven piston that might be found in a simple steam engine or a steam hammer. The piston has a diameter of 6.31 cm, and steam is introduced in the left-hand chamber at a gauge pressure of 27.3 kg$_f$/cm². The right-hand chamber is open to the atmosphere. What is the effective force pushing the piston to the right?
    Answer: 3410 kg$_f$

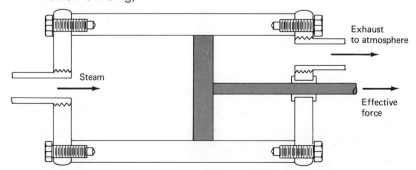

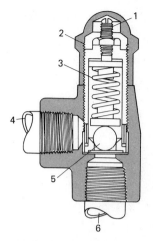

1. Pressure adjusting screw
2. Adjusting screwcap
3. Compression spring
4. Return port
5. Ball
6. Pressure port

Fig. 6-10 The spring-operated pressure relief valve. The device is easily adjustable and can be mounted in any position.

6-5 Fluid Density

In Chap. 5, we defined a solid's density as the ratio of its mass to its volume. Since fluids also have mass and take up space, we can express fluid density the same way:

$$\text{Density} = \frac{\text{mass}}{\text{volume}} \tag{5-1}$$

For instance, 1 cm³ of pure water at room temperature has a mass of 0.998 23 g. The density of this water is then 0.998 23 g/cm³. Table 6-2 lists the densities of some other common liquids.

Liquids usually expand slightly when they are heated, and this makes their density decrease. Under pressure, liquids compress slightly and their density increases. These density changes are very small, however, which means that the values in Table 6-2 are accurate under a wide range of conditions.

Gases are another story, as we see in Fig. 6-11. A pressure of 210 kg_f/cm^2, gauge [204 atmospheres (atm)], is needed to compress water by 1 percent of its volume. But the same pressure compresses air into 1/200 of its original volume! Thus the density of gases depends a great deal on the pressure. Table 6-3 lists the densities of some common gases at atmospheric pressure and room temperature (20°C, or 68°F). We should keep in mind that these gas densities are quite different at other temperatures and pressures.

TABLE 6-2   DENSITIES OF SOME COMMON LIQUIDS*

Liquid	g/cm³	lb$_m$/ft³
Alcohol, denatured	0.792	49.4
Carbon tetrachloride	1.63	102
Ethylene glycol	1.109	69.11
Fuel oil	0.93	58
Gasoline	0.66–0.69	41–43
Kerosene	0.82	51.2
Mercury, 0°C	13.596	848.8
Mercury, 20°C	13.546	845.6
Sulfuric acid, concentrated	1.834	114.5
Water, pure, 4.00°C	1.0000	62.428
Water, pure, 20.0°C	0.99823	62.317
Water, sea	1.025	63.99

*The temperature is 20°C unless stated otherwise.
Note: 1 g/cm³ = 1 kg/L = 1000 kg/m³

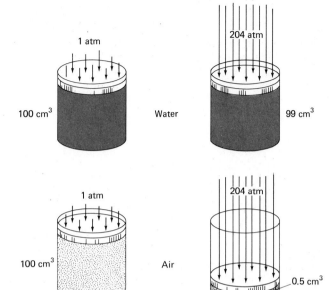

Fig. 6-11 Compressibility of fluids: gases are easily compressed, while liquids are not.

TABLE 6-3   DENSITIES OF SOME COMMON GASES AT 20°C [68°F] AND
1 ATM PRESSURE

Gas	g/cm³	lb_m/ft³
Acetylene	0.001093	0.06823
Air, dry	0.001205	0.07523
Ammonia	0.0007184	0.04485
Argon	0.001662	0.1038
Carbon dioxide	0.001842	0.11499
Carbon monoxide	0.001165	0.07274
Chlorine	0.002995	0.18692
Helium	0.0001663	0.01038
Hydrogen	0.0000837	0.005228
Methane	0.0006679	0.04170
Neon	0.0008389	0.05238
Nitrogen	0.001165	0.07274
Oxygen	0.001332	0.08312
Propane	0.001873	0.1168
Steam (110°C)	0.000533	0.0332
Sulfur dioxide	0.002727	0.1702

Note: 1 g/cm³ = 1 kg/L = 1000 kg/m³

The tremendous elasticity of gases leads many people to think that air and other gases have very little mass. In fact, this is not true. The simple demonstration in Fig. 6-12 shows that the air in an inflated football has a measurable mass.

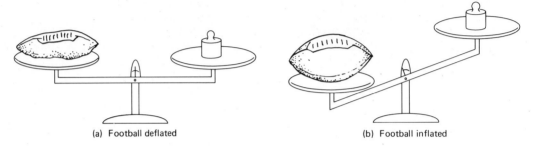

(a) Football deflated   (b) Football inflated

Fig. 6-12 Demonstrating that air has mass.

In the standard U.S. oxygen cylinders used for welding, 220 ft³ of oxygen has been compressed into just 1.3 ft³ of cylinder volume. The filled cylinders hold the gas at a pressure of 2500 psig (at 70°F). The actual mass of the oxygen can be found from Eq. (5-2):

$$\text{Mass} = (\text{density})(\text{volume})$$

Using the normal density of oxygen from Table 6-3, and the normal volume of this oxygen, we have

$$\text{Mass} = 0.083\ 12\ \frac{\text{lb}_m}{\text{ft}^3}\ (220\ \text{ft}^3)$$

or about 18 lb$_m$. In other words, the filled oxygen cylinder weighs about 18 lb more than an empty one.

Notice that it is quite incorrect to say that there is 2500 "pounds" of oxygen in the cylinder. In fact, the weight of the oxygen is around 18 lb while its pressure is 2500 *pounds per square inch* (psi, or lb/in³) gauge.

We used old USCS units in this discussion to point out the common error of confusing pressure with weight and mass. The same comments hold for the metric kilogram-force per square centimeter: The number of kilograms of gas in a cylinder is quite different from the gas pressure in kilogram-force per square centimeter. Of course, this kind of mistake cannot be made when the kilopascal is used as the pressure unit.

Exercises

11. A large tank truck holds 33 000 L of gasoline [8700 gal]. What is the gasoline's mass? There is 1000 cm³ in 1 L.
     Answer: 22 000 to 23 000 kg
12. A rectangular swimming pool measures 18.6 ft by 32.1 ft, with an average depth of 5.0 ft. (*a*) Find the mass of the water it holds in pound-mass. (*b*) Find the weight of this water in pounds.
     Answer: (*a*) 186 000 lb$_m$, (*b*) 186 000 lb
13. The cylinder of an engine holds 330 cm³ of air when the piston is at bottom dead center. What is this air's mass?
     Answer: 0.40 g
14. A kerosene tank holds $100\overline{0}$ L and weighs 122 kg$_f$ when empty. How much weight must its legs support when it is full?
     Answer: 942 kg$_f$
15. Express the normal density of air in kilograms per cubic meter.
     Answer: 1.205 kg/m³

6-6 Weight Density

As we saw in Chap. 4, an object's weight depends on its mass and on the earth's gravitational attraction. But if we stay close to the earth's surface, gravity doesn't change very much, and we can say that a 1-kg mass weighs about 1 kg$_f$, or 9.81 N. Similarly, at points close to the earth's surface, 1 lb$_m$ weighs about 1 lb.

We have seen that a very important physical property of any substance is its density, defined as its mass per unit volume. By referring to the density tables, we were able to calculate the mass of any material, solid or fluid, provided that we knew its volume.

Often we are interested in a substance's weight rather than its mass. Accordingly, we can define the *weight density* to be the weight per unit volume.

Definition: A substance's *weight density* is the ratio of the substance's weight to the volume that this weight occupies:

weight density

$$\text{Weight density} = \frac{\text{weight}}{\text{volume}} \qquad (6\text{-}3)$$

Since an object's weight depends on where it is, its weight density also varies from place to place. But again, as long as we stay near the earth's surface, this variation is very small.

Weight density has units of weight (force) divided by a unit of volume: newtons per cubic centimeter, newtons per liter, kilogram-force

Application: The Atmospheric Engine

Four hundred years ago, laws were written in England against burning wood as a fuel. This had nothing to do with more modern concerns about pollution. Since England is on an island, its defense depended on a strong navy. And in those days, ships were built of wood.

Naturally enough, this was a tremendous boon to the coal mining industry. But as years passed and the easy-to-get coal was stripped away, the open-pit mines had to be dug deeper and deeper. Water seepage became a big problem. Every mine had to keep teams of horses and oxen on hand to drive the pumps. And the animals had to be fed during dry periods when the water table fell and the pumps weren't needed.

This was the situation in 1705 when Thomas Newcomen invented the atmospheric engine. Newcomen was a blacksmith, and in working with heat and water (for quenching and tempering) he was familiar with the phenomenon shown in Fig. 6-6. Why not use the atmosphere to drive the mine pumps? He built a huge iron cylinder nearly 3 m [10 ft] high and 1 m [3 ft] in diameter. In this cylinder he fit a movable piston for the atmosphere to act on. Steam was introduced in the cylinder from a low-pressure boiler (at about 0.2 kg_f/cm^2, gauge), and with the help of a counterweight the piston rose to the top of the cylinder. At this point, the steam valve was manually closed and cold water was splashed on and into the hot cylinder to condense the steam. Atmospheric pressure then pushed the piston into the partial vacuum in the cylinder. With the help of a "walking beam," this arrangement could operate a mine's lift pumps. The engine could be fueled with the same coal the mine was producing, and it didn't have to be "fed" when the pumps weren't needed. No wonder Newcomen found it hard to keep pace with the orders.

One essential feature of the engine was the "valve tender," a young boy whose job it was to open and close the steam and water valves at the right times. This wasn't as hard as it might sound, since the engine could make only about eight cycles each minute. Even so, it must have been tedious: open, close, open, close, hour after weary hour for a 14-h working day. One young valve tender named Humphrey Potter didn't care for this. One day he rigged up some ropes and pulleys between the

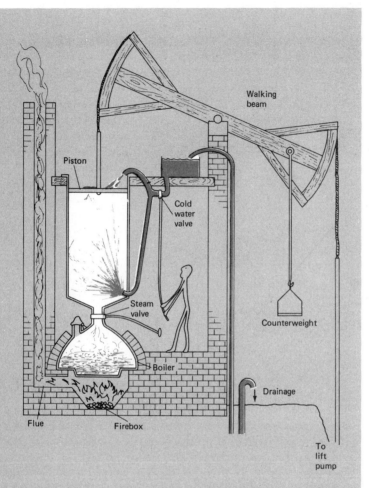

Walking beam

Piston

Cold water valve

Steam valve

Counterweight

Boiler

Drainage

Flue   Firebox

To lift pump

walking beam and the valve handles, then lay back in the grass to watch the engine do its own thing. From the mine pit, the owner noticed the engine pumping faster than he'd ever seen. Running up the hill to congratulate his valve tender, he found the boy loafing. Humphrey Potter, inventor of the automatic valve gear, was also the first person in history to automate himself out of a job.

As large as the atmospheric engine was, it could develop only around 20 horsepower (hp), and it needed rather large quantities of fuel to do this. Some 70 years after its invention, a Scot named James Watt made vast improvements on the device and transformed it into the steam engine.

per cubic meter, pounds per cubic foot, and so on. The official SI unit is newtons per cubic meter. But by using the gram-force per cubic centimeter, we make things easy on ourselves. Notice that a 1-g mass weighs about 1 g_f. The density of kerosene, from Table 6-2, is 0.82 g/cm³. This means that the weight density of kerosene is about 0.82 g_f/cm³. Similarly, the weight density of seawater is about 1.025 g_f/cm³.

When we use the USCS units, density is normally given in pound-mass per cubic foot. Then the weight density in pounds per cubic foot has the same numerical value. Referring to Table 6-2, we see that the weight density of pure water is about 62.3 lb/ft³, and the weight density of gasoline is between 41 and 43 lb/ft³.

6-7 Variation of Pressure with Depth

We already mentioned that a fluid's pressure may be due to gravity. Let's examine this more closely. At the top surface of a lake, the gauge pressure is zero. The deeper we go, the greater the weight of the water above us, and the greater the pressure. We can find the gauge pressure at any depth by using the following formula:

$$\text{Gauge pressure due to gravity} = (\text{depth})(\text{weight density}) \qquad (6\text{-}4)$$

Example 6-4 shows the use of this formula.

Example 6-4 Pressure behind a Dam.

The water behind the Grand Coulee Dam on the Columbia River has a depth of $55\bar{0}$ ft. Let's calculate the pressure exerted by this water on the deepest part of the upstream side of the dam.

Equation (6-4) requires the weight density of the water. From Table 6-2, we see that the water's density at 20.0°C is 62.317 lb_m/ft³. Therefore its weight density is about 62.3 lb/ft³. Putting this value and the water's depth into the equation gives us

$$\text{Gauge pressure} = (\text{depth})(\text{weight density})$$

$$= 550 \text{ ft} \left(62.3 \, \frac{\text{lb}}{\text{ft}^3} \right)$$

$$= 34\ 300 \text{ lb/ft}^2$$

Since there is 144 in² in 1 ft², this result may also be expressed as 238 psi, gauge. This is approximately the same as the pressure developed in the cylinder of a car engine during ignition.

Notice that this pressure of 238 psig occurs only near the bottom of the dam. Halfway up the pressure is only half as great, and near the top the gauge pressure is practically zero. ◄

One very important thing to note about Eq. (6-4) is that the shape of the fluid's reservoir or container is not mentioned. This is because the shape doesn't matter. Figure 6-13 shows an arrangement of glass tubing known as Pascal's vases. The tubing must be more than a few millimeters in diameter, but other than that it can include any complicated shapes we might imagine. When a liquid is poured into one vase, it flows through the connecting sections at the bottom and rises to the same level in all the vases. This means that the pressure at the bottom of each vase is the same. Why? The answer is that if it weren't, the liquid would flow from the high-pressure vases to the low-pressure vases and make the levels unequal.

Fig. 6-13 Pascal's vases. Liquids seek their own level regardless of the shape of the conduit. This means that the pressure exerted by a column of the liquid at rest depends on the column's vertical height rather than its shape.

So the pressure under a liquid depends only on the depth rather than on the amount of liquid. Figure 6-14 shows two dams of equal height. One has a very long lake backed up behind it, while the other, because of the terrain, forms only a short lake. Which dam has to be stronger? Both lakes are of equal depth. The pressure against the bottom of each dam is therefore the same, and the dams have to be of equal strength.

Example 6-5 The Construction Caisson.

Figure 6-15 is a diagram of a construction *caisson,* used for building bridge piers underwater. The caisson is first lowered into the water until its bottom sinks into the mud. The water is then pumped out of the inside, and the top is sealed. Workers may now climb inside and use power

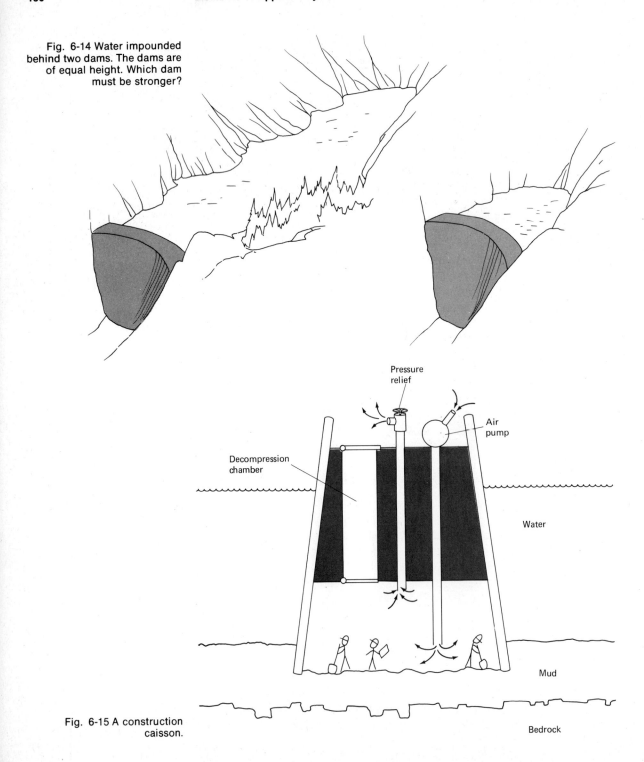

Fig. 6-14 Water impounded behind two dams. The dams are of equal height. Which dam must be stronger?

Pressure relief

Air pump

Decompression chamber

Water

Mud

Bedrock

Fig. 6-15 A construction caisson.

shovels to dig out the mud. This allows the caisson to settle down to the bedrock. Now the workers leave and the caisson is filled with concrete. When the concrete has cured, the caisson is removed, leaving the finished pier resting on bedrock.

The problem is that water seeks it own level and therefore tries to seep through the mud and into the caisson while the workers are digging. To prevent this, the air in the caisson must be pressurized. Let's say that the water is 5.5 m [18 ft] deep. What air pressure is needed to keep the water out?

Since water flows toward any region at a lower pressure, the inside air pressure must be at least as great as the pressure under 5.5 m of water. Using Eq. (6-4) and the fact that 5.5 m = 550 cm, we have

$$\text{Gauge pressure} = (\text{depth})(\text{weight density})$$

$$= 550 \text{ cm} \left(1.0 \, \frac{g_f}{cm^3} \right)$$

$$= 550 \text{ } g_f/cm^2$$

Since there is 1000 g_f in 1 kg_f, we can also write

$$\text{Gauge pressure} = 550 \, \frac{g_r}{cm^2} \left(\frac{1 \text{ } kg_f}{1000 \text{ } g_f} \right)$$

$$= 0.55 \text{ } kg_f/cm^2$$

Of course, the caisson can't simply be pumped up to this pressure and then left alone. This would soon asphyxiate the workers. Instead a separate exhaust is equipped with a pressure relief valve adjusted to a pressure slightly higher than 0.55 kg_f/cm^2, and the compressor pump runs continuously. This gives a continuous change of fresh air while maintaining the required pressure.   ◀

Exercises

16. Find the gauge pressure, in kilogram-force per square centimeter, under (a) 267 m of seawater, (b) 5.23 m of kerosene, (c) 12.5 cm of mercury.
    Answers: (a) 27.4 kg_f/cm^2, gauge, (b) 0.43 kg_f/cm^2, gauge, (c) 0.169 kg_f/cm^2, gauge

17. Find the gauge pressure, in pounds per square inch, at a depth of 35.6 ft in fresh water.
    Answer: 15.4 psig

18. Find the gauge pressure, in kilopascals, at the bottom of a tank of denatured alcohol 2.73 m in height.
Answer: 2120 kPa

6-8 Other Units of Pressure

So far, our units of pressure have been the same as the units of stress listed in Table 5-2. There are, however, some other pressure units we should be familiar with. These are based on the *mercury barometer,* which is basically an instrument for measuring atmospheric pressure.

The device is shown in Fig. 6-16. The tall glass tube is about 800 mm long and is sealed at one end. With the open end up, the tube is filled to the top with mercury, then covered (a finger works fine) and inverted in an open bowl of mercury. The column of mercury in the tube drops only slightly, stabilizing at about 760 mm [29.9 in]. In fact, if the atmospheric pressure is at its standard value of 1.033 227 kg$_f$/cm², the mercury column is exactly 760 mm high.

It is, of course, the atmosphere that is supporting the column of mercury. The fluid pressure produced under this depth of mercury at 0°C is

$$\text{Pressure} = (\text{depth})(\text{weight density})$$

$$= 76.000 \text{ cm} \left(13.596 \ \frac{g_f}{\text{cm}^3} \right)$$

$$= 1033.3 \ \frac{g_f}{\text{cm}^2} \left(\frac{1 \text{ kg}_f}{1000 \text{ g}_f} \right)$$

$$= 1.033 \ 3 \text{ kg}_f/\text{cm}^2$$

Notice that this answer agrees well with our earlier value for standard atmospheric pressure.

It is common practice to say that standard atmospheric pressure is just 760 mm of mercury (mmHg), or 29.92 in of mercury (inHg). In other words, the height of the mercury column is used as a pressure unit. As mentioned earlier, atmospheric pressure does not always have this exact value, even at sea level. Instead, it fluctuates with weather conditions, and meteorologists use this fact to predict the weather. If a barometer is sealed inside a closed container, it may be used to measure pressures well below atmospheric. A mechanical vacuum pump, for instance, can produce a pressure as low as 0.001 mmHg. Even lower pressures can be achieved by other means, such as oil diffusion pumps.

Table 6-4 lists conversion factors between millimeters of mercury, inches of mercury, and some of the other pressure units we've discussed.

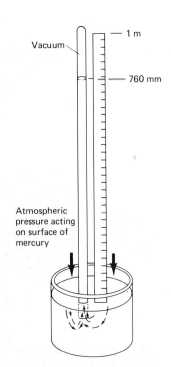

Vacuum

— 1 m

— 760 mm

Atmospheric pressure acting on surface of mercury

Fig. 6-16 The mercury barometer.

TABLE 6-4   CONVERSION FACTORS FOR UNITS OF PRESSURE

	kPa	$\dfrac{kg_f}{cm^2}$	mmHg	inHg	atm
1 kilopascal =	1	0.010 197	7.500 6	0.295 30	$9.869\ 2 \times 10^{-3}$
1 kilogram-force per square centimeter =	98.067	1	735.56	28.959	0.967 84
1 millimeter of mercury* =	0.133 32	$1.359\ 5 \times 10^{-3}$	1	0.039 370	$1.315\ 8 \times 10^{-3}$
1 inch of mercury* =	3.386 4	0.034 532	25.4	1	0.033 421
1 standard atmosphere =	101.33	1.033 2	760	29.921	1

*At 0°C [32°F].
1 bar = 100 kPa
1 in of water at 60°F = 0.248 84 kPa
1 torr = 1 mmHg = 0.133 32 kPa
See Table 5-2 for other units of pressure.

6-9 Surface Tension

The surface of a liquid is always in tension, as if it were a stretched rubber membrane. This is true whether the liquid is in contact with a gas or with another liquid with which it doesn't mix (as oil on water). This *surface tension* supports small objects like needles or razor blades, even though they are denser than the liquid (Fig. 6-17). It also makes life very easy for water spiders.

We are all familiar with the way water and other liquids form drops when they are splashed or otherwise broken up. Again, surface tension is responsible. Liquid drops tend to be nearly spherical, although they may distort as they fall (Fig. 6-18). This shape results from the tendency of the surface to contract to as small an area as possible. A spherical drop can easily be seen if a small amount of water is poured into a jar of kerosene.

Surface tension is closely related to cohesion. The molecules of a liquid attract one another fairly strongly, so they try to group with as few molecules at the surface as possible. To penetrate the surface, we need to overcome this cohesive force. We have seen that as a liquid is heated, the rapid molecular motion decreases the effect of such cohesive forces. And, just as you might expect, a liquid's surface tension is found to decrease at higher temperatures.

Surface tension not only resists penetration of the surface by other objects, but also makes it difficult for the liquid to flow through small openings. Hot water, for instance, is better for washing than cold water

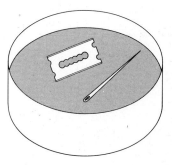

Fig. 6-17 A razor blade or a needle will remain on the water's surface if placed there very gently.

because of its lower surface tension. Where liquid penetration is important, surface tension can also be reduced by adding *detergents*. Soap detergents help water flow into the small openings in fabrics, and the additives in high-detergent motor oil allow the oil to flow into the small openings and channels in an engine's bearings, valve guides, and other critical areas.

6-10 Capillary Action

If a liquid is able to "wet" a solid, this means that the adhesive forces between the solid and liquid molecules are greater than the cohesive forces holding the liquid itself together. Liquid molecules then attach themselves to the solid. Notice that a given liquid may wet only some solids. Water wets clean glass, but it won't wet wax or fresh asphalt. Oil, which has a fairly low surface tension, wets most solids to some extent. Mercury wets clean copper or silver, but beads up on most other surfaces.

The adhesion of a liquid for a solid, combined with the liquid's surface tension, gives rise to a phenomenon known as *capillary action*. Figure 6-19a shows a series of small-bore glass tubes placed in a pan of water. The tubes are open at both ends. The water wets the glass and is drawn toward the walls of the tubing. Inside, the water's surface tension pulls the surface as tightly as it can (again, as if it were a stretched rubber membrane), and this draws a certain amount of water up the tube. The water rises to a height where these adhesive and surface-tension forces are balanced by the downward pull of gravity. The smaller the tubing, the higher the column of water. In a glass tube with a bore the size of a human hair, the water rises about 30 cm.

As shown in Fig. 6-19b, different liquids rise to different heights in the same size tube. If the liquid does not wet the tube (as with mercury in glass), the capillary action is negative. But in all cases, capillary action is noticeable only when the tubing diameter is very small—typically less than a few millimeters. With larger diameters, a liquid always seeks its own level, as we pointed out in the discussion of Pascal's vases.

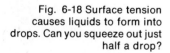

Fig. 6-18 Surface tension causes liquids to form into drops. Can you squeeze out just half a drop?

Capillary action is responsible for many common phenomena. It causes the sap to rise in plants and trees, and limits the maximum height to which the plant can grow (the uppermost leaves have to get nourishment). It makes water flow into blotters and towels and allows us to clean up oil spills with sawdust. The circulation of blood through the small blood vessels (capillaries), the rise of kerosene in a lamp wick, and the flow of lubricants in journal bearings are all helped along by the capillary action just described.

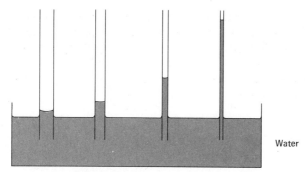

(a) A liquid climbs higher in smaller–bore tubing.

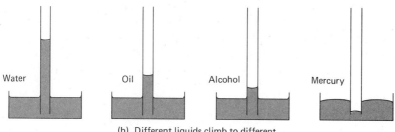

(b) Different liquids climb to different
heights in tubes of the same size.

Fig. 6-19 Capillary action in
small-bore glass tubes.

6-11 Viscosity

We said that fluids tend to flow from regions of high pressure to regions
of low pressure. Except for the capillary action that takes place with
liquids in small conduits, fluids also tend to flow downhill. Yet not all
fluids can be poured with equal ease. The ability of a fluid to flow under
its own weight is described by its viscosity.

> Definition: *viscosity* (more technically, *kinetic viscosity*)
> is a measure of a fluid's resistance to being poured. It is
> also related to the fluid's resistance to flow under
> pressure.

A fluid with a high viscosity is very sluggish and pours very
slowly—molasses, for instance. A fluid with a low viscosity pours very
easily; gasoline is an example. The viscosity of liquids is strongly
dependent on temperature. As a liquid is heated, its viscosity drops and
it flows more easily. It is for this reason that crude oil is often warmed
before it is pumped through pipelines in the winter. Gases, on the other
hand, increase in viscosity as they get warmer. Stated another way, a
cold gas exhibits less fluid friction than a warm one. This is one factor
that makes aircraft perform better in colder weather.

Application: Soldering of Pipes

One important application of capillary action is the soldering of copper pipe and fittings. Molten solder wets copper, brass, or some other metals only if they are clean. Thus the parts to be joined must be first cleaned with fine steel wool and then coated with a rosin flux to prevent oxidation on heating.

The solder is not heated directly. Instead, the torch flame is directed at the pipe and then the fitting. When the flux begins to boil, the metal is probably hot enough to melt the plumber's solder. The solder is then touched to the joint. As it melts, capillary action draws the solder into the narrow gap between the pipe and the fitting—even if it has to run uphill.

Care should be taken not to disturb the joint while the solder is freezing. But once solid, the solder holds the pipe and fitting together with a rigid, watertight bond.

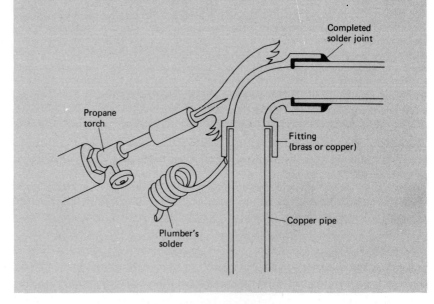

There are many units for measuring viscosity, but we would create a great deal of unnecessary confusion by discussing them here. One metric unit, the *centistoke* (cSt), is used in the graph in Fig. 6-20. This graph shows how the viscosity of various fluids varies with temperature. Notice that the viscosity scale is logarithmic. (From the graph, the viscosity of crude oil at 0°C is more than 10 times the viscosity of water at 0°C.)

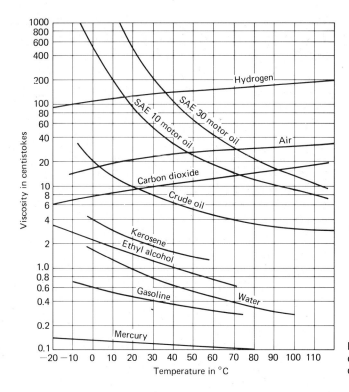

Fig. 6-20 Effect on temperature on the viscosity of some common fluids.

The SAE viscosity index, commonly used to label motor oils, is a number that indicates the oil's ability to flow over a range of temperatures. From Fig. 6-20, we see that SAE 30 oil always has a higher viscosity than SAE 10 oil at the same temperature. We also see that SAE 30 oil at 20°C [68°F] flows just as well as SAE 10 oil at 0°C [32°F]. This is why the higher SAE numbers are used for engine lubricants in warm climates, and the lower numbers are used in cold climates. If SAE 30 oil were used in an engine during an Alaskan winter, it would be very difficult to turn over the cold engine and the oil wouldn't circulate properly if the engine did start. If we used SAE 10 oil in an engine during an Arizona summer, it would get too thin and leak past the piston rings and other oil seals in the engine. The "multiviscosity" motor oils have a flatter viscosity curve and can be used under a wider range of conditions than the single-numbered oils.

At this point, the reader may have noticed something strange about Fig. 6-20: How is it that air has a greater viscosity than water? The answer is that this viscosity is an index of the fluid's ability to flow under its own weight. Mercury flows (or pours) more easily than gasoline, which flows more easily than water, and so on up the graph. Air simply does not flow very readily unless it is pushed by some other agent. This, in fact, is a very fortunate thing for us. If air had a lower viscosity, we would probably have constant and terrible windstorms on our planet.

Summary   Fluids are all around us: the air we breathe, the water we drink, and even our blood, perspiration, and teardrops. Some fluids drive our engines while others lubricate them. We use fluids as coolants, and we use them in heating systems. In fact, there is scarcely a device or technical application that doesn't somehow depend on the properties of fluids. Throughout the rest of this book we see many instances where an understanding of the properties of fluids is essential to an understanding of how things work.

Terms You Should Know

fluid	weight density
fluid pressure	caisson
fluid force	mercury barometer
atmospheric pressure	surface tension
vacuum	detergent
gauge pressure	capillary action
absolute pressure	viscosity
pressure relief valve	

Problems

1. The diagram shows an "inverted siphon," used to transfer water across a valley from one reservoir to another. Its operation is based on the principle that a liquid seeks its own level. What gauge pressure, in kilogram-force per square centimeter, must the pipe be capable of holding?
2. The tank has a bursting strength of 2.65 kg_f/cm^2. How high, in meters, can the column of water be in the pipe before the drum ruptures?
3. Air is being pumped into a diving bell that has been lowered to a depth of 38 ft in sea-water. Calculate the gauge pressure the

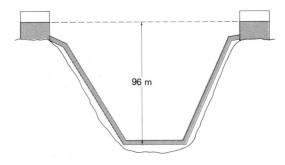

96 m

compressor must supply in (a) pounds per square inch, (b) kilogram-force per square centimeter, (c) kilopascals.

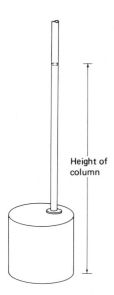

Height of
column

4. A building has a flat roof that measures 9.62 m by 12.13 m and weighs 12 200 kg$_f$. It rests on the foundation walls with airtight caulking but no vertical ties. At the warning of an approaching storm, the building is shut up tight. By how many millimeters of mercury must the barometer drop to allow the normal inside air pressure to blow the roof off the building?*

5. A ship sinks in 23 m of seawater. Salvage divers find that a watertight door is closed, sealing one compartment full of air. The door measures 2.2 m [7.2 ft] by 0.92 m [3.0 ft] wide. What force would be needed to pull the door open against the pressure of the water?

6. A "plumber's helper" is 12.7 cm in diameter. (a) What is the largest upward pressure it can generate on a badly clogged drain? (b) What upward force must be applied to generate this pressure?

7. Water is to be piped to the top ceiling of a 14-story building, where each story is 3.0 m high. What minimum pressure must be applied to the system at ground level to do this?

8. A lift pump is used to pump kerosene. What is the maximum height the kerosene can be lifted when the atmospheric pressure is at its standard valve?

9. One liter of water is completely boiled away, producing steam at atmospheric pressure. What is the volume of this steam, in liters?

Note: This actually happens; in a tornado, it is the inside air pressure that blows the building apart.

7
Force
and
Acceleration

Introduction   We have already mentioned forces a number of times. In Chap. 4 we said that weight is a force, and we talked about the electric force that holds atoms together and gives rise to cohesion and adhesion. Later, in Chap. 5, we mentioned elastic forces, and we defined *stress* as force divided by area. In Chap. 6 we extended this idea and said that the *pressure* on a confined fluid can be calculated in the same way. We also mentioned a force called surface tension. So we already know a certain amount about forces.

In this chapter, we look in greater detail at the kinds of forces, what they have in common, and how they contribute to an object's motion.

7-1 Origin of Forces   In Chap. 4 we defined force as "a push or a pull in a particular direction." Now this may sound a bit silly, because how can you push something without automatically doing it in some direction? But we have to be careful here. Sometimes we have pushes that act in all directions at once. One example is a submarine that is underwater. In this case, the water pushes in on the submarine's hull from all sides at the same time. We speak not of the *force* of the water, but of the water's *pressure*. Of course, it's entirely possible to exert a *force* by using a fluid. This is done in automotive brake systems, aircraft landing gear, industrial presses, and a large number of other devices. In these cases, a particular direction is singled out by building the devices so that one part (usually a piston) can move while the rest of the device can't. The thing to remember is that we use the word "force" only when a direction is being singled out.

There are a number of different kinds of forces. Table 7-1 lists the important ones. Every mechanical device is subject to three of these: gravity, friction, and elastic forces. Every electric device involves electric and magnetic forces (which, because they usually occur together, are sometimes referred to as *electromagnetic* forces). Only the nuclear force is beyond our everyday experience, but even this is now finding application in nuclear power generating facilities.

We said that the direction of a force is important. One reason is that often several forces act on an object at the same time. Figure 7-1 shows an oil-drilling platform being towed by five ships. Obviously, the ships must

7-2 Combinations of Forces

TABLE 7-1   PRINCIPAL KINDS OF FORCES

1. *Gravity*
   Gravity is the force of attraction between matter and the earth and, to a much lesser extent, between every piece of matter and every other piece of matter. The force of gravity on an object is the same as the object's *weight*. The direction of this force is the same as the direction taken by a plumb line.
2. *Electric forces*
   Forces of attraction or repulsion arise between electrically charged objects that are separated in space. Electric forces are responsible for chemical reactions and for the various forms and properties of matter, including adhesion, cohesion, and surface tension. They also cause electricity to flow through conductors.
3. *Magnetic forces*
   Closely related to electric forces, magnetic forces are caused by charges in motion and act to deflect the paths of moving charges. They also attract certain materials that are naturally magnetic (iron, for instance), and they may be produced by permanent magnets made of such materials. Magnetism is unrelated to gravity, although the two forces may act together (permanent magnets are heavy, and the earth itself has a weak magnetic field).
4. *Elastic forces*
   Elastic forces allow solids to "remember" their original shape after being deformed. They are also responsible for the expansion of hot gases. Although they actually have electrical origins, it is easier to speak of them as a separate type of force. Forces produced by the expansion and contraction of muscles also fall into this category.
4. *Friction*
   Arising through the surface contact between two objects, friction slows the movement of one object over another or prevents motion completely.
6. *Nuclear force*
   A very powerful and short-range attraction that holds the nucleus of an atom together, nuclear force takes part in processes like radioactivity and nuclear fission and fusion.

Fig. 7-1 Five ships towing an oil-drilling platform. The total towing force depends not only on the force in each towline, but also on the directions of the towlines as well. (Courtesy Standard Oil of California)

pull in slightly different directions to keep from interfering with one another. The total towing force depends not only on how hard each ship pulls, but also on the directions of all the towlines. At the same time, there are other forces on the platform: its weight (gravity), an upward buoyant force (resulting from gravity acting on the water around the platform), and a force of fluid friction holding the platform back as it is pulled forward. How the platform moves depends on how all these forces combine. The platform may speed up, or slow down, or move at a steady speed in a straight line, or move in a curved path. It may also spin, twisting up the towlines and creating a real problem. Or, in the worst possible case, it may sink.

But this is a very complicated problem to start out with. Let's look at some simpler combinations of forces.

We all understand the tug-of-war principle: there are two teams, each pulling like mad, but no one goes much of anywhere. Each team might pull with a total force of, say, 250 kg$_f$. But because the two 250-kg$_f$ forces oppose each other, the overall effect is just as if there were no force. Yet if we take these same two 250-kg$_f$ forces and point them in the same direction, as in Fig. 7-2, the result is the same as a 500-kg$_f$ force. Now these forces can do something.

What happens if we take the same forces and combine them in some other direction, like the bottom sketch in Fig. 7-2? Well, for

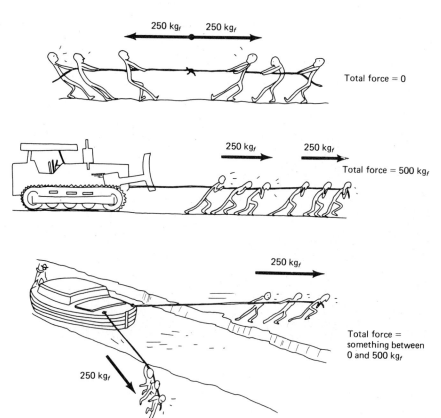

250 kg$_f$ 250 kg$_f$

Total force = 0

250 kg$_f$ 250 kg$_f$

Total force = 500 kg$_f$

250 kg$_f$

Total force =
something between
0 and 500 kg$_f$

250 kg$_f$

Fig. 7-2 The total of two forces
acting together depends on the
relative directions of the forces.

starters we can say that the total force is somewhere between 0 and 500
kg$_f$. But to be more accurate, we need the angle between the two
forces (which is the same as the *direction* of one force relative to the
other). Suppose that we know the angle, and suppose that it is 75°.
Then we can find the total force by making a scale drawing, where we
represent our forces by little arrows. This is shown in Fig. 7-3, using a
scale of 1 cm = 50 kg$_f$. On this scale, both 250-kg$_f$ forces are
represented by arrows 5.0 cm long. The two arrows are drawn with the
75° angle between them. To find the total force (sometimes called the
resultant force), we use the *parallelogram method*: we draw in sides
parallel to the two arrows as shown, and then we draw in the diagonal.
The length of this diagonal gives us a measure of the total force. In Fig. 7-3,
for instance, the diagonal is 7.9 cm in length. The total force is then

$$7.9 \text{ cm} \left(\frac{50 \text{ kg}_f}{1 \text{ cm}} \right) = 400 \text{ kg}_f$$

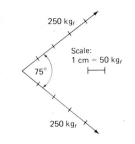

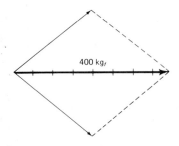

Fig. 7-3 The parallelogram method for adding forces on a scale drawing.

Example 7-1 uses this type calculation.

Example 7-1 Mobile Home in a Windstorm.

A typical mobile home measures 4.27 m by 21.3 m [14 ft by 70 ft] and has an exterior height of 3.43 m [135 in]. It weighs about 7700 kg$_f$ [17 000 lb]. In a 100-kmph [62-mph] windstorm, the wind load from Fig. 5-13 is around 220 kg$_f$/m². If this wind blows directly against the long face of the home, which has an area of 73 m², it produces a force of

$$\text{force} = (\text{stress}) (\text{area})$$

$$= 200 \ \frac{kg_f}{m^2} \ (73 \ m^2)$$

$$= 14 \ 600 \ kg_f$$

Now the mobile home is acted on by two forces: the vertical force of gravity and the horizontal force of the wind. In Fig. 7-4, these forces are drawn to a scale of 1 cm = 4000 kg$_f$. The length of the wind-load arrow is then

$$14 \ 600 \ kg_f \left(\frac{1 \ cm}{4000 \ kg_f} \right) = 3.65 \ cm$$

and the length of the gravity arrow is

$$7700 \ kg_f \left(\frac{1 cm}{4000 \ kg_f} \right) = 1.93 \ cm$$

The arrows are drawn with a 90° angle between them, since one force is horizontal while the other is vertical. The diagonal of the parallelogram has a measured length of about 4.13 cm. This gives a total, or resultant, force of

$$4.13 \ cm \left(\frac{4000 \ kg_f}{1 \ cm} \right) = 16 \ 500 \ kg_f$$

Will this force tip over the home? Not completely. As one side begins to lift, it presents a smaller frontal area to the wind, which reduces the wind load and deflects the airstream. The home will rock, but will probably remain upright in this storm. With higher winds, however, this rocking may be considerable, and wind catching the underside of the floor may roll the home over on its side. It makes

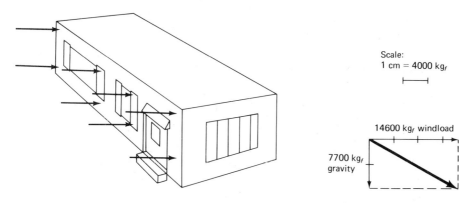

Scale:
1 cm = 4000 kg_f

14600 kg_f windload

7700 kg_f
gravity

Fig. 7-4 Using the
parallelogram method to find
the resultant force on a mobile
home in high wind.

sense, then, to plant shrubs and trees around such a home to prevent
excessive wind loading. ◀

There is another way of adding forces that looks a bit different
but really amounts to the same thing. In Fig. 7-5, two 250-kg$_f$ forces
have been drawn head to tail on a scale of 1 cm = 50 kg$_f$. These forces
are the same as those in Fig. 7-3, and there is the same 75° angle
between the directions they act in. The total force is then found from the
length of an arrow drawn from the tail of the first arrow to the tip of the
second. As the drawing shows, this length is still 7.9 cm, which
converts to a force of $4\overline{0}0$ kg$_f$.

Some people get upset when they see this second method, the
head-to-tail method, because the arrows are drawn head to tail when
the forces actually act at the same point. But there should be no cause
for alarm. The procedure is simply a way to add forces when their
directions aren't the same. It is a way of getting an answer. And it
works.

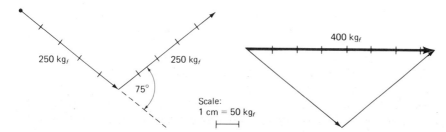

250 kg$_f$

250 kg$_f$

75°

Scale:
1 cm = 50 kg$_f$

400 kg$_f$

Fig. 7-5 The head-to-tail
method of adding forces on a
scale drawing. The same forces
are being added as in Fig. 7-3.

One advantage of this second method is that it can be adapted to any number of forces, simply by stringing them all together head to tail on a sketch. We see a case of this in Example 7-2.

Example 7-2 Total Force on an Airplane.

A certain light plane weighs 2100 kg_f [4600 lb] fully loaded. As it takes off, its wings develop a vertical lifting force of 2500 kg_f. Its engine is pulling it forward with a force of 600 kg_f, while air friction (*drag*) is holding it back with a force of 200 kg_f. What is the total force on the plane?

Figure 7-6 shows the situation, with the forces represented by arrows. To add these forces, we may draw the arrows head to tail as in Fig. 7-7. It doesn't matter in what order we string the arrows together, as long as each arrow points in the right direction and its length is based on the scale factor. The total force, or resultant, is represented by an arrow drawn from the tail end of the first force to the arrowhead on the last one. This arrow has a measured length of 2.8 cm. The total force is then

$$2.8 \text{ cm} \left(\frac{200 \text{ kg}_f}{1 \text{ cm}} \right) = 560 \text{ kg}_f$$

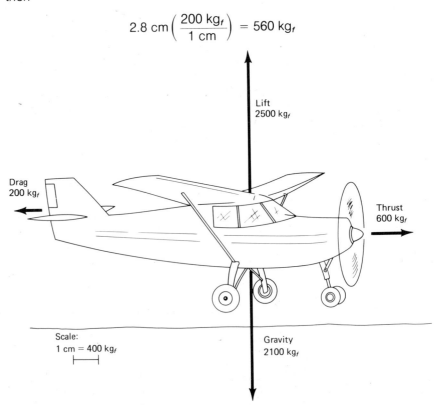

Fig. 7-6 Forces on an airplane taking off from a runway.

Notice that this resultant force is pulling the plane forward and upward at an angle 45° to the horizontal. ◀

Exercises

1. The following forces are to be represented at a scale of 1 cm = 80 kg$_f$. What are the proper lengths for the arrows? (a) 8$\overline{0}$0 kg$_f$, (b) 430 kg$_f$, (c) 981 kg$_f$.
   Answers: (a) 10.0 cm, (b) 5.4 cm, (c) 12.3 cm

2. The following forces are to be represented at a scale of 1 in = 50 lb. What are the proper lengths for the arrows? (a) 75.0 lb, (b) 305 lb, (c) 38 lb
   Answers: (a) 1.50 in, (b) 6.10 in, (c) 0.76 in

3. Two forces act perpendicular to each other. One measures 1200 kg$_f$ while the other measures 800 kg$_f$. (a) What is the resultant force? (b) In what direction does the resultant force act, as measured from the 1200-kg$_f$ force? Use a ruler and protractor, and base your answer on an accurate scale drawing.
   Answers: (a) 1400 kg$_f$, (b) 34°

4. Two 750-kg$_f$ forces act with a 30° angle between them. How big is the total force?
   Answer: 1450 kg$_f$

5. Find the resultant of a 100-kg$_f$ force, a 120-kg$_f$ force, and an 80-kg$_f$ force if they act in the directions shown.
   Answer: 150 kg$_f$ in a direction 20° below that of the 1$\overline{0}$0-kg$_f$ force.

6. Find the resultant of a 9$\overline{0}$0-kg$_f$ force, a 780-kg$_f$ force, and a 450-kg$_f$ force that act in the directions shown.
   Answer: Resultant is zero; these three forces completely balance one another.

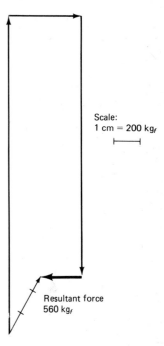

Scale:
1 cm = 200 kg$_f$

Resultant force
560 kg$_f$

Fig. 7-7 Finding the total force on the airplane in Fig. 7-6.

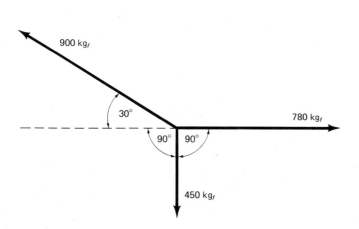

900 kg$_f$

30°

90° | 90°

780 kg$_f$

450 kg$_f$

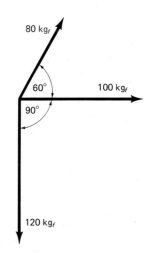

80 kg$_f$

60°

100 kg$_f$

90°

120 kg$_f$

7-3 Newton's First Law of Motion

We have seen that forces can originate in many ways. But no matter how they originate, the rule for combining forces is always the same as described in the last section.

Now let's look at what forces *do*. There are three basic rules here (scientists and engineers call them *laws*). These force laws were discovered by a scientist named Isaac Newton over 300 years ago. Even today they are still called Newton's laws.

Newton's first law of motion

> *Newton's First Law of Motion*: An object at rest stays at rest, and an object in motion stays in motion in a straight line and with no change in speed unless some unbalanced force acts on it.

Newton's first law is sometimes called the law of inertia. Let's examine what it says more closely. The first part is obvious: things just don't start moving all by themselves. If we go into the parking lot and discover that our car is missing, we can safely assume that someone or something has moved it. But suppose that something is already moving. Then it continues to move straight ahead at a constant speed if it is left alone. A force is needed to slow it down, and a force is needed to speed it up. A force is also needed to make it move in a curved path.

Think about your experiences in driving a car. You're driving along a level stretch of four-lane road, and you shift into neutral. It takes a very long time for the car to coast to a stop. But does the car stop all by itself? No, forces are needed to stop it—forces of air friction over the body and mechanical friction in the wheel bearings. There is also an elastic force sometimes referred to as "rolling friction." The tires flatten slightly under the weight of the car, and this offers some resistance to the rolling of the tire. Together, these forces don't amount to very much. But they do all act in opposition to the direction of the motion, and so they account for the gradual loss of speed. The point is that if these forces *weren't* present, we would expect the car to coast for as long as the road stayed level.

As a result, when a car is at cruising speed, its engine is loafing. It may be turning over a large number of rpm's, but it has to generate only enough thrust to overcome the total force trying to slow the car down. In terms of power, this may be only 30 to 50 hp. As we will see in Chap. 17, an engine generates power on *demand*. The rated horsepower is the *maximum* that the engine can generate; most of the time the engine is turning out quite a bit less. This is one reason that a car with a 200-hp engine may get nearly the same gas mileage as the same car with a 100-hp engine.*

Now let's suppose that our car is cruising along at a constant speed (Fig. 7-8). This means that the engine, through the drive train, is generating just enough force to compensate for the total force trying to slow down the car. It's just like the tug-of-war game; the two forces exactly cancel. As far as Newton's first law is concerned, the *net force* is zero. The *unbalanced force* is zero, which is the same thing. And because there is no unbalanced force, the car continues along with no change in speed.

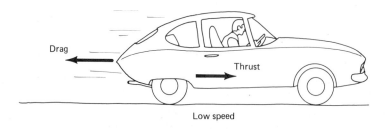

Low speed

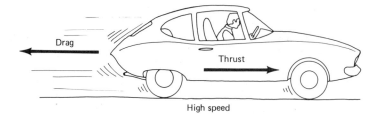

High speed

Fig. 7-8 In a car that is cruising at a steady speed on a level road, the forward thrust is exactly balanced by the total "drag." The total force is zero. At high speeds, it takes more thrust to balance the increased drag, but the total force is still zero.

Give it a little more gas. The car speeds up a bit, then levels off at a new cruising speed. There is more force pushing the car forward. But friction depends on speed, and this is particularly true with air friction. As a result, there is now more force holding the car back. The car levels off at the new speed where the total forward force is just balanced by the total backward force.

* I hope I don't start any arguments here. The point is that the 100-hp engine does not give anywhere near twice the gas mileage of a 200-hp engine in the same car. If the car is cruising at 50 mph, either engine would adequately provide the 50 or so hp to do this. If there is any significant difference in fuel consumption, it is due mainly to how the car is driven and what accessories it has on it. Of course, if we compare two different cars and one is lighter or more streamlined than the other, this will also contribute to a reduced fuel consumption, and we have muddied the comparison.

equilibrium   There is a word used to describe situations like this:

> Definition: *Equilibrium* is the state where all the forces acting on an object cancel one another.

The term "equilibrium" applies not only to objects in motion. Bridges, buildings, and other structures are designed so that the forces always stay in equilibrium. If the forces ever get out of equilibrium, the structure fails.

What about turns? Again, a car can't be turned unless there is some lateral force to push it into the turn. On an icy road, there may be very little friction between the tires and the road. As a result, the car may fail to steer when the front wheels are turned. The car has a *natural tendency* to want to go straight, and this tendency must be overcome in order to change the direction of the motion.

7-4 Acceleration

If an object's speed changes or if the direction of its motion changes, we say that it has *accelerated*. We can accelerate a car by stepping on the gas, stepping on the brake, or by turning the steering wheel. (Possibly we could do a few other maneuvers, too.) We are using the word "acceleration" in its engineering sense here, and this is a bit broader than the way the average person uses the word.

Like forces, accelerations have *direction*. This is how the direction is given:

object is:	acceleration is:
speeding up	in direction of motion
slowing down	opposite to direction of motion
turning	toward center of turn
moving at constant speed in a straight line	zero

Some people get upset that the word "deceleration" doesn't fit into this scheme. If necessary, we can regard deceleration as a negative acceleration. In other words, an acceleration opposite to the direction of the motion is what most people call a deceleration.

If an object is speeding up or slowing down (rather than turning), we can define its acceleration quantitatively as follows:

acceleration   Definition: The *acceleration* of an object in straight-line motion is given by

$$\text{Acceleration} = \frac{\text{change in speed}}{\text{time}} \qquad (7\text{-}1)$$

Thus, if a car speeds up from rest to 60 kmph [36 mph] in 5.0 s, the car has an acceleration of 12 kmph/s. The unit here is read as "kilometers per hour, per second." Similarly, if a boat has an acceleration of 2.0 kmph/s, this means that at each second the boat picks up another 2.0 kmph of speed. At this rate, then, it speeds up by 10 kmph in 5.0 s.

Example 7-3 Acceleration of a Car.

A certain car speeds up from 50 to 85 kmph in 4.5 s. What is its acceleration?

Using Eq. (7-1), we have

$$\text{Acceleration} = \frac{\text{change in speed}}{\text{time}}$$

$$= \frac{85 \text{ km/h} - 50 \text{ km/h}}{4.5 \text{ s}}$$

$$= \frac{35 \text{ km/h}}{4.5 \text{ s}}$$

$$= 7.8 \text{ km/(h} \cdot \text{s)}$$

Again, the unit here is read as "kilometers per hour, per second." ◀

Example 7-4 Speed of a Rocket.

A certain rocket accelerates at 25 kmph/s. If it starts from rest, how fast is it going after 6.0 s?

The rocket's speed is increasing by 25 kmph *each second*. In a time of 6.0 s, this builds up a speed of

$$\text{Speed} = (\text{acceleration}) (\text{time})$$

$$= 25 \frac{\text{km}}{\text{h} \cdot \text{s}} (6.0 \text{ s})$$

$$= 150 \text{ km/h}$$

or about 90 mph. ◀

Example 7-5 Acceleration of a Cruising Airplane.

A certain airplane cruises due east at a steady speed of 240 kmph for 2.0 h. What is its acceleration?

Let's be careful here. The speed of the plane doesn't change at all. It isn't speeding up, and it isn't slowing down. It isn't changing direction, either. Its acceleration is therefore zero. ◄

In engineering and trades applications, it is common to express accelerations in g's. The rate at which an object picks up speed when it is in free fall is $1g$ of acceleration. Since gravity changes from place to place, the g of acceleration depends slightly on where we are, too. In fact, we have already listed the local values of this quantity in Table 4-3. But because the variation in g is relatively small, we don't lose much accuracy by using g_0, the *standard free-fall acceleration.* To keep things simple, we will do this from now on. If an object is accelerating in a straight line at $1g$, its speed changes by 35.304 kmph each second [21.937 mph each second]. We may therefore write

$$1g = 35.304 \ \frac{km}{h \cdot s}$$

Figure 7-9 shows how an object behaves under an acceleration of $1g$. The graph gives the actual vertical speed of a typical person jumping out of an airplane without a parachute. The person's speed increases by about 35 kmph during each of the first 5 s, which is what we mean by an acceleration of $1g$. If it weren't for the viscosity of air and the friction it generates, the person would continue to speed up by this amount each second until he or she hit the ground. Instead, the speed levels off at about 190 kmph [120 mph]. After this speed is reached, there is no further acceleration. In this case, 190 kmph is the *terminal*

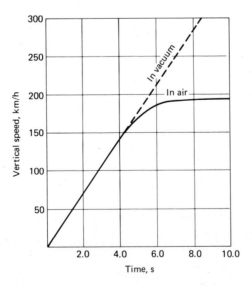

Fig. 7-9 The speed gained by a falling object depends on how long it falls.

speed. Different objects falling through different fluids reach different terminal speeds. The terminal speed of a parachutist, for instance, is around 22 kmph [14 mph].

Any unit of speed divided by any unit of time gives a valid unit of acceleration. This leads to many possibilities, the most common of which are listed in Table 7-2.

TABLE 7-2   *CONVERSION FACTORS FOR UNITS OF ACCELERATION*

	$\dfrac{m}{s^2}$	$\dfrac{ft}{s^2}$	$\dfrac{km}{h \cdot s}$	$\dfrac{mi}{h \cdot s}$	g_0
1 meter per second, per second =	1	3.280 8	3.600	2.236 9	0.101 97
1 foot per second, per second =	0.304 8	1	1.097 3	0.681 82	0.031 081
1 kilometer per hour, per second =	0.277 78	0.911 34	1	0.621 37	0.028 325
1 mile per hour, per second =	0.447 04	1.466 7	1.609 3	1	0.045 586
1 standard free-fall acceleration =	9.806 65	32.174	35.304	21.937	1

Note: $1 \dfrac{M}{s^2} = 1 \dfrac{N}{kg}$

$1 \dfrac{ft}{s^2} = 1 \dfrac{lb}{slug}$

Exercises

7. A car traveling at 76 kmph brakes to a stop in 5.2 s. What is its acceleration in (a) kilometers per hour, per second; (b) miles per hour, per second; (c) g's?
   Answers: (a) −15 kmph/s, (b) −9.1 mph/s, (c) −0.41 g
8. A bullet is shot from a gun with a muzzle velocity of 488 m/s. It reaches this speed after spending only 3.22 ms in the barrel. What is the bullet's acceleration in (a) meters per square second; (b) kilometers per hour, per second; (c) miles per hour, per second; (d) g's?
   Answers: (a) 152 000 m/s², (b) 546 000 kmph/s, (c) 339 000 mph/s, (d) 15 500 g's
9. A freight train accelerates at 0.351 m/s², starting from rest. (a) What is its speed in kilometers per hour after 30.0 s? (b) How long would it take to reach a speed of 60 kmph if this acceleration continued?
   Answers: (a) 37.91 kmph, (b) 47.5 s

7-5 Newton's Second Law of Motion

Accelerations occur only as the result of forces. Newton's second law, which is sometimes called the *law of acceleration,* tells us how this happens.

Newton's second law of motion

> *Newton's Second Law of Motion*: An object accelerates in the direction of the total (or resultant) force on it, and the amount of this acceleration is greater when the object's mass is less.

This law may be written as a formula:

$$\text{Force, in kg}_f = (\text{mass, in kg})(\text{acceleration, in } g\text{'s}) \qquad (7\text{-}2)$$

or in USCS units.

$$\text{Force, in lb} = (\text{mass, in lb}_m)(\text{acceleration, in } g\text{'s}) \qquad (7\text{-}3)$$

Thus it is harder to accelerate a dump truck than a drag racer, because the truck has more mass. The dragster in Fig. 7-10 is made mainly of aluminum, which keeps its mass to just 613 kg [1350 lb$_m$]. So its acceleration would be large even with a normal-size engine. With the oversized power plant, it can actually maintain an average acceleration greater than 1g over a ¼-mi track.

Let's look at a numerical example of Newton's second law.

Example 7-6 Performance of a Sports Car.

A certain sports car has a mass of 1380 kg [3040 lb$_m$]. On a straight track, we want it to accelerate from rest to 100 kmph [62.1 mph] in 9.81 s. What total forward force is needed to do this?

Fig. 7-10 A dragster achieves its high acceleration because of the large thrust accelerating such a small mass. (Courtesy ALCOA)

To apply Eq. (7-3), we first need to calculate the acceleration. This is

$$\text{Acceleration} = \frac{\text{change in speed}}{\text{time}}$$

$$= \frac{10\overline{0} \text{ km/h}}{9.81 \text{ s}}$$

$$= 10.2 \frac{\text{km}}{\text{h·s}} \left[\frac{1 \, g}{35.304 \text{ km/(h·s)}} \right]$$

Notice that the second factor here is the conversion from kilometers per hour, per second to g's. This gives

$$\text{Acceleration} = 0.289_g$$

Then using Eq. (7-2), we have

$$\text{Force, in kg}_f = (\text{mass, in kg})(\text{acceleration, in } g\text{'s})$$

$$= 1380 \text{ kg} (0.289 \, g)$$

$$= 399 \text{ kg}_f$$

This 399 kg$_f$ [879 lb] is the total, or resultant, forward force needed to give the car this performance. It is the difference between the thrust and the drag. If the drag is known, it can be added to this 399 kg$_f$ to get the total traction needed at the drive wheels. Later, we see how this force can be related to the torque and power requirements of the engine. ◀

We may also use Newton's second law to find the acceleration when we know the force and the object's mass. In this case, the formulas are rewritten as

$$\textit{Acceleration, in g's} = \frac{\text{force, in kg}_f}{\text{mass, in kg}} \qquad (7\text{-}4)$$

or in old USCS units,

$$\text{Acceleration, in g's} = \frac{\text{force, in lb}}{\text{mass, in lb}_m} \qquad (7\text{-}5)$$

Let's look at an example.

Example 7-7 Acceleration of a Rocket.

The rocket in Fig. 7-11 has a mass of 7500 kg. Its engine develops an upward thrust of 16 000 kg$_f$. Let's calculate the rocket's acceleration.

Notice that the rocket has a weight of around 7500 kg$_f$. This much thrust is needed just to lift it above the ground, with no acceleration. The resultant force on the rocket, which is the force available to accelerate it, is what is left over after this tug-of-war between the thrust and gravity. From the picture, we see that this force is 8500 kg$_f$.

Equation (7-4) then tells us that

$$\text{Acceleration, in } g\text{'s} = \frac{\text{force, in kg}_f}{\text{mass, in kg}}$$

$$= \frac{8500}{7500} g$$

$$= 1.1 g$$

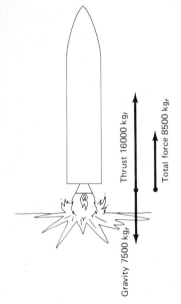

Fig. 7-11 The total, or resultant, force on the rocket is the difference between its upward thrust and its weight. This resultant force determines the rocket's acceleration.

Thrust 16000 kg$_f$

Total force 8500 kg$_f$

Gravity 7500 kg$_f$

Since 1 g amounts to a speed gain of 35.304 kmph each second, this rocket is gaining speed at a rate of (1.1)(35.304 kmph/s), or 39 kmph each second. This gives it a speed of around 390 kmph [230 mph] in just 10 s. Of course, higher speeds increase the drag, which acts against the accelerating force. But at the same time, the rocket's engine is consuming fuel, which decreases the mass that is accelerated by this lower force. The two effects tend to cancel each other.

When the rocket reaches extremely high altitudes, the air gets too thin to provide much drag; at the same time, the decrease in gravity lowers the remaining weight of the rocket. Under these conditions, the rocket may be accelerated to very high speeds before its fuel runs out. The typical speed of an orbiting artificial satellite is around 30 000 kmph. Space shots are based on an analysis similar to this one, with the detailed calculations done automatically by computer. ◀

In Example 7-7 we had to be careful to base the acceleration on the total force, which meant that gravity had to be included. More commonly, objects will accelerate on the level and gravity does not have to be considered. Example 7-8 shows such a case. At the same time, we demonstrate the use of old USCS units.

Example 7-8 Acceleration of a Truck.

A certain van has a mass of 4500 lb$_m$. Traction in the drive wheels produces a forward force of 950 lb. Rolling friction holding back the van is 100 lb. Let's estimate the truck's acceleration on level ground and the time to reach a speed of 30 mph.

The situation is shown in Fig. 7-12. Although the truck weighs about 4500 lb, this force is balanced by another 4500-lb upward force supporting the wheels. If this upward force did not exist, the truck would sink into the road until it hit something strong enough to provide a 4500-lb supporting force. The resultant force along a vertical axis is therefore zero, and we don't have to worry about it. The resultant forward force is 850 lb: the 950-lb thrust minus the 100-lb rolling friction. Using Eq. (7-5) gives us

$$\text{Acceleration, in } g\text{'s} = \frac{\text{force, in lb}}{\text{mass, in lb}_m}$$

$$= \frac{850}{4500}g$$

$$= 0.189g$$

Since 1 g amounts to a speed change of 21,937 mph each second, in 1 s our van gains a speed of

$$0.189 \, (21.937 \, \frac{\text{mi}}{\text{h}}) = 4.1 \, \frac{\text{mi}}{\text{h}}$$

We have rounded off here to the original two-digit accuracy. At this rate, the van accelerates to 41 mph in $\overline{1}0$ s.

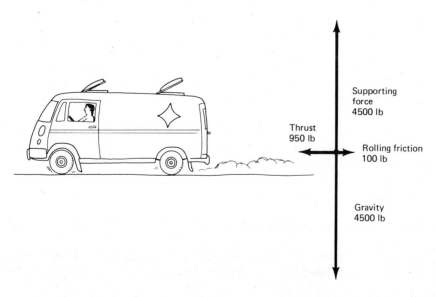

Thrust
950 lb

Supporting
force
4500 lb

Rolling friction
100 lb

Gravity
4500 lb

Fig. 7-12 On a level road, the force supporting a vehicle balances the force of gravity. The vehicle's performance is then affected only by the horizontal forces.

The question, however, was to find the time needed to speed up to 30 mph. Since the van gains 4.1 mph each second, the time required is

$$\text{Time} = \frac{30 \text{ mi/h}}{4.1 \text{ mi/(h·s)}}$$

$$= 7.3 \text{ s}$$

We should point out again that this is an *approximate* calculation. This is not because of any shortcoming in Newton's laws; in fact, they are perfectly accurate. The difficulty is in calculating with quantities that are inherently variable. The force on a van could never stay at a constant 850 lb as the van speeds up. Instead, it changes as the engine's rpm changes. And, of course, we already mentioned the variable effect of air drag. This type of analysis, then, leads only to "ballpark" performance predictions. The actual performance of a vehicle must be found through time and speed trials. ◀

Exercises

10. A truck has a mass of 6200 kg. What net forward force is needed to give it an acceleration of 0.088 g on a level road?
   Answer: 550 kg$_f$

11. When its engine is shut off, a small 880-kg boat slows down from 30 to 25 kmph in 3.3 s. (*a*) What is the acceleration? (*b*) What is the average force producing this acceleration?
   Answers: (*a*) −1.5 kmph/s, or −0.043 g, (*b*) 38 kg$_f$

12. A car has a mass of 3200 lb$_m$. What net force is needed to give it an acceleration of (*a*) 0 g, (*b*) 0.5 g, (*c*) 1.0 g?
   Answers: (*a*) 0 lb, (*b*) 1600 lb, (*c*) 3200 lb

13. What is the acceleration if a 2.72-kg mass is acted on by a net force of (*a*) 0 kg$_f$, (*b*) 2.72 kg$_f$, (*c*) 10.5 kg$_f$?
   Answers: (*a*) 0 g, (*b*) 1.00 g, (*c*) 3.86 g

14. A certain car traveling at 70 kmph experiences a total drag force of 45 kg$_f$. The car's mass is 1250 kg. What is the car's acceleration if the forward thrust is (*a*) 0 kg$_f$, (*b*) 45 kg$_f$, (*c*) 120 kg$_f$?
   Answers: (*a*) −0.036 g, (*b*) 0 g, (*c*) 0.060 g

7-6 Newton's Third Law of Motion

We've all heard about action and reaction. A jet engine gets forward thrust by blowing gases out its exhaust at high speed. A gun tends to "recoil" when it's shot. Step off a small boat onto a dock and you may land in the water if the boat isn't moored. These effects all result from Newton's third law.

Newton's Third Law of Motion: Forces always occur in pairs. If one object exerts a force on another, then the second object always exerts an equal and opposite force back on the first.

For instance, the airplane propeller in Fig. 7-13 is just a giant fan that blows air back over the plane. The *reaction force*—the air pushing forward on the propeller—pushes the plane forward. The action of a car's weight on its coil springs produces a reaction that holds the car up. In the same way, the elastic reaction of a bridge pier supports the bridge's weight. Action and reaction forces are always exactly equal and exactly opposite in direction. It really doesn't make much difference, then, which force we call the action and which the reaction. In the examples just given, however, it is customary to label the forces as we have done here.

This brings us to an important question. If forces *always* exist in pairs, and if each pair of forces is *always* equal and opposite, why don't these forces just cancel? Why can we sometimes produce resultant forces that accelerate things? The key is that action and reaction act on different things. The rocket accelerates in one direction while its

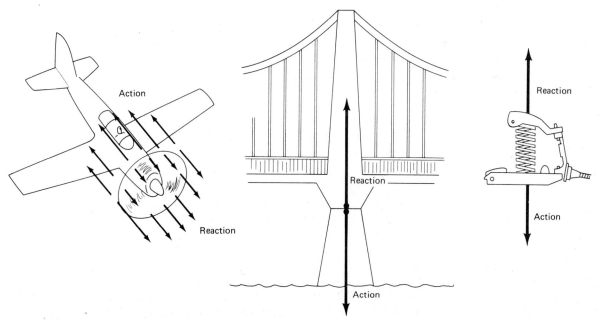

Fig. 7-13 Action and reaction. Forces always occur in pairs.

Application: Interior Ballistics

Although the Chinese were the first to discover gunpowder, they seem to have used it only in rockets and fireworks. In Europe, Roger Bacon wrote about the explosive mixture in the late 1200s, and by the early 1300s people were already using it to fire projectiles out of tubes (at one another). In those early days, history records that it was nearly as dangerous to do the shooting as to be shot at.

Although there have been many mechanical refinements in guns through the years, the basic principle is still the same: Ignite a relatively slow-burning explosive in a tube so the high-pressure gases accelerate a bullet. Once it leaves the muzzle, the bullet is on its own. The study of the bullet's motion within the bore is called *interior ballistics*; *exterior ballistics* is concerned with its continuing motion toward a target.

The *caliber* of a bullet is the same as the bore diameter of the gun barrel that it fits. This may be expressed in millimeters or inches. When expressed in inches, the unit is customarily omitted. Thus a .22-caliber gun has a bore diameter of 0.22 in, and a .22-caliber bullet has about this same diameter (at least after it is shot).

A typical .30-caliber bullet has a mass of 110 gr, or 7.1 g. It may be accelerated to a speed of $6\bar{0}0$ m/s [$2\bar{0}00$ ft/s] in a barrel 58 cm long. This happens in a very short time: about 1.9 ms. Thus the average acceleration is

$$\frac{6\bar{0}0 \text{ m/s}}{0.001\ 9 \text{ s}} \left(\frac{1\ g}{9.81 \text{ m/s}^2} \right) = 32\ 000\ g$$

What average force is needed to produce this tremendous acceleration? Using Newton's second law and expressing the bullet's mass as 0.0071 kg, we get

$$\text{Force} = 0.007\ 1\ (32\ 000)\ \text{kg}_f$$

$$= 230\ \text{kg}_f$$

or about 500 lb.

Now as this force propels the bullet forward, its reaction (equal and opposite) propels the gun backward. Why doesn't it knock over the shooter? Well, if the gun is very light, it just might do that. But if the gun has a large mass, its acceleration under this reaction force is small. Suppose that the gun has a mass of 10 kg [22 lb$_m$]. Then its average acceleration is

$$\text{Acceleration} = \frac{\text{force}}{\text{mass}}$$

$$= \frac{230}{10}\,g$$

$$= 23\,g$$

This is the same as 230 m/s². Since this acceleration lasts only 0.001 9 s (the time it takes the bullet to leave the barrel), the gun's recoil velocity is

$$230\,\frac{m}{s^2}\,(0.001\ 9\ s) = 0.44\,\frac{m}{s}$$

This is only 1.6 kmph, or 1.0 mph. Of course, this is a fairly heavy gun. If the gun is made lighter, it will be easier to carry around but more painful to shoot.

We may also estimate the average gas pressure in the bore. The bore diameter is 0.30 in, or 0.76 cm. The cross-sectional area of the bore is therefore π (0.38 cm)², or 0.45 cm². The 230-kg$_f$ force accelerating the bullet acts on this area, so the average pressure must be

$$\text{Pressure} = \frac{\text{force}}{\text{area}}$$

$$= \frac{230\ \text{kg}_f}{0.45\ \text{cm}^2}$$

$$= 510\ \text{kg}_f/\text{cm}^2$$

This is nearly 4 times the pressure in a fully charged commercial oxygen cylinder. No wonder so many of the early guns blew themselves apart.

Interior ballistics is also concerned with problems of rifling the barrel, lubricating the bore, shell and chamber design, properties of the explosive itself, and so on.

exhaust gases are forced the other way. If you shoot a rifle, you may become painfully aware that the rifle kicks backward as the bullet is accelerated forward. An accelerating car may throw up mud or gravel behind it. As for our bridge, the action-reaction pair is shown for the bearing surface between the superstructure and the pier. If it can, the pier moves downward under the action force. But if it rests on bedrock, the forces *on the pier* are balanced, and it won't go anywhere. Meanwhile, the superstructure is attracted to the earth through gravity. If this were the only force *on the superstructure,* it would fall and accelerate as it plunged into the water. Instead, the elastic reaction of the pier balances this downward force, and the superstructure goes nowhere.

    We see this same principle in Fig. 7-14. A stalled truck is to be pushed. Pushing from the truck's bed accomplishes nothing, because all the forces *on the truck* remain balanced. But pushing from the ground puts the action on the truck and transmits the reaction to the ground. The truck goes one way while the ground is pushed the other way. (You don't believe that the ground is pushed anywhere? Take a closer look. Where did those footprints come from?)

7-7 Circular Motion   We said that a moving object normally wants to go straight, and to make its path curve is to accelerate it (even though its speed may not

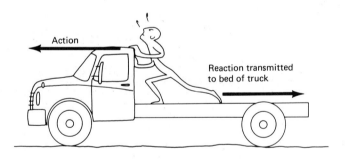

Action

Reaction transmitted to bed of truck

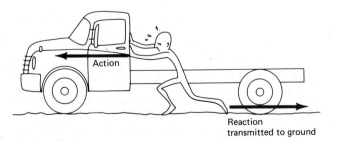

Action

Reaction transmitted to ground

Fig. 7-14 Pushing a truck. It takes an unbalanced force to start the truck moving.

change). The direction of this acceleration is toward the center of the turn. An acceleration of this type is called a centripetal acceleration.

Definition: A *centripetal acceleration* is experienced by any moving body that is making a turn. The direction of the acceleration is toward the center of the turn. The centripetal acceleration is greater at high speeds, and is also greater if the turning radius is decreased.

centripetal acceleration

Figure 7-15 shows the top view of a car making a turn at a steady speed. Notice that although the speed doesn't change, the motion *does* change toward the right in a right-hand turn. Since the change in motion always points toward the center of curvature of the turn, this is the direction of the centripetal acceleration. Obviously, the motion changes more rapidly at high speeds or in "tight" turns. In these cases, the centripetal acceleration is high. At low speeds and/or gradual turns, the centripetal acceleration is low.

Figure 7-16 shows how centripetal acceleration depends on the speed and turning radius. A well-designed racing car may achieve a

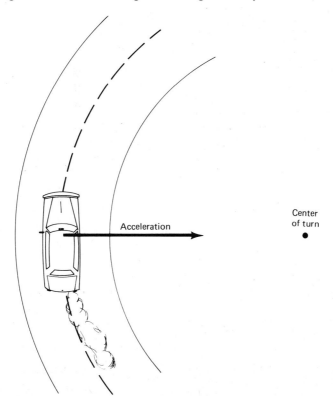

Acceleration

Center
of turn

Fig. 7-15 Centripetal acceleration. Any object making a turn is accelerating toward the center of the turn.

Application: Measuring the Drag on a Moving Car

Aerodynamic and friction drag affect a vehicle's performance, because these forces oppose any forward thrust accelerating the vehicle. For an accurate analysis of vehicle performance, then, drag must be taken into account.

A measurement of total drag can be based on Newton's second law. The car is taken onto a flat, straight track on a calm day (a track is better than a highway because of the current national speed limit, and also for safety reasons). An observer in the passenger seat carries a stopwatch and a note pad. The driver begins cruising at a steady speed, say $1\overline{0}0$ kmph. The driver then disengages the clutch, at the same time signaling to the observer, who starts the stopwatch. When the speed has dropped to 90 kmph, the driver again signals the observer, who stops the watch and records the measurements. Notice that the speed measurement should be made from the driver's seat to avoid the parallax error in reading a speedometer from an angle. The acceleration (deceleration) is then calculated, from which Eq. (7-2) gives the average drag causing this acceleration. For instance, if a 1600-kg vehicle slows from $1\overline{0}0$ to 90 kmph in 9.5 s, the acceleration is 0.030 g and the drag force is $(1600)(0.030)$ kg$_f$, or 48 kg$_f$ [106 lb].

Since the drag force depends on vehicle speed, a series of trials can be made. Actually, the drag varies during the trial

centripetal acceleration as high as 0.8g in a turn, but most conventional passenger cars spin out of control long before this.

Of course, it takes force to produce accelerations, and centripetal accelerations are no exception. No moving object will turn unless it is pushed or pulled into the turn. The pushing or pulling agent is called a centripetal force.

centripetal force

Definition: A *centripetal force* is any force that gives rise to a centripetal acceleration. The magnitude of a centripetal force may be found from Newton's second law.

Notice that any of the forces we've studied can be centripetal if they cause circular motion. For a car in a flat turn, the centripetal force

itself (the drag at 100 kmph is higher than at 90 kmph). Still, it is reasonably accurate to quote the result for the midpoint of the speed range: in the case just given, the drag of 48 kg$_f$ occurs at about 95 kmph.

With enough data at different speeds, a graph of vehicle drag versus speed can be drawn. Such a graph is shown for the author's 1972 Mustang convertible.

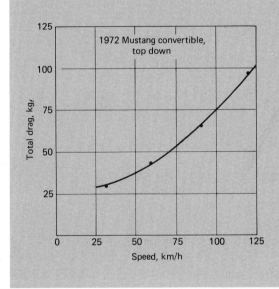

is the lateral friction between the tires and the road; for a car in a banked turn, it is an elastic reaction to gravity; for a rotating tire tread, it is the elastic force in the radial plies; for an airplane in a banked turn, it is the force produced by excess air pressure on the underside of the tilted wing; for a rotating crankshaft, it is the elasticity of the steel; and so on. In fact, no circular motion can ever take place without a centripetal force to produce it.

Example 7-9 Centripetal Force on a Car in a Turn.

A certain car has a mass of 1830 kg [4250 lb$_m$]. It enters a curve with a turning radius of 48 m [157 ft] at a speed of 50 kmph [31 mph]. Let's find the lateral force needed to successfully negotiate the turn.

From Fig. 7-16 we see that the centripetal acceleration in a 48-m turn at 50 kmph is 0.42g. Newton's second law then tells us how much

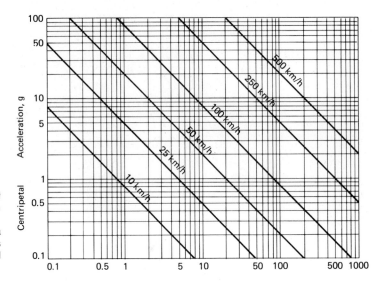

Fig. 7-16 Centripetal acceleration depends on an object's speed and turning radius. A high speed and/or a small turning radius contributes to a large centripetal acceleration.

force is needed to produce this acceleration. Using Eq. (7-2), we have

$$\text{Force, in kg}_f = (\text{mass, in kg})(\text{acceleration, in } g\text{'s})$$

$$= 1930 \text{ kg } (0.42 \text{ g})$$

$$= 810 \text{ kg}_f$$

Notice that this is a tremendous force (almost 1800 lb, in old USCS units). And the car isn't even going all that fast! This 810 kg$_f$ must come from friction between the four tire "patches" and the road. If there is an icy spot or an oil slick, the friction will be considerably less than this critical amount, and the car will skid forward in a nearly straight line. Depending on the vehicle's steering characteristics, whether the front tires or the rear tires break away first, and whether the driver is braking or accelerating in the turn, the car may also spin as it skids forward. ◀

Exercises

15. A large truck tire has a radius of 61 cm. What is the centripetal acceleration on the tread when the tire is rolling at (a) 25 kmph, (b) 5$\overline{0}$ kmph?
    Answers: (a) 8.1 g, (b) 32 g

16. A car tire has a typical tread diameter of 28 in. What is the centripetal acceleration when it rolls at (a) 15.5 mph, (b) 31 mph, (c) 62 mph?
    Answers: (a) 2.2 g, (b) 14 g, (c) 56 g

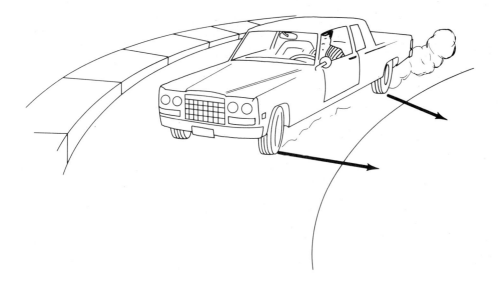

17. Assume that a passenger car can withstand a centripetal acceleration of 0.5 g without loss of control. What should the minimum turning radius be on a flat road if the speed limit is (a) 50 kmph, (b) 100 kmph?
    Answers: (a) 40 m, (b) 160 m
18. A plane flying at 500 kmph does a vertical "loop." If the maximum centripetal acceleration is 6g, what is the approximate diameter of the loop?
    Answer: Twice the radius, or 680 m
19. A railroad boxcar has a mass of 58 000 kg. What lateral force must the rails supply to get it to make a 50-m turn at 25 kmph?
    Answer: 5800 kg$_f$

It always takes a force to produce an acceleration. According to Newton's second law, the acceleration and the accelerating force both have the same direction. But Newton's third law tells us that every force has an equal and opposite reaction force. So whenever we see something accelerate, we can also expect to find a second force that acts *opposite* to the direction of the acceleration.

Since this may sound a bit confusing, let's follow an imaginary car trip and watch what these action and reaction forces do. The driver is a 50-kg woman. She sits in the driver's seat, buckles the seat belt, and starts the engine. So far, there is no acceleration, no accelerating force, and, of course, no reaction. Then she releases the brake, shifts into gear, and opens the throttle. The car accelerates forward, and the

7-8 Reaction Forces during Accelerations

woman's body presses her seat backward. The seat experiences a rearward reaction to the forward force which it must apply to accelerate the driver with the car. Suppose that the car accelerates at 0.3 g. Then everything in it also accelerates at 0.3 g. The seat is pushing the woman's body forward with a force of (50 kg)(0.3 g), or 15 kg$_f$. At the same time, the woman's body is pressing the seat *backward* by this same amount, 15 kg$_f$.

The trip continues. As the car's speed increases, the driver gradually eases off the gas pedal. Eventually the car levels off at a steady speed of, say, 89 kmph [55 mph]. Now there is no acceleration. With no accelerating force, the rearward reaction on the seat has also disappeared.

Then suddenly a deer darts across the road. The woman jams on the brakes and flies forward against the seat belt (or harness). Since the car is now slowing down, its acceleration is backward (opposite to the direction of the motion). The force on the woman's body is also backward, slowing her down along with the car. But the *reaction* of the woman's body on the harness is forward. If the car has a negative acceleration of 1g (the approximate limit on most road surfaces), the 50-kg woman pulls forward against the restraining device with a force of 50 kg$_f$. When the car has stopped, this reaction force vanishes along with the decelerating force.

Suppose that the car is steered into a left-hand turn with a centripetal acceleration of 0.5 g. The woman's body has a natural tendency to want to travel straight as the car turns out from under her. To make the turn along with the car, our driver must have a force of (50 kg)(0.5 g), or 25 kg$_f$, acting on her toward the left. This force comes from friction in the seat, the tension in the seat belt or harness, and her grip on the steering wheel. At the same time, the woman also pulls all these things toward the right (the outside of the turn) with an equal force of 25 kg$_f$. This reaction force toward the outside of a turn is called a centrifugal force.

centrifugal force

Definition: A *centrifugal force* is the outward reaction to the inward centripetal force that acts on an object moving in a curved path.

Notice that the centrifugal and centripetal forces are equal and opposite. Notice also that they act on different things. Figure 7-17 shows a railroad train rounding a shallow curve. The centripetal force acts on the train's wheels, pushing them toward the inside of the curve. The centrifugal force is the reaction acting on the train's rails, pushing them outward. Cases have been reported of trains ripping out sections of rails by taking a bend too fast. The centrifugal force is responsible.

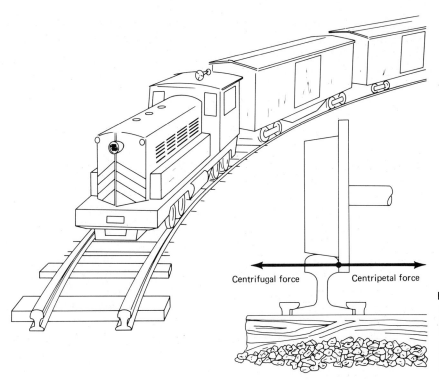

Centrifugal force   Centripetal force

Fig. 7-17 A railroad train rounds a shallow curve. The centripetal force is an inward force acting on the train's wheels. The centrifugal force is the outward reaction acting on the rails.

When a car is cruising at 89 kmph [55 mph], the treads on the tires are executing a very rapid circular motion. The centripetal force needed to keep the tread from flying off is around 3000 kg$_f$ [6600 lb]. We may therefore say that the centrifugal force pulling the radial plies outward is 3000 kg$_f$. We might note here that doubling the car's speed to 178 kmph [110 mph] increases both forces by a factor of 4, not 2! And unless the tire was designed for racing, the resulting 1200-kg$_f$ centrifugal force will probably pull it apart.

We see, then, that important reaction forces can arise during all three types of accelerations: speeding up, slowing down, or turning. In one interesting case, unexpectedly large reactions arose from a combination of two of these. There was a sharp turn leading to a certain movable bridge, and this caused many drivers to brake suddenly on the bridge approach. Eventually the reaction forces shifted the bridge pier so much that it became impossible to close the bridge properly.

A collision between two moving vehicles leads to large forces and large accelerations. So does the impact of a hammer on a nail or a falling brick striking a sidewalk. In all these cases, a moving object is stopped

7-9 Collisions and Impacts

Application: Banking of Turns

When a moving object leans to one side or the other, it has a natural tendency to go into a turn. This can be seen by rolling a coin across a hard floor. If the coin begins to lean to the right, it also curves toward the right, tracing a tighter and tighter turn until it falls over. Bicycles and motorcycles are conventionally steered by leaning. If an airplane is banked, it slips forward for awhile, eventually begins to turn even without any rudder action. The same effect can be seen with a speedboat. And we all know that it is easier to steer a car around a banked curve than around a flat one.

Why does this happen? Let's look at the airplane. In straight and level flight, the lift balances the weight and the resultant force is zero. (For convenience, we have represented the total lift as a single arrow, even though the lift actually acts over the entire wing surface. Similarly, the weight is also represented by a single arrow.) But since the lift force acts perpendicular to the wing, it pulls not just up but sideways as well when the plane is banked. Now the resultant force is no longer zero; as shown in the second sketch, it acts sideways and down. This pulls the plane into a descending turn, or spiral.

If the pilot wants to execute a level turn, he or she must increase the lift while banking the aircraft. As shown in the third sketch, this produces a sideways (horizontal) resultant, which is the centripetal force needed for the turn.

If the bank is too shallow for the speed and turning radius, the plane skids off the outside of the turn. If the plane is banked too steeply for its speed and turning radius, it *slips* off the inside of the turn and loses altitude. An intentional slip is a useful maneuver for losing altitude, particularly when landing in a crosswind.

A car is supported by the road surface on which it travels. In a banked curve, this supporting force performs the same

very quickly. And the *quickness* of the stop, or the acceleration, gives rise to the large forces.

You are already familiar with this if you have ever tried to catch a raw egg. But if you haven't had this experience, think about how you might do it. Would you catch it stiff-armed? This would stop it very

function as the airplane's lift. The car can also depend on lateral friction to hold it on the road in a turn. But when the curve is banked properly for the turning radius and the car's speed, the road itself pushes the car around the turn and friction has nothing to do with it.

Lift

Gravity

Straight and level flight:
zero resultant force

Bank:
resultant force causes
downward spiral

Bank with increased lift:
resultant force causes
level turn

quickly, and the large acceleration would require a force that could easily break the shell. More than likely, you would catch it with a broad follow-through that stopped it very gradually. This would keep the force small.

Suppose that we have a 1400-kg car that runs head-on into a brick wall at a speed of 30 kmph. Can we say what the force of the impact is? Not without more information. The car might have a very stiff frame and rigid bumpers. In that case, the metal may deform very little, and the collision may shorten the car by only, say, 10 cm. This leads to a very high acceleration of 16g. The collision force is then 1400 kg, multiplied by 16, or 22 000 kg$_f$ [nearly 50 000 lb]! But suppose that the car has a very light frame and a collapsible bumper. Between the movement of the bumper and some compression or bending in the frame, the car may travel 40 cm during the collision. In this case, the acceleration is only around 4g, and the force of the collision is 5600 kg$_f$ [12 000 lb].

The point is that nothing ever "stops on a dime." There is always a certain amount of travel during the stop, and the shorter the travel, the larger the forces involved. A hammer, for instance, can easily be swung at a speed of 90 kmph. If it hits a finger, the finger isn't likely to give more than 5 mm. This means that the acceleration of the hammerhead is around 640g during the impact. If the hammer's mass is 1 kg, the force on the finger is 640 kg$_f$ [1400 lb]. The author has one finger that will never be the same.

Figure 7-18 shows the acceleration of objects that are stopped from four different speeds. In all cases, the acceleration is greater when the object is stopped in a shorter distance.

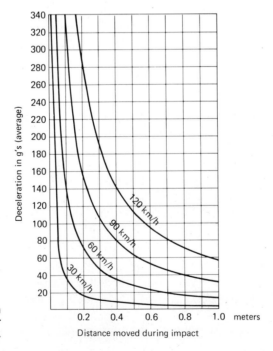

Fig. 7-18 Acceleration developed during impact at four different speeds.

What is the effect of large accelerations on the human body? Well, most people black out under $5g$ or $6g$ if it lasts around 10 s or more, simply because the heart cannot pump with the necessary force. But for very short exposures, the acceleration can go as high as $40g$ with no bad effects. Beyond this, some internal parts of the body continue to move after others have stopped, and there is both internal and external injury. The human limits are shown in Fig. 7-19. Passenger restraint systems in cars are designed with these limits in mind.

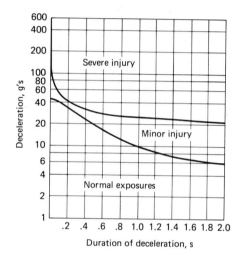

Fig. 7-19 Human acceleration limits, perpendicular to spine.

Example 7-10 Passenger Restraints in Collisions.

A certain car strikes a wall at 90 kmph and moves 0.8 m during the impact. Can the accident be survived? (Notice that the car has crumpled considerably.)

From Fig. 7-18, we see that the acceleration during impact is nearly $40g$. Figure 7-19 shows us that this is survivable if it doesn't last too long. But how long does the impact take?

We can reason this way: The car slowed down from 90 to 0 kmph. This means its average speed during impact was around 45 kmph. Expressing this in meters per second, we get 12.5 m/s. The car traveled 0.8 m at this average speed. The time to do this was

$$\text{Time} = \frac{0.8 \text{ m}}{12.5 \text{ m/s}}$$

$$= 0.06 \text{ s}$$

So even though the car bent itself up considerably, the whole process still took just a small fraction of a second. From Fig. 7-19 we see that $40g$ for 0.06 s is a "normal exposure." The passenger will probably, however, have bruises from the seat belt and/or shoulder harness.

What happens if the driver and passenger aren't restrained? They keep right on going when the car stops. Some people think they could hold themselves against the dashboard in such an instance, but this is folly. Even if their reactions were quick enough, a $40g$ acceleration with a body mass of 60 kg [130 lb$_m$] would require a force of 2400 kg$_f$ [5300 lb]. No one is this strong. ◄

Exercises

20. A plane in a level turn at a 60° bank will accelerate at $2.0g$. What is the centripetal force on a 90-kg passenger in the plane?
    Answer: 180 kg$_f$
21. A 1500-kg car traveling at 60 kmph comes to a stop in 50 cm. (a) What is the acceleration? (b) What is the force during impact?
    Answers: (a) $28g$, (b) 42 000 kg
22. A falling brick has a mass of 3.1 kg. It hits some soft ground at a speed of 90 kmph and embeds itself to a depth of 10 cm. What is the force of the impact?
    Answer: 992 kg$_f$

Summary Forces may arise in many different ways. Regardless of their sources, however, the forces on an object combine according to a tug-of-war principle, in which their directions as well as their magnitudes are important. And the total, or resultant, force on an object determines its behavior.

Newton's three laws of motion tell us how forces interact with matter. It is impossible to push on an object without having the object push back, or to pull on an object without having the object pull back. Things can move without the action of a resultant force, but it takes a resultant force to *change* the motion. Such changes in motion are called *accelerations*.

If an object moves in a curved path, the acceleration is called a centripetal acceleration. It takes a centripetal force, acting toward the center of the curve, to produce such an acceleration. The reaction to such a force is called a centrifugal force. The centrifugal force acts not on the turning object itself, but on the things it is in contact with.

Terms You Should Know

resultant force

parallelogram method

head-to-tail method

Newton's first law of motion

drag

equilibrium

acceleration

standard free-fall acceleration (g_0)

terminal speed

Newton's second law of motion

Newton's third law of motion

reaction force

centripetal acceleration

centripetal force

centrifugal force

Problems

1. A seat belt is to restrain a 100-kg passenger under an acceleration of 50g. The belt is to have a width of 5.0 cm and a thickness of 0.3 cm. (a) What total force must the belt provide? (b) What is the minimum allowable ultimate strength of the belt in kilogram-force per square centimeter?

2. When its engine is shut off while cruising, a certain boat slows down from $4\bar{0}$ to $3\bar{0}$ kmph in 4.6 s. The boat's mass is 2560 kg. Use this information to estimate the actual forward thrust needed to keep the boat cruising at a speed of 35 kmph.

3. On an icy road, the centripetal force available (resulting from friction) may be only 10 percent of a vehicle's weight. At what speed will a car skid on a flat, icy turn when the radius of curvature is 50 m?

4. A jetliner has four engines that can each produce a thrust of 8200 kg$_f$. The plane's total mass at takeoff, including passengers and fuel, is 150 000 kg. (a) What is the plane's maximum acceleration at takeoff? (b) Approximately how long would it take the plane to accelerate from rest to its takeoff speed of 240 kmph?

5. A pickup truck has a mass of 1830 kg. When empty, it can accelerate from rest to 80 kmph in 9.7 s (on a level road). Suppose, now, that it is loaded with 1 t of sand. (a) Approximately how long does it take the truck to accelerate from rest to 80 kmph? (b) Approximately how much speed could it gain in 9.7 s, starting from rest on a level road?

6. An empty railroad flatcar has a mass of 15 t. It is part of a train in which the cars behind it have a gross mass of 1020 t. The flatcar's frame has an elastic limit in compression of 300 000 kg$_f$. The entire train is moving on a straight and level track. (a) What deceleration in g's can the flatcar withstand before it begins to crumple? (b) With this deceleration, how long does it take the train to stop from a speed of 90 kmph? (c) Assuming that the train's *average* speed as it stops is 45 kmph, what (approximately) is the minimum stopping distance that will not damage the flatcar?

8

Static
Equilibrium

We have seen that an object experiences no acceleration when all the forces on it balance out. If the object is already moving when this happens, it continues to drift along at a steady speed in a straight line. Or if the object is at rest, it stays at rest.

In this chapter, we examine the case of structures and other systems that are designed to remain at rest. Such systems are said to be in static equilibrium.

static equilibrium

Definition: A system is in a state of *static equilibrium* when it is at rest and shows no tendency to begin moving.

Bridges, crane booms, houses, water towers, and a host of other structures are designed to be in static equilibrium. They are also designed to remain in this state under a variety of changing conditions.

8-1 Concurrent and Nonconcurrent Forces

We have to begin with a distinction between two kinds of force systems. In Fig. 8-1a we see a nail being driven into a board. The hammer transmits a force to the head of the nail, while friction and elastic compression in the wood apply a resisting force to the tip of the nail. Ideally, if the nail is being driven straight, these two forces act along

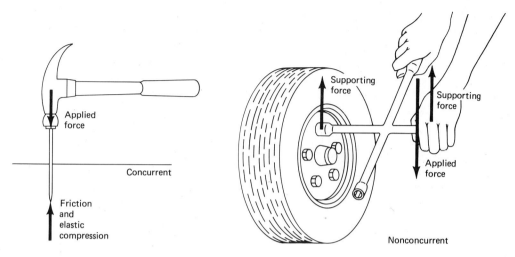

Fig. 8-1 Concurrent- and noncurrent-force systems.

lines passing through the nail's center. Forces that act through the same point are called *concurrent forces*. Concurrent forces act to move an object in a straight line. Notice the difference between this and Fig. 8-1*b*. Here a lug wrench is being used to tighten a wheel nut. Three forces are shown. If we add the two upward forces, we find that they are balanced by the downward force. But these forces do not act through the same point, and we say that the force system is *nonconcurrent*. Nonconcurrent forces act to produce a twisting action, or a rotation.

We can summarize this difference in the following definitions:

> Definition: *Concurrent forces* act at or through the same point on an object. They combine to produce motion along a line.

concurrent forces

> Definition: *Nonconcurrent forces* act on or through different points on an object. They combine to produce rotations, possibly in addition to motion along a line.

nonconcurrent forces

Notice that a single force is neither concurrent nor nonconcurrent. These terms describe properties of a *system* of two or more forces.

A force system is seldom perfectly concurrent, and so there is usually at least a slight tendency for a structure to bend or twist under load. However, in many structures and machines, the *main* forces are concurrent. In these cases, we can analyze the system with the help of the following principle:

8-2 The First Condition for Equilibrium

First Condition for Equilibrium: A system of concurrent forces produces equilibrium if and only if the total, or resultant, force is zero.

Let's use an example to show how this principle is applied.

Example 8-1 Tension in a Guy Wire.

A light, horizontal strut is supported by a guy wire as shown. Hanging from the end of the strut is a mass of 450 kg. The problem is to find the tension in the guy wire. We will neglect the weight of the strut itself and the small amount of friction in the pivot.

    This system is in static equilibrium. To apply the first condition for equilibrium, we look for a point where the important forces are concurrent. This point is at the tip of the horizontal strut. The forces here can be

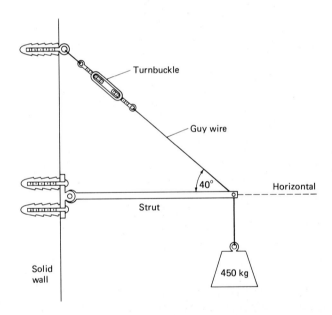

sketched as shown (not to scale). Notice that the 450-kg_f load acts downward (the direction of gravity). The tension in the guy wire acts up and to the left; it is pulling the system in this direction to keep it from falling. The third main force is the elastic compression in the strut. Since the strut acts to keep the load away from the wall, this means that the strut is *pushing* toward the right on our point of concurrency. The sketch showing the force vectors is called a *free-body diagram*.

    The next step is to make a scale drawing in which these three forces add to zero. If we use the head-to-tail method, the head of the last

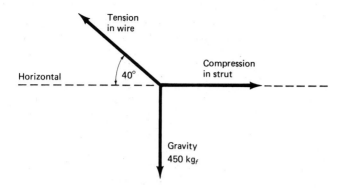

arrow must come back to the tail of the first one. At the same time, the direction of each arrow must be drawn correctly. Using a scale of 1 cm = 200 kg$_f$, we get the following scale drawing.

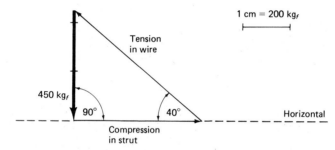

We may now measure the length of the tension arrow. This comes out as 3.5 cm. The tension we are looking for is then

$$\text{Tension} = 3.5 \text{ cm} \left(\frac{200 \text{ kg}_f}{1 \text{ cm}} \right)$$

$$= 7\overline{0}0 \text{ kg}_f$$

So in this case, the tension in the wire is much greater than the load supported by the system. Unless the wire has breaking strength greater than 7$\overline{0}$0 kg$_f$, the system will immediately fail when it is loaded. Ideally, the cable should have an elastic limit greater than 7$\overline{0}$0 kg$_f$, so there is no permanent deformation under the load.

The scale drawing also gives us the compression in the strut, if we are interested. This arrow measures 2.7 cm in length, giving a compression of 540 kg$_f$. ◀

Making the scale drawing is often easier if we use complementary angles. Since the vertical is always 90° from the horizontal, a force that acts at an angle 30° from the horizontal is also acting at 60° from the vertical. Figure 8-2 shows the construction of a scale drawing from a certain free-body diagram. First, the force that is already known is drawn to scale. Then from the tip of this arrow, a line is drawn 30° to the horizontal to represent the direction of force *A*. This can easily be done by measuring 60° from the vertical. The direction of force *B* is 50° from the horizontal, which places it 40° from the vertical. A line is then drawn with this direction. The completed sketch is a scale drawing of three forces added to give a total force of zero. A measurement from this drawing gives us the necessary sizes of forces *A* and *B*.

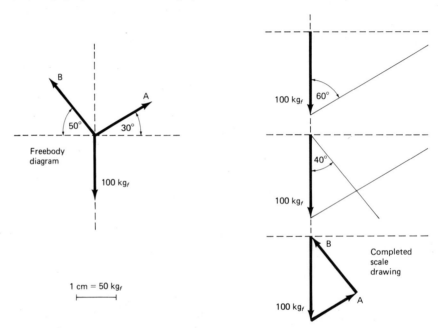

Fig. 8-2 Constructing a scale drawing to find the forces producing equilibrium. If a force acts at 30° from the horizontal, it may be easier to draw it by measuring 60° from the vertical.

1 cm = 50 kg$_f$

Of course, the drawing need not be done this way. Any technique that gets the force directions right leads to the right answer. Let's look at another example.

Example 8-2 Forces in a Simple Truss.

The simple truss shown supports a small bridge. The vertical strut (sometimes called a king post) supports a total weight of 1600 kg$_f$. This weight includes about half the weight of the bridge (called the *dead load*) plus half the weight of a vehicle centered on the bridge (called the *live load*). Why half? The reason is simply that the bridge will have two of these trusses, one on each side of the road. Anyway, the problem is to find the forces in the diagonals.

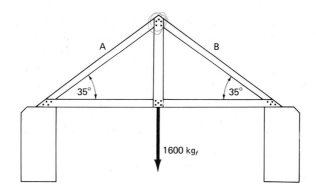

The truss, of course, has to be in equilibrium. There are many points on the truss where forces act, but the point we are interested in is at the top. Why here? At the top, the forces we are trying to find act concurrently with the force we know. The free-body diagram for this point looks like this:

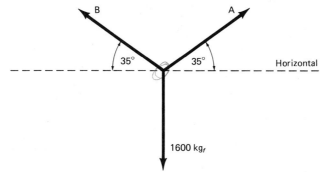

The strut on the left pushes up and to the right, while the strut on the right pushes up and to the left. For simplicity, we have labeled the forces A and B.

Now we can make the scale drawing of the forces. Using a scale of 1 cm = 400 kg$_f$, we get the following:

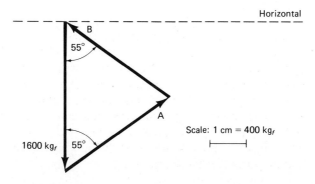

From this, we can measure arrows A and B. We find that each arrow is 3.5 cm long. Thus both forces are equal to

$$3.5 \text{ cm} \left(\frac{400 \text{ kg}_f}{1 \text{ cm}} \right) = 1400 \text{ kg}_f$$

The diagonal braces must each be capable of withstanding about 1400 kg$_f$ of compression. ◀

Let's summarize this procedure for solving equilibrium problems:

1. Find a point where the known force (or forces) is concurrent with the unknown force (or forces).
2. Decide on the direction of each force at this point.
   a. Gravity always acts vertically downward.
   b. Flexible cables always pull rather than push.
   c. Rigid bars or beams may either push or pull (cases of bending or twisting do not lend themselves to this procedure). Cover the beam with one hand and ask yourself in which direction the system would move if this beam weren't there. The force always acts opposite to this direction.
3. Sketch a free-body diagram showing all the principal forces at the point picked in step 1.
4. Make a scale drawing in which the forces are added to give a zero resultant. Be very careful with the directions on this drawing. (Use a protractor.)
5. Measure the lengths of the arrows representing the unknown forces. Use the scale factor to convert these lengths to the actual forces.

Exercises

1. Solve Example 8-1 with the 40° angle replaced by a 25° angle between the guy wire and the strut.
   Answer: Tension in wire = 1060 kg$_f$, compression in strut = 970 kg$_f$
2. Assuming that neither strut bends under the load, find the two forces. Ignore the weight of the struts themselves. *Hint*: Force A acts horizontally to the left, while B acts up and to the right at 30° from the horizontal, or 60° from the vertical.
   Answer: A = 640 kg$_f$, B = 740 kg$_f$
3. Solve Example 8-2 with the 35° angles replaced by 25° angles.
   Answer: Compression in each diagonal is 1900 kg$_f$

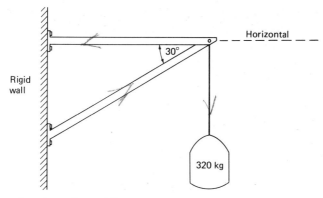

4. Find the forces in wires *A* and *B*.
    Answer: A = 135 kg$_f$, *B* = 110 kg$_f$

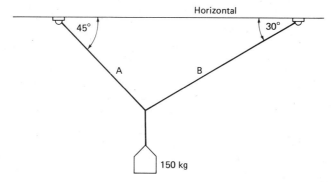

5. Find the forces in struts *A* and *B*. Assume that neither strut bends or
    twists. Ignore the weight of the struts themselves.
    Answer: A = 2600 kg$_f$, *B* = 3200 kg$_f$

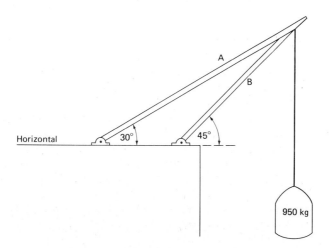

8-3 Torque

Torque is the technical term for a twisting action. Torques are closely related to forces. However, it's possible for a small force to produce a large torque, or for a large force to produce little or no torque. It all depends on how the force is applied, as we see in Fig. 8-3.

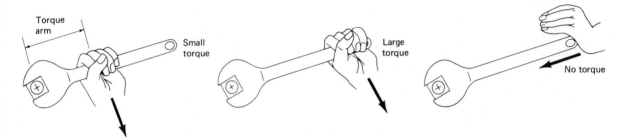

Fig. 8-3 The torque produced by a force depends on the direction of the force and the length of the torque arm.

Notice that two factors affect how much torque a force produces:

1. The length of the *torque arm* between the force and the center of rotation*
2. The direction of the force

When we're interested in getting a large torque out of a given force, we usually apply the force at a right angle to the torque arm and as far as possible from the center of rotation. In trying to remove a rusted nut, we may even resort to a "cheater bar," as shown in Fig. 8-4. This is simply a section of pipe slipped over the wrench handle to increase the length of the torque arm.

The quantitative definition of torque is as follows:

torque　　　Definition: *Torque* is the twisting action produced by a force, found by multiplying the force times the effective torque arm between the force and a center of rotation:

$$\text{Torque} = (\text{force})(\text{effective torque arm}) \qquad (8\text{-}1)$$

Torque is sometimes also referred to as the "moment of a force." In this book, we stay with the term "torque."

Notice that the torque depends on something called the effective

---

* The torque arm may also be referred to as a lever arm. We use the term "torque arm" here because in many cases (a belt-and-pulley system, for instance) it may not be obvious that there is a lever at work.

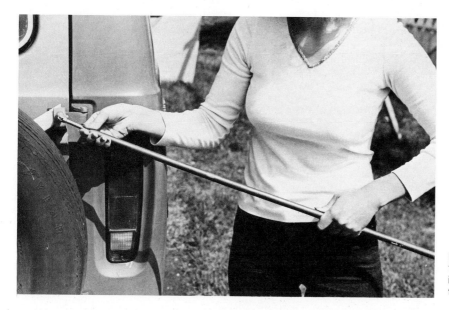

Fig. 8-4 A "cheater bar," used to remove a rusted nut. Increasing the torque arm increases the torque.

torque arm. If a force is applied at a right angle to a wrench handle, the entire distance from the pivot is effective. But if the force is applied at some other angle, the torque arm is effectively shorter. The *effective* torque arm in such cases is the perpendicular distance from the line of action of the force, as shown in Fig. 8-5.

The unit of torque is a unit of force times a unit of length. This leads to many possibilities: newton-meters, kilogram-meters, kilogram-centimeters, pound-inches, and ton-feet, to name a few. Table 8-1 lists the conversion factors between some of these units. The torque wrenches

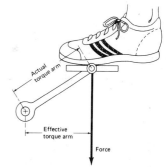

Fig. 8-5 The effective torque arm for torque calculations is the perpendicular distance between the line of aciton of the force and the center of rotation.

TABLE 8-1 CONVERSION FACTORS FOR TORQUE UNITS

	N·m	**kg$_f$·m**	**kg$_f$·cm**	**lb·in**	**lb·ft**
1 newton-meter =	1	0.101 97	10.197	8.850 7	0.737 56
1 kilogram-meter =	9.806 65	1	100	86.796	7.233 0
1 kilogram-centimeter =	0.098 067	0.01	1	0.867 96	0.072 330
1 pound-inch =	0.112 98	0.011 521	1.152 1	1	0.083 333
1 pound-foot =	1.355 8	0.138 26	13.826	12	1

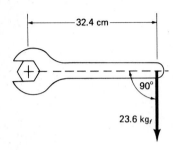

used by automobile mechanics are usually calibrated in units of kilo-gram-meters ($kg_f \cdot m$) or pound-feet ($lb \cdot ft$). Occasionally, we also see the newton-meter ($N \cdot m$) used on such tools. The newton-meter is the official SI unit for torque, and it is almost always used in scientific work.

Example 8-3 Torque on a Bolt.

A bolt is tightened by a wrench as shown. How much torque is being applied?
　　The effective torque arm here is the entire 32.4 cm between the point of application of the force and the center of rotation of the bolt. Why? The answer is that the force is acting at a right angle to the wrench handle. Using Eq. (8-1), we get

$$\text{Torque} = (\text{force})(\text{effective torque arm})$$

$$= 23.6 \text{ kg}_f \text{ (32.4 cm)}$$

$$= 765 \text{ kg}_f \cdot \text{cm}$$

This result may also be expressed as 7.65 $kg_f \cdot m$, 75.0 $N \cdot m$, 55.4 $lb \cdot ft$, or 664 $lb \cdot in$. ◄

Example 8-4 Torque on a Crankshaft.

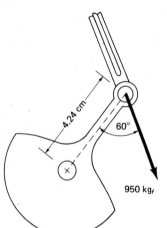

Both the size and the direction of the force on the crankshaft change as the shaft rotates. This, in turn, leads to a variable torque. Let's find the torque developed when the shaft is as shown.
　　In this position, the force is not perpendicular to the torque arm. The effective torque arm is therefore less than 4.24 cm. To find out what it is, we can make a scale drawing. From this drawing, we can measure the effective torque arm as 3.7 cm. (If you know how to do a trigonometric calculation, this is more direct and gives an effective torque arm of 3.67 cm.) Then using Eq. (8-1), we have

$$\text{Torque} = (\text{force})(\text{effective torque arm})$$

$$= 950 \text{ kg}_f \text{ (3.7 cm)}$$

$$= 3500 \text{ kg}_f \cdot \text{cm}$$

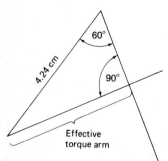

Effective torque arm

1 cm

This is better expressed as 35 $kg_f \cdot m$ [250 $lb \cdot ft$]. This is approximately the peak torque developed during the power stroke in one cylinder of a large engine. But because the torque varies so greatly, and the engine has a four-stroke cycle, the steady torque output of an eight-cylinder engine will probably be somewhat less than this figure. ◄

As with all our formulas, we can write Eq. (8-1) in other ways. If we are interested in the force needed to produce a certain torque, we can write

$$\text{Force} = \frac{\text{torque}}{\text{effective torque arm}} \qquad (8\text{-}2)$$

Or if we want to know the torque arm needed to develop a certain torque from a certain force, we can write

$$\text{Effective torque arm} = \frac{\text{torque}}{\text{force}} \qquad (8\text{-}3)$$

Thus, if an electric motor produces a torque of 3.0 $kg_f \cdot m$ and turns a circular saw blade 0.2 m in radius, the cutting force at the blade can be found from Eq. (8-2). Dividing the torque by the 0.2-m effective torque arm gives a cutting force of 15 kg_f [33 lb].

If the same motor drives a sanding belt and the belt must transmit a frictional force of 36 kg_f, Eq. (8-3) can be used to get the effective torque arm: 3.0 $kg_f \cdot m$ divided by 36 kg_f, or 0.083 m [3.3 in]. This is also the radius of the rollers that should be used to drive the belt.

Exercises

6. A force of 18.6 kg_f acts perpendicular to a torque arm of 0.175 m. Find the torque in (a) kilogram-meters, (b) newton-meters, (c) pound-feet.
   Answers: (a) 3.26 $kg_f \cdot m$, (b) 31.9 N·m, (c) 23.5 lb·ft
7. A force of 58.3 lb acts perpendicular to a torque arm of 18.7 in. Find the torque in (a) pound-inches, (b) pound-feet, (c) kilogram-meters.
   Answers: (a) 1090 lb·in, (b) 90.0 lb·ft, (c) 12.6 $kg_f \cdot m$
8. Find the torque, in kilogram-meters, developed in the system shown.
   Answer: 13 $kg_f \cdot m$
9. Find the torque in Exercise 3 when the angle is changed to (a) 20°, (b) 70°, (c) 90°.
   Answers: (a) 5.9 $kg_f \cdot m$, (b) 15 $kg_f \cdot m$, (c) 17 $kg_f \cdot m$
10. What force is needed to develop a torque of 12.5 $kg_f \cdot m$ when the torque arm is 18.5 cm? Assume that the force acts perpendicular to the torque arm.
    Answer: 67.6 kg_f
11. What effective torque arm is needed to produce a torque of 9.82 $kg_f \cdot m$ from a force of 24.5 kg_f?
    Answer: 0.401 m

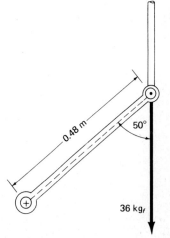

0.48 m

50°

36 kg_f

8-4 The Second Condition for Equilibrium

Suppose that you are driving on a sandy road and you suddenly step on the brake pedal very hard. The wheels lock, and the car begins to skid. There is a force of friction at each tire tread which is attempting to rotate the wheels. But there is another force on each brake disk (or drum) which develops a countertorque to oppose the rotation. The wheels lock because this countertorque balances the torque produced by road friction.

Notice that we did not say that the *forces* balance. In fact, they don't balance at all in this case (friction on the brakes is greater than on the tires). To keep a tire or anything else from rotating, the *torques* must cancel. This is the idea behind the second condition for equilibrium.

second condition for equilibrium

Second Condition for Equilibrium: A system of nonconcurrent forces combines to produce no rotations if the sum of the clockwise torques equals the sum of the counterclockwise torques.

A simple application is the equal-arm balance shown in Fig. 8-6. An unknown mass is placed in one pan, and known masses are added to the other pan until the system balances. The right-hand weight produces a torque acting to produce a clockwise rotation. At the same time, the left-hand weight produces a torque that tries to rotate the system

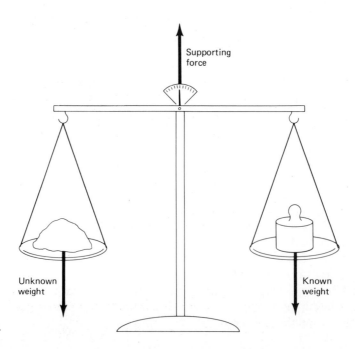

Fig. 8-6 The equal-arm balance.

Static Equilibrium **209**

counterclockwise. Equilibrium is achieved when these two torques are equal. Since the two torque arms are also equal, the system balances when the weight in one pan equals the weight in the other pan. At the same time, the upward force supporting the balance must equal the sum of all the downward forces.

The equal-arm balance is useful when the weights (or masses) being compared are small. For larger masses, like a person or a vehicle, the unequal-arm balance is used. Example 8-5 shows how this device works.

Example 8-5 The Unequal-Arm Balance.

Figure 8-7 shows the basic features of this device. The vehicle to be weighed drives onto a platform which is free to move vertically a very small amount (typically a few centimeters). The vehicle's weight is transmitted to a very short torque arm on the balance's beam. The other end of the beam is a long torque arm on which a weight can be slid back and forth. When the weight is properly positioned, its torque lifts the weighing platform.

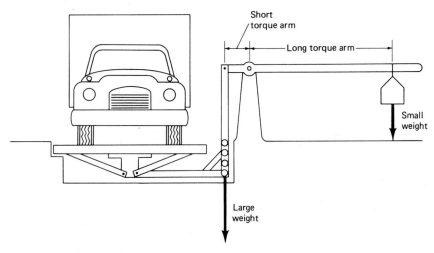

Fig. 8-7 Principle of the unequal-arm balance. A large weight on a short torque arm can be balanced by a smaller weight on a long torque arm.

Suppose that the short torque arm measures 0.91 cm and a sliding weight of 125 kg$_f$ lifts the platform when the long torque arm measures 0.753 m. What is the vehicle's weight?

Application: Bridges

The earliest bridges were probably trees that had fallen across streams. Today's viaducts use the same principle: stiff horizontal girders supporting a road over a series of short spans. This is the cheapest way to build a bridge.

For intermediate spans, trusses or arches may be used. In crossing deep valleys, the structure is commonly built below the road surface; when the clearance is lower, it is usually on top. In either case, the uppermost structural members in a truss or an arch are always in compression.

The strength of a truss bridge is based on the rigidity of triangles. Still, a triangle is rigid only as long as no side is stressed beyond its elastic limit. The bridge must be designed so that this never happens under any possible combination of dead and live loads. Because stronger structural members also increase the dead load, this can become a very complicated design problem.

A cantilever is a rigid structural member that projects horizontally from its point of support, so that the top is in tension and the bottom in compression. A cantilever bridge is usually built as two separate cantilevers, which are then joined in the center of the span. (Sometimes these cantilevers are joined by a short truss.) Although a cantilever may look something like a truss (both are made of triangles), it is held up on a different principle. If the center of a cantilever bridge is removed, the rest of the bridge is still stable.

The principle of the suspension bridge is simple: String a number of cables between two towers, and then hang the road from these cables. In practice this arrangement is not very stable; changing loads cause it to swing from side to side and maybe even up and down. Thus all but the most primitive suspension bridges use trusses to stiffen the span.

Although the world's longest truss bridge spans 376 m [1230 ft], most span less than 200 m. Steel arches do better. One in New York City spans 504 m [1650 ft], and many are longer

than 250 m. Cantilever bridges can be built to support very
heavy loads over long spans. One built in Quebec in 1917
carries a railway over a span of 549 m [1800 ft]. The cantilever
design was very popular in the last century because it is so easy
to build, fairly small structural members can be used, and these
are positioned in sequence beginning at the ends and working
toward the middle. The Brooklyn Bridge, built in 1883 and still in
use, was the world's first large suspension bridge. The main
span is 486 m [1596 ft] long. The record for a single span is now
held by the Humber River suspension bridge in England: 1410 m
[4626 ft].

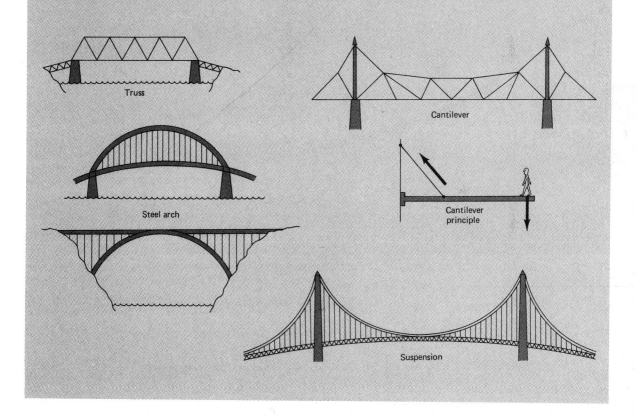

Truss

Cantilever

Steel arch

Cantilever principle

Suspension

A sketch of the forces on the balance beam looks like this (not to scale):

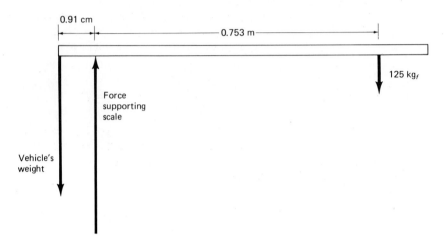

The clockwise torque, from Eq. (8-1), is

$$\text{Clockwise torque} = 125 \text{ kg}_f \,(0.753 \text{ m})$$

$$= 94.1 \text{ kg}_f \cdot \text{m}$$

According to the second condition for equilibrium, this is balanced by an equal counterclockwise torque of 94.1 kg$_f$·m. To generate this much torque at an effective lever arm of just 0.91 cm [0.009 1 m], we need a fairly large force. Using Eq. (8-2),

$$\text{Force} = \frac{\text{torque}}{\text{effective torque arm}}$$

$$= \frac{94.1 \text{ kg}_f \cdot \text{m}}{0.009\ 1 \text{ m}}$$

$$= 10\ 300 \text{ kg}_f$$

The device will be adjusted so it takes no weight to balance it when there is no load on the platform. Thus our result here is the weight of the vehicle itself: 10 300 kg$_f$ [22 700 lb]. Notice that the balance support holds a total of 10 300 + 150 kg$_f$, plus the weight of the balance beam and platform assembly.

When the balance is designed to measure very heavy loads, it may be a bit more complicated than we have indicated in this example.

In particular, it may use two or even three beams with unequal torque arms arranged so that one beam trips the next. Such a device is called a *compound* unequal-arm balance. It allows very large loads to be balanced by relatively small weights. The basic principle, however, is the same as in this example: A small force with a large torque arm can balance a large force with a short torque arm. ◀

Exercises

12. The drill shown can develop a torque of 3.2 $kg_f \cdot m$. The work must be clamped or otherwise restrained to keep the bit from "grabbing" and spinning it. Find the force necessary to keep the work from spinning if it is clamped (*a*) 75 cm, (*b*) 7.5 cm, (*c*) 2.0 cm from the hole.
    Answers: (*a*) 4.3 kg_f, (*b*) 43 kg_f, (*c*) 160 kg_f

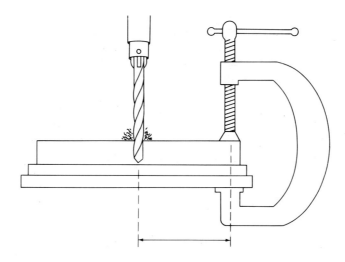

13. Find the force *F* needed to balance the weight as shown.
    Answer: 18.8 kg_f

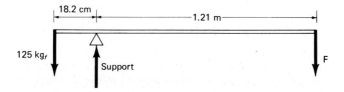

8-5 Center of Gravity

Suppose that you have to lift a 3-m [10-ft] section of pipe. Where do you grab it? Probably very near the center, 1.5 m from either end. Certainly you would never try to lift the entire pipe by grabbing it at one end. The center is a natural balancing point, and holding it there means that your wrist has to apply little or no torque to keep the entire pipe above the ground.

Suppose, now, that you go to lift a chain saw with a 60-cm [24-in] cutting bar. Let's say that the entire saw, including the engine, measures 81 cm [32 in] tip to tip. Do you lift the saw from the center of this 81 cm? No. You lift it by the handle, which is just forward of the engine. Unlike the pipe, the natural balancing point of the saw is closer to one end.

Every object, regardless of its size or shape, has a natural balancing point somewhere. This point is called the object's center of gravity.

center of gravity

Definition: The *center of gravity* of an object is its natural balancing point. The entire weight of the object can be considered to act at this point.

Figure 8-8 shows our pipe and saw, each suspended by a rope. In both cases, the center of gravity rests below the point of suspension. If an object has a uniform shape and composition from one end to the other, its center of gravity is the same as its geometrical center. This is the case with boards, solid doors, oil drums, bricks, and even a balanced tire (whose center of gravity is in the middle of the hole). For objects with nonuniform shape or composition, the center of gravity is closer to the

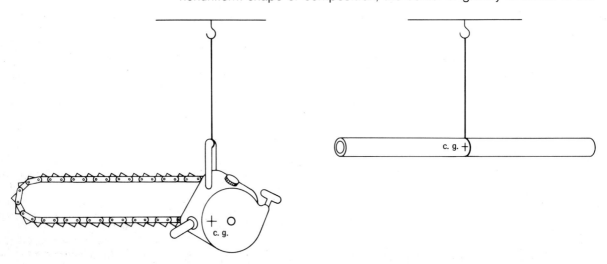

Fig. 8-8 An object's center of gravity is its natural balancing point.

heavier end. Thus a car with a front engine has its center of gravity closer to the front than to the rear.

The force of gravity on an object can be considered to act at the object's center of gravity. In Fig. 8-9, we see a rectangular sign supported by two cables. The center of gravity has been abbreviated c.g. In the top sketch, this system is in equilibrium, with each cable supporting half the weight of the sign, and with the right-hand cable supplying a counterclockwise torque about the c.g. that is balanced by the clockwise torque of the left-hand cable. In the second sketch, one cable has come loose. The system is no longer in equilibrium. There is an unbalanced torque that rotates the sign about the remaining point of suspension. When the sign swings so the c.g. passes to the left of the point of suspension, the unbalanced torque acts to swing it back toward the right. The sign will swing back and forth for awhile before it comes to rest. Notice that when it does finally stop swinging, the sign's c.g. is directly below the point of suspension. In this position, the two remaining forces act along the same line, and no torques are produced.

Regardless of what point an object is suspended from, it will finally come to rest with its c.g. below the point of suspension. This principle can be used to find the c.g. of an irregular shape (Fig. 8-10). A cutout of the shape is made from cardboard, sheet metal, wood, or some other convenient material. Three small holes are made at different places along the edge. The shape is suspended by a wire through each hole in turn, and a plumb bob and chalk line are used to mark the vertical in each case. The three lines will intersect at a single point, which is the c.g. In fact, the shape will balance horizontally on a fingertip placed beneath this point.

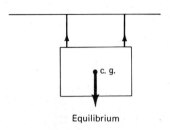

Equilibrium

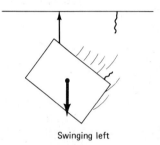

Swinging left

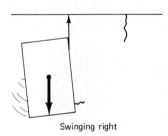

Swinging right

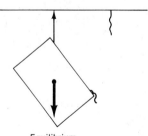

Equilibrium

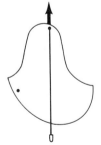 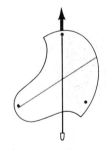 

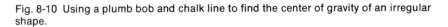

Fig. 8-10 Using a plumb bob and chalk line to find the center of gravity of an irregular shape.

Fig. 8-9 Motion of a sign when one of the suspending cables snaps. The sign finally comes to rest with its center of gravity directly below the remaining point of suspension.

The c.g. of a vehicle is affected by the placement of the loads it carries. This is a very important consideration in loading airplanes. Since an airplane's lift is developed between the leading and trailing edge of the wing, the plane cannot sustain level flight if the c.g. falls outside this range. If the c.g. is too far forward, the plane will dive; if it is too far to the rear, the plane will stall. The plane's c.g. may change in flight as fuel is consumed.

As we mentioned earlier, the c.g. of a tire, or a complete wheel, is at its geometrical center. Thus a wheel resting on a level surface will be supported at a point directly below its c.g., as in Fig. 8-11. This wheel will not roll by itself. Suppose now that the surface is inclined a small amount. The wheel still may not roll, since its bottom is flattened slightly by its weight. But if the surface is inclined enough, the wheel's weight will no longer act on the contact point (remember, weights always act downward). In this case, the wheel's weight produces a torque that causes it to roll downhill.

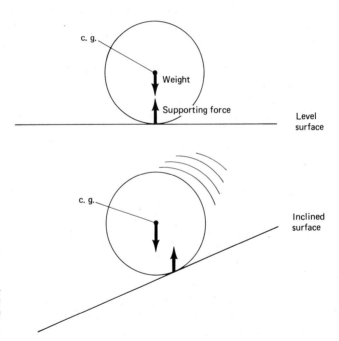

Fig. 8-11 A wheel will roll under its own weight if its center of gravity is ahead of the point of support.

As long as the wheel's c.g. is at its center, the wheel will probably roll smoothly. But if the c.g. is off-center, the torque will vary and the wheel will roll with a series of jerks. This can easily be seen by placing some chewing gum in a jar lid and rolling the lid. At high speeds, the centrifugal force of the off-balance weight causes a tremendous pounding as the wheel rolls. For this reason, car and truck wheels must be balanced. This amounts to fastening small weights at various places on the wheel rim until the c.g. is at the wheel's center. Since this is done when the wheel is at rest, it is called a *static balance*.

But sometimes a wheel can be statically balanced and still wobble when it rolls at high speed. Let's examine why this might happen. Figure 8-12 shows a tire with a small imperfection in the casing that has been statically balanced with a rim weight. These two weights exert centrifugal forces on the wheel as it rolls. The weights are not equal, and neither are their speeds on their turning radii. Still, it turns out that the two centrifugal forces are exactly equal and in opposite directions when the tire has been statically balanced. Then where does the wobble come from? From the fact that the wheel has thickness. Suppose that one weight is on the inside of the tire but the other is on the outside. The two centrifugal forces now act along different lines of action (they are nonconcurrent). Together, they twist the wheel, and the direction of this twist changes from inward to outward and back again during each revolution. The effect is greater at high speeds because the centrifugal reaction forces increase with speed.

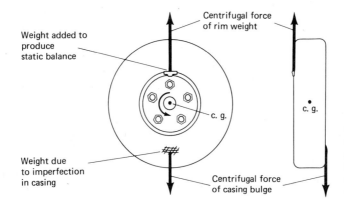

Fig. 8-12 A statically balanced wheel may still wobble at high speeds if the centrifugal reaction forces combine to produce a net torque about the wheel's center of gravity.

In practice, the problem is corrected by performing a *dynamic balance*. A roller drives the wheel at high speed, and a sensor detects any wobble, at the same time recording the placement of the weights needed to correct it. The wheel is then stopped and the weights are fastened to the rim, on the inside if necessary.

The concept of center of gravity is useful for analyzing systems where the weight contributes to equilibrium. Example 8-6 shows a familiar case.

Example 8-6 The Crane.

Figure 8-13 shows a mobile crane unit used to position a platform. We want to know how far behind the rear wheel the platform can be lowered. The truck's front wheels are not to leave the ground.

Let's suppose that the entire crane unit weighs 8.8 t [8800 kg$_f$] and that the platform weighs 1.9 t [1900 kg$_f$]. The crane's center of gravity is 4.3 m in front of the rear axle.

We may then make the following sketch, which is the free-body diagram of the entire system.

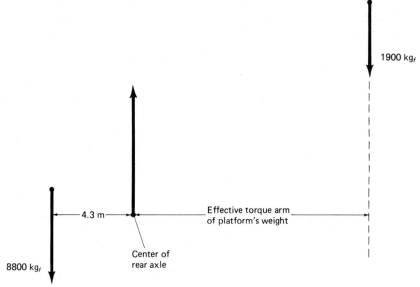

This diagram shows the positions of the forces just as the truck's front wheels are about to leave the ground. At this point, all the weight is supported by the rear set of wheels.

The crane's weight, acting at its c.g., produces a counter-clockwise torque about the rear axle. This torque is

$$\text{Counterclockwise torque} = (\text{force})(\text{effective torque arm})$$

$$= 8800 \text{ kg}_f \, (4.3 \text{ m})$$

$$= 38\,000 \text{ kg}_f \cdot \text{m}$$

If the platform produces a clockwise torque exceeding this value, the entire system will topple to the right. Thus, 38,000 kg$_f$·m is the upper limit on the clockwise torque. To get this much torque from a 1900-kg$_f$ weight, we need an effective torque arm of

$$\text{Effective torque arm} = \frac{38\,000 \text{ kg}_f \cdot \text{m}}{1900 \text{ kg}_f}$$

$$= 20 \text{ m}$$

This is our answer. We can lower the platform a maximum of 20 m behind the rear wheels. If we swing the boom out farther than this, the platform comes crashing down as the entire system topples backward.

Notice that we didn't worry about the supporting force at the rear set of wheels. Although this force is about 10,700 kg$_f$ (8800 kg$_f$ + 1900 kg$_f$), it produces no torque about the rear axle. Just like the wrench in Fig. 8-4, the 10 000 kg$_f$ cannot produce a torque about the point on which it acts. ◀

Exercises

14. A cylindrical drum filled with oil measures 64 cm in diameter and 106 cm in height. Where is its center of gravity?
    Answer: 53 cm up from the bottom and 32 cm in from the sides
15. A flyweight has a mass of 5.71 kg and a c.g. 6.41 cm from its center of rotation. It is free to rotate in a vertical plane as shown. (a) What torque applied to the shaft will hold the c.g. level with the center of rotation? (b) What shaft torque is necessary to hold the c.g. directly below the center of rotation?
    Answers: (a) 36.6 kg$_f$·cm, or 0.366 kg$_f$·m, (b) 0
16. A car has a wheelbase of 3.15 m [124 in] and a mass of 1750 kg [3850 lb$_m$]. Its center of gravity is 1.02 m behind the front wheels. (a) How much force is needed to lift the car at the front wheels, leaving the rear wheels on the ground? (b) How much force is needed to jack up the rear axle, leaving the front wheels on the ground?
    Answers: (a) 1180 kg$_f$, (b) 567 kg$_f$

Vertical

8-6 Stability

We have all seen cases where something is balanced and left to stand, but then it topples over "by itself." A golf ball on a tee, a canoe with two people standing in it, or a bicycle that is moving very slowly all show this tendency toward toppling over. Such systems are said to be *unstable*.

The *stability* of a system refers to how it responds to small disturbing forces or torques. For instance, we can stand a penny on edge on a flat table without too much trouble. Since it is now in equilibrium, the penny should stay that way indefinitely if there are no other forces or torques. But, in practice, vibrations in the table and wind currents in the room will eventually combine to knock the penny over. Thus the penny is unstable when standing on edge.

It is not enough to design a structure to be in equilibrium. The structure must also be able to respond to various disturbing influences without permanent damage. It must be, in other words, in stable equilibrium.

Application: The Grab

The grab is a device for picking up bulk materials like coal and iron ore. It has two heavy steel jaws, which are connected and pivoted at a point above their centers of gravity. A closing cable is attached to this pivot, and pulling up on this cable causes the jaws to snap shut. This happens because the c.g. of each jaw tries to move under their common point of suspension.

The filled grab is raised and moved around with the closing cable. A second cable is used to open the grab and lower it or raise it in the open position. This opening cable is connected through a splice to each jaw. With tension in the opening cable and the closing cable slack, each jaw moves its separate way to align its c.g. below the new points of suspension.

The operation of the grab, both in closing and in opening, depends on the principle that an object's c.g. tries to align vertically with the point of suspension.

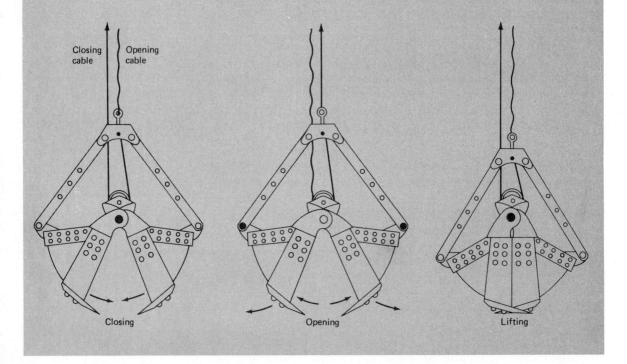

Closing cable　Opening cable

Closing　　　Opening　　　Lifting

Definition: *Stable equilibrium* is the state of a system where, if it is disturbed, forces or torques are produced which return the system to its original state.

stable equilibrium

Definition: *Unstable equilibrium* is the state of a system where, if it is disturbed, forces or torques are produced which move the system farther away from its original state.

unstable equilibrium

Definition: *Neutral equilibrium* is the state of a system where, if it is disturbed, no additional forces or torques are produced.

neutral equilibrium

Figure 8-14 shows simple examples of these three kinds of equilibrium. Actually, any number of forces and torques can combine to produce stability. In many of the important cases, however, the center of gravity determines the stability: The lower the c.g., the more stable the system. Any system has a natural tendency to go into a state that lowers its c.g., if it can. Thus if a truck has a lower c.g. when lying on its side than when standing on its wheels, we should not be too surprised if we find the truck lying on its side someday.

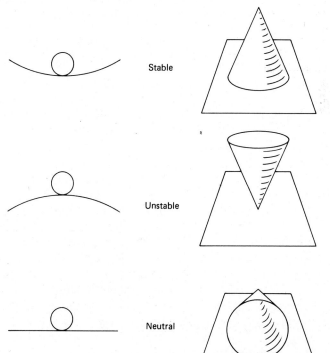

Stable

Unstable

Neutral

Fig. 8-14 Stable, unstable, and neutral equilibrium.

In Fig. 8-15, we see a crate with a high c.g. It is standing upright, and it is stable. Why? The reason is that if we tip it slightly, the supporting force acts wholly at the leading edge. The weight acts on a line of action behind this point and produces a torque that acts to stand the crate back up. But if we tip the crate too far, gravity now acts to the right of the supporting force and the torque pulls the crate over. How far is too far? As we see in the figure, a plumb line from the c.g. must intersect the base if the crate is to remain upright.

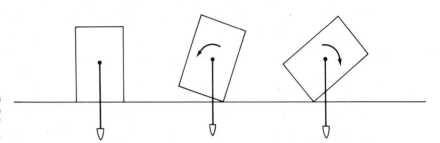

Fig 8-15 An object will remain upright as long as a plumb line dropped from the center of gravity falls within the base.

The crate in Fig. 8-15 is stable, but it would be more stable if its c.g. were lower. In that case, it would have to be tipped farther to get the plumb line from the c.g. to fall outside the base.

Summary A structure may be held in static equilibrium by the sheer weight of its parts; masonry walls, concrete piers, and earth dams are examples. But most modern constructions achieve equilibrium through a complicated balance of many opposing forces and torques. Of course, the weight of the structure still plays a part, and this is why center of gravity is such an important concept. But there are also elastic forces of tension and compression in the structural members, wind load and other "live loads," and the torques produced by these different forces acting on different points in the structure. All these forces and torques must cancel one another over a wide range of conditions. As a result, the force in a structural member can often be much greater than the weight it seems to support. If a structure supports a weight of 1 t, for instance, some of its parts may have to withstand forces of many metric tons.

The failure of a seemingly minor structural member can begin a domino effect that leads to the failure of the entire structure. This is why bridges and airframes must be given periodic, detailed inspections. A weld that seems to be an insignificant part of a bridge may bring down the whole edifice if it fails.

Terms You Should Know

static equilibrium	torque arm
concurrent forces	second condition for equilibrium
nonconcurrent forces	center of gravity
first condition for equilibrium	stable equilibrium
free-body diagram	unstable equilibrium
torque	neutral equilibrium

Problems

1. A roof is loaded with a 250-kg$_f$ vertical force, as shown. (a) What force do the rafters transmit to the walls? (b) If the walls are tied together with a horizontal joist, what is the tension in the joist?

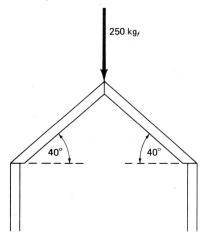

2. Find the weight supported by piers A and B. Hint: Find the force needed at B to keep the system from rotating around the top of pier A.

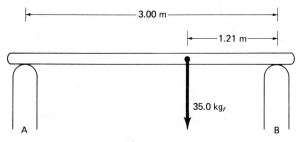

3. Find the tension in wires A and B.

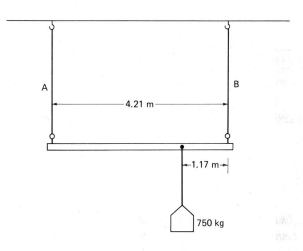

4. A small drawbridge has a span of 12 m and a mass of 14 t. It lifts in two halves, pivoting at the ends and separating in the middle. (a) Approximately what torque is needed to open the bridge? (b) Approximately what torque would be needed if the entire span were pivoted at one end? Hint: Make a sketch, and assume that the bridge's mass is uniformly distributed across the span.

5. An airplane has two wing-mounted engines, each of which produces a thrust of 1650 kg$_f$. The engines are centered 4.52 m from the plane's centerline. One of the engines fails in flight. (a) What torque is needed to keep the plane flying straight when the remaining

engine is at full throttle? (*b*) If the rudder is located 14.2 m behind the plane's c.g., what lateral force on the rudder is needed to produce this torque?

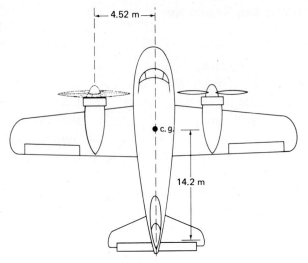

9
Energy and Power

The term "energy" has been much in the news in recent years. We have become painfully aware of the energy it takes to move our cars, heat or cool our homes, and light our lights. The world has suddenly discovered that energy costs money.

In this chapter, we look into the technical aspects of energy: where it comes from, where it goes, and what it can do along the way. We also see that energy and power are actually different things, even though they are routinely confused with each other in newspapers and magazines.

Introduction

We have seen that forces sometimes cause motion, and sometimes they don't. We can kick a football and send it through the air, but if we kick a dump truck, the only likely result is a sore foot. There is an obvious physical difference between these two cases: kicking the football causes motion, while kicking the dump truck doesn't.

When a force causes an object to move, we say that the force does *work* on the object. The amount of work depends on the size of the force and the distance the object moves in the direction of the force.

9-1 Work

> Definition: The *work* done by a force is the product of the force and the distance that it moves an object in the direction it acts.

work

Work = (force)(distance moved in direction of force)　　(9-1)

Suppose that we try to push a car that has two flat tires (Fig. 9-1). We struggle and groan and shed a lot of perspiration, but the car doesn't budge. How much work has been done? Exactly none. Without motion, there is no work.

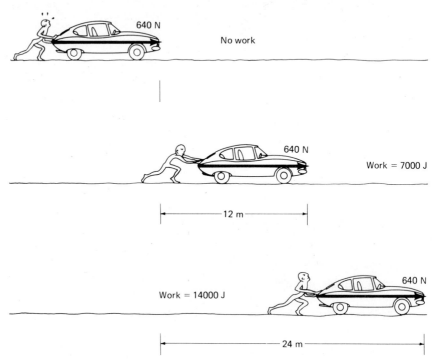

640 N

No work

640 N

Work = 7000 J

|—————— 12 m ——————|

640 N

Work = 14000 J

Fig. 9-1 Work depends on the distance that a force moves an object.

|—————————— 24 m ——————————|

Suppose, now, that we push a car which does move. Let's say that the force is 640 N [65 kg$_f$, or 144 lb], and the car moves forward 11 m [39 ft]. According to Eq. (9-1), the work done is

$$\text{Work} = 640\,\text{N} \, (11\,\text{m})$$

$$= 7000\,\text{N} \cdot \text{m}$$

What if we push the car twice as far? This still takes a force of 640 N, but we are using the force to accomplish more. Since the car goes twice as far, the force now does twice as much work: 14 000 N·m.

In the International System, the unit of work is the *joule* (J). This is the work done by a 1-N force acting through a distance of 1 m:

$$1 \text{ J} = 1 \text{ N} \cdot \text{m}$$

Thus our previous answers may be written as 7000 J and 14 000 J. Since the joule is an SI unit, the SI multipliers may also be used. Thus, 7000 J is the same as 7.0 kilojoules (kJ).

If a force of 1 kg_f acts through a distance of 1 m, then 1 $\text{kg}_f \cdot$m of work has been done. This unit is related to the joule by

$$1 \text{ kg}_f \cdot \text{m} = 9.806 \ 65 \text{ J}$$

In the old USCS units, force is measured in pounds and distance in feet. The unit of work is then the foot-pound (ft·lb). This is slightly larger than the joule:

$$1 \text{ ft} \cdot \text{lb} = 1.355 \ 8 \text{ J}$$

Example 9-1 Work in a Hydraulic Log Splitter.

Figure 9-2 shows the important features of this device. A cylinder and a wedge are rigidly fastened to a frame. A piston moves a ram that slowly pushes the log against the wedge. The top bar is latched in place as a safety precaution to prevent a misaligned log from "bucking" out of the device.

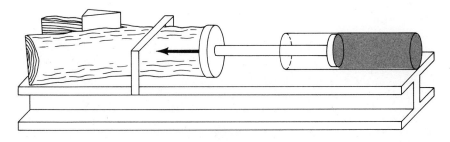

Fig. 9-2 A hydraulic log splitter.

Suppose that the cylinder has an inside diameter of 4.20 cm, and the operating pressure is 17.8 kg_f/cm², gauge. The piston moves 0.65 m in splitting a log of this length. What work is done?

The piston's area is 13.9 cm², so the force on the ram is

$$\text{Force} = (\text{pressure})(\text{area})$$

$$= 17.8 \; \frac{\text{kg}_f}{\text{cm}^2} \; (13.9 \text{ cm}^2)$$

$$= 247 \text{ kg}_f$$

The work done by this force is found from Eq. (9-1):

$$\text{Work} = (\text{force})(\text{distance moved})$$

$$= 247 \text{ kg}_f \; (0.65 \text{ m})$$

$$= 160 \text{ kg}_f \cdot \text{m}$$

Ordinarily, we express the final result in *joules*.

$$\text{Work} = 160 \text{ kg}_f \cdot \text{m} \; \left(\frac{9.806 \; 65 \text{ J}}{1 \text{ kg}_f \cdot m} \right)$$

$$= 1600 \text{ J}$$

The reader can verify that this is equivalent to 1200 ft·lb.

At this point, this example might seem like simply a "fun with figures" exercise. However, as we will soon see, this calculation can lead us to the required size of the pump motor. It can also tell us how fast the job can be done if we already know something about the pump and motor. Work, in fact, is a very useful quantity for describing the operation of power equipment. ◀

Exercises

1. A force of 2500 N moves a boat a distance of 90.0 m. What work is done, in joules?
   Answer: 225 000 J, or 225 kJ
2. Find the work done in lifting a 780-kg elevator through a vertical distance of 22 m.
   Answer: 17 000 kg$_f$·m, or 170 000 J, or 170 kJ
3. A force of 270 lb pulls a railroad car. Find the work done in pulling the car 1 mi in (a) foot-pounds, (b) megajoules.
   Answers: (a) 1 400 000 ft·lb, (b) 1.9 MJ
4. A piston has a stroke of 12.5 cm. In moving this distance, it is to do

982 J of work. Find the required force on the piston in (a) newtons, (b) kilogram-force.
Answers: (a) 7860 N, *(b)* 801 kg$_f$

Figure 9-3 shows an archer shooting an arrow from a bow. First she draws the bowstring. Since this requires a force moving through a distance, work is being done. Then with the bow "cocked," she takes aim. There is no work done here since there is no motion. If this had been a crossbow, it could stay loaded indefinitely. Finally, she releases the arrow. The elastic force in the bowstring propels the arrow forward.

9-2 Potential Energy

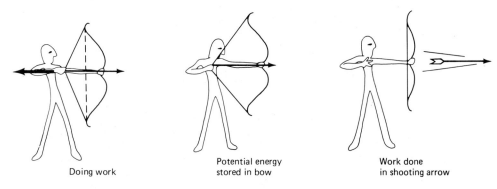

Doing work

Potential energy
stored in bow

Work done
in shooting arrow

Fig. 9-3 Work done in stretching a bowstring is stored as potential energy. When the string is released, this same work propels the arrow.

How much work does the bowstring do in shooting the arrow? The same work, it turns out, as the archer did in stretching the string in the first place. The bow just stores this work for awhile.

There are many systems that can store work and then release it on demand. Sometimes the stored work is generated by natural processes, like the water held behind a dam or the electric charge separation between a storm cloud and the ground. Water behind the dam does work when it flows through a water turbine; if the dam breaks, the water will do work in a devastating way. When the storm cloud discharges in a lightning bolt, the work may splinter a tree. Chemical fuels formed long ago by natural processes can do work if they are burned in an engine. In cases like these, the "stored work" is called potential energy.

Definition: *Potential energy* is work that has been stored by changing the positions of the parts of a system in such a way that this work can later be recovered.

potential energy

GRAVITATIONAL

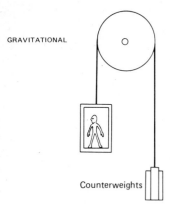

Counterweights

ELASTIC

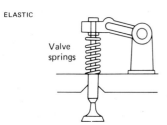

Valve
springs

CHEMICAL

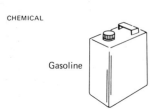

Gasoline

ELECTRICAL

Charged
capacitor
in electronic
photoflash

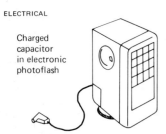

NUCLEAR

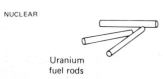

Uranium
fuel rods

Fig. 9-4 Some sources of potential energy.

Figure 9-4 shows some familiar potential-energy sources. When the source releases its work through the action of gravity, we say that the potential energy is gravitational. When elastic forces do the work, as in closing an engine valve, the potential energy is elastic. If the work is performed by a chemical reaction, the potential energy is chemical. The capacitor in a photoflash stores work by separating positive and negative electric charges that try very hard to get back together again; here the potential energy is electric. And a reactor's fuel rods do work through the electrical repulsion of the positive charges in an atom's nucleus—hence nuclear potential energy.

Example 9-2 The Weight-Driven Mechanical Clock.

The movement of a mechanical clock has three elements: a source of potential energy, an escapement to convert the energy to work just a little at a time, and a regulator to keep the escapement ticking at a constant rate. The regulator may be a balance wheel on small clocks or a

swinging pendulum on larger ones. The potential-energy source is usually a battery, a spring, or a weight (although mechanical clocks have also been driven by falling water drops, changes in barometric pressure, radioactive sources, and other exotic means).

Let's suppose that it takes 6.3 J of work to drive a certain mechanical movement for one day. The potential energy is to be supplied by a weight that is raised once a week. To be on the safe side, we will want enough potential energy for eight days: $(8)(6.3 \text{ J}) = 5\bar{0}$ J.

If we use a weight of $1\bar{0}$ N, raising it 1 m stores $1\bar{0}$ N·m of work, or $1\bar{0}$ J. To store $5\bar{0}$ J, such a weight would have to be raised a distance of $5\bar{0}/1\bar{0}$ m, or 5.0 m [16 ft]. Unfortunately, this is a bit high to be practical—particularly if the clock is to be in a home.

There is a simple alternative: Increase the weight. If we double the weight, it will take twice as much work to lift it 1 m, and we will get twice as much work back out when it drops again.

Let's say that we have only a 1.1-m drop available. What weight do we need to store $5\bar{0}$ J? One way to answer this is to rewrite Eq. (9-1):

$$\text{Work} = (\text{force})(\text{distance})$$

$$\text{Force} = \frac{\text{work}}{\text{distance}}$$

$$= \frac{5\bar{0} \text{ J}}{1.1 \text{ m}}$$

If we recall that 1 J = 1 N·m, this gives

$$\text{Force} = \frac{5\bar{0} \text{ N·m}}{1.1 \text{ m}}$$

$$= 45 \text{ N}$$

This is the same as about 10 lb, or 4.6 kg_f. The mass hanging from the string driving the time mechanism is therefore about 10 lb_m, or 4.6 kg. Often there are two such weights: one to operate the clock and the other to operate the strike or chime mechanism.

Now the design of a clock actually involves more factors than we've considered here. Our point has been to show how gravitational potential energy can be harnessed to do useful work. ◀

9-3 Kinetic Energy

We have seen that work can come from the direct action of a force, or it can be released from a source of potential energy. Now let's talk about what work can do when it isn't being stored as potential energy.

We already saw one example—the bow and arrow. The released work propels the arrow forward at great speed (close to 300 kmph). A similar thing happens when a release of chemical potential energy shoots a bullet from a gun; here the speed may be 1600 kmph or greater. An airplane in a dive will gain speed very quickly as the gravitational potential energy decreases. We get the same effect with a falling elevator or with a truck that loses its brakes on a hill. If the main valve is knocked loose from a compressed-gas cylinder, the cylinder will shoot around like a rocket. (This has led to serious industrial accidents and is a good reason for securely harnessing such cylinders.) In cases like these, the work, or the potential energy, goes into increasing an object's motion. We say that the object has gained kinetic energy.

> Definition: An object's *kinetic energy* is work stored in its motion. It may be calculated from the formula

kinetic energy

$$\text{Kinetic energy, in joules} = \tfrac{1}{2}(\text{mass, in kg})(\text{speed, in m/s})^2 \qquad (9\text{-}2)$$

Let's look closely at Eq. (9-2). If two objects are moving at the same speed but one has more mass, then the more massive object has more kinetic energy. Thus a truck cruising at 80 kmph has more kinetic energy than a car at 80 kmph. This should make sense; after all, it takes more work to get the truck up to this speed. The formula also tells us that kinetic energy increases very quickly with increases in speed. If a car accelerates from 40 to 80 kmph (doubling its speed), its kinetic energy increases by a factor of *4*. Why? The answer is that the formula tells us to square the speed. If we double something and then square it, we get 2^2, or 4, times as much. Similarly, if we triple the speed, the kinetic energy increases by a factor of 9.

Example 9-3 Kinetic Energy of a Moving Car.

A certain car has a mass of 1560 kg [3440 lb_m]. Let's find its kinetic energy at 45.0 and 90.0 kmph.

Equation (9-2) calls for speed in the SI unit, meters per second. Making this conversion, we have

$$45.0\ \frac{km}{h} = \left(45.0\ \frac{km}{h}\right)\left(\frac{1000\ m}{1\ km}\right)\left(\frac{1\ h}{3600\ s}\right)$$

$$= 12.5\ m/s$$

Similarly, we find that 90.0 kmph is the same as 25.0 m/s. At 12.5 m/s, the kinetic energy is

$$\text{Kinetic energy} = \tfrac{1}{2}(1560 \text{ kg})(12.5 \text{ m/s})^2$$

$$= \tfrac{1}{2}(1560)(156 \text{ J})$$

$$= 122\ 000 \text{ J}$$

We may write this with an SI multiplier as 122 kJ.

At a speed of 25.0 m/s, we should expect to get 4 times the kinetic energy. For the nonbelievers in the audience, let's try it:

$$\text{Kinetic energy} = \tfrac{1}{2}(1560 \text{ kg})(25.0 \text{ m/s})^2$$

$$= \tfrac{1}{2}(1560)(625 \text{ J})$$

$$= 488\ 000 \text{ J}$$

This is the same as 488 kJ. Notice that it *is* just 4 times the previous answer.

Suppose now that this car accelerates from 45.0 to 90.0 kmph. This means a kinetic energy gain of 488 kJ − 122 kJ, or 366 kJ. This 366 kJ is the work we must do on the car to increase its speed by this amount. The work can come directly from the car's engine, or it can come from gravitational potential energy by coasting the car down a steep hill, or it can come from a combination of the two. ◄

Okay, so now we know how to calculate kinetic energy. What significance does this quantity have? As we said in the definition, kinetic energy is work that is stored in an object's motion. A moving object can stop only by transferring this work to its surroundings, that is, by producing a force that acts through some distance. If the stopping distance is small, as in a car collision, this force can become very large. This is exactly the same idea that we talked about in Sec. 7-9. Figure 9-5 shows a test car intentionally being driven into a barrier to measure the collision force and its effects.

As of this writing, there is a national 55-mph speed limit in the United States. Earlier, speed limits of 65 mph and higher were common. Although 65 mph is only 18 percent faster than 55 mph, a car's kinetic energy at 65 mph is 40 percent greater than at 55 mph. Thus 40 percent more work is done in stopping the faster-moving car, and any collision will be significantly more violent. In fact, statistics show that there have been fewer traffic fatalities since the reduction in speed limit.

Fig. 9-5 This test car is being towed into a barrier at 30 mph to measure the collision force and its effects. This force arises when the car's kinetic energy is converted to work during the collision. (Courtesy Chrysler Corporation)

If we know an object's mass and its kinetic energy, we can find its speed by rewriting Eq. (9-2):

$$\text{Speed, in m/s} = \sqrt{\frac{(2)(\text{kinetic energy, in J})}{\text{mass, in kg}}} \qquad (9\text{-}3)$$

Let's look at an example.

Example 9-4 The Pile Driver.

When a heavy structure is to be built on soft ground, it should be supported on pilings resting on bedrock. The pilings, usually made of steel, are pounded in with a pile driver. There are several versions of this device; a simple one is shown here. This is basically a big hammer with guides to keep it on target.

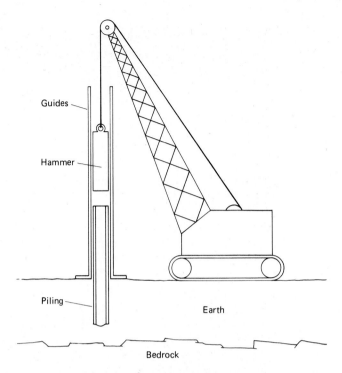

The hammer has a typical mass of 8200 kg [18 000 lb$_m$], and it is dropped perhaps 6.5 m [21 ft]. Lifting the hammer takes a force of about 8200 kg$_f$. The work done in lifting it 6.5 m is

$$\text{Work} = (\text{force})(\text{distance})$$

$$= (8200 \text{ kg}_f)(6.5 \text{ m}) \left(\frac{9.806\ 65 \text{ J}}{1 \text{ kg}_f \cdot \text{m}} \right)$$

$$= 520\ 000 \text{ J}$$

This work is stored as gravitational potential energy. When the hammer is released, this potential energy changes into kinetic energy. Thus the hammer has 520 000 J of kinetic energy when it hits the piling.

How fast is the hammer going? Using Eq. (9-3), we have

$$\text{Speed, in m/s} = \sqrt{\frac{2 \text{ (kinetic energy, in J)}}{\text{mass, in kg}}}$$

$$= \sqrt{\frac{(2)(520\ 000)}{8200}} \ \frac{\text{m}}{\text{s}}$$

$$= 11 \text{ m/s}$$

This is the same as 40 kmph [25 mph]. If the pile driver pushes the piling in 0.25 m when it strikes, the force it exerts is about (520 000)/(0.25 m), or 2 100 000 N. This, of course, is much larger than the weight of the hammer itself.

Pile drivers may also be steam-operated, which eliminates the need for the crane. ◀

The kinetic energy we have been talking about is mechanical. A moving object may carry this energy from one place to another. When we shoot a gun, the work done on the bullet goes into the bullet's kinetic energy; when the bullet hits something at some distance away, this kinetic energy is changed back into work. Thanks to the bullet, we can kill a rat without having to get too close.

But there are ways to transfer energy from one place to another without using moving objects. This happens every morning when the sun comes up: the sun's energy crosses 150 million km [93 million mi] of empty space and warms the earth. Light from a car's headlamps illuminates the road ahead of it. Sound from a speaker can be heard some distance away. Are light and sound and radiant heat really energy? Yes. The test is whether they can do work. Radiant heat can be used to drive solar engines, light can be used to generate electricity which will drive a motor, and a loud enough sound (as from a "sonic boom") can break a window.

Although it is not customary to refer to sound and light as "kinetic energy," they do have a certain similarity to the moving bullet. These forms of energy all originate in one place and travel somewhere else. Table 9-1 summarizes the different kinds of "traveling energy."

Exercises

5.　What is the kinetic energy, in megajoules, of a 55 000-kg tank car moving at 12 m/s?
Answer: 4.0 MJ

6.　A plane has a mass of 1850 kg and cruises at a speed of 192 kmph. (a) What is its kinetic energy in megajoules? (b) If the same plane lands at a speed of 96 kmph, what is its kinetic energy in megajoules?
Answers: (a) 2.63 MJ, (b) 0.658 MJ

7.　A truck has a mass of 11.5 t. (a) How much work, in megajoules, is needed to accelerate it from rest to a speed of 40.0 mph? (b) How much work is needed to accelerate it from rest to 80.0 kmph? (c) How much work is needed to accelerate it from 40 to 80 kmph?
Answers: (a) 0.710 MJ, (b) 2.84 MJ, (c) 2.13 MJ

8.　An 8.5-kg object is initially at rest. What is its speed after 2.7 kJ of work is done on it?
Answer: 25 m/s

TABLE 9-1 FORMS OF TRAVELING ENERGY

Type of energy	Characteristics
Sound	Generated by any moving object surrounded by a fluid; travels through solids, liquids, and gases, but not a vacuum; speed is determined by the density and elasticity of the substance it travels through; speed in air is about 331 m/s [1090 ft/s].
Electromagnetic (EM) waves—radio waves, radiant heat, light, ultraviolet, X-rays, gamma rays	Generated by accelerating charges; travel through vacuum and gases; depending on the frequency, may also travel through certain solids and liquids; speed in vacuum is $2.997\ 9 \times 10^8$ m/s [186 280 mi/s].
Electricity	Consists of a stream of electrons or other charges pushed by an electric force; flows through any substance and even a vacuum, but generally flows best through metals.
Heat	Normally flows from high-temperature regions to low-temperature regions; flows through any substance and even a vacuum, but best through materials known as conductors.
Water waves	Generated by wind or other agents that disturb a liquid's surface; travel horizontally while the motion of the liquid is elliptical or circular.
Mechanical kinetic energy	Associated with moving solids, liquids, and gases.

9-4 Conservation of Energy

We have looked at cases where work is changed into potential energy, and others where potential energy is changed into mechanical kinetic energy. Actually, there are many other possibilities. For instance, we can change heat energy into electricity with a device known as a thermocouple. We can change nuclear potential energy into heat in a fission reactor. We can change work into electricity with a generator (or alternator). We can change electricity into chemical potential energy through the process of electrolysis (which will separate water into hydrogen and oxygen that can then be used as a fuel). In fact, there is a way to change any form of energy into any other form of energy. And every physical event—everything that happens—can be described as an energy transformation.

This brings us to a very important principle: We can never actually create energy, and we can never actually use it up. All we can do, if we are clever enough, is change it from one form to another. We may state the principle this way:

law of conservation of energy

> *Law of Conservation of Energy*: Energy can neither be created nor destroyed. It can only be changed from one form to another.

We could write this as a formula, but that would make it more complicated that it has to be. The idea is simply that all the energy can be completely accounted for in any process. In this scheme, work is counted as a type of energy.

Suppose that 20 kJ of electric energy goes into a motor, but the motor does only 18 kJ of work (Fig. 9-6). The law of conservation of energy guarantees us that the other 2 kJ is not lost; it has to have gone somewhere. Where? For one thing, the motor makes a great deal of noise when it operates, so part of the 2 kJ has gone into sound energy. We will also notice the motor getting hot as it runs; this is because the rest of the 2 kJ has gone into heat. Altogether, the motor has converted 20 kJ of electric energy into 20 kJ of other forms of energy.

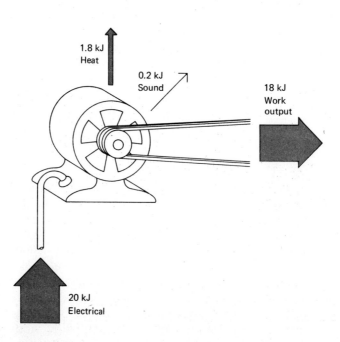

1.8 kJ
Heat

0.2 kJ
Sound

18 kJ
Work
output

20 kJ
Electrical

Fig. 9-6 Typical energy transformation in a small electric motor.

Exercises

9. Name two devices that convert chemical potential energy into work.
   Answer: Rocket engine, dynamite
10. Name a device that converts electric energy into sound. Name a device that converts sound into electric energy.
    Answers: Loudspeaker, microphone
11. What energy transformation takes place when you turn on the room lights?
    Answer: Electric energy is converted to light and heat.
12. What energy transformation takes place when coal is burned in a furnace?
    Answer: Chemical potential energy is converted to heat, light, some sound, and some waste chemical potential energy in the unburned gases and soot.

9-5 Efficiency

If energy is never destroyed, why do we hear so much concern in the news about conserving energy? Isn't energy *always* conserved 100 percent? Yes, it is. We always have just as much energy left at the end of a process as we started with. Using energy never destroys it; using energy just converts it to a different form. But the problem is this: There are very few energy transformations where all the energy goes the same way. We can transform electric energy to heat in a resistance heater, and we get just as much heat energy out as we put electric energy in. But this is an unusual case. If we use heat to drive a steam-turbine engine, the work we get out is only around 35 percent of the heat energy we supply. Most of the rest is unrecoverable heat that must be exhausted to the environment.

Furthermore, it's very easy to change other forms of energy into heat, but it's very difficult to go the other way—heat to electricity or chemical potential, for instance. It's as if heat were the lowest form of energy and all other energy forms downgrade themselves until they become heat. If you turn off the lights at night, the room immediately gets dark. Where has the light energy gone? It's been absorbed by the walls, the floor, the ceiling, and yourself, and it has heated everything a little. Turn off the stereo, and the sound quickly stops. Again, the energy has been absorbed and transformed to heat. Brake a moving car to a stop. Where has the kinetic energy gone? To heat in the brake disks and pads. In fact, we are often told to "pump" the brakes when going downhill to allow air to circulate over the brake pads and carry away this heat.

Every energy transformation, then, produces at least some heat. If we are trying to get work from an engine but we also get a lot of heat, this heat is undesirable and represents "wasted" energy. There is probably nothing useful we can do with it. In fact, we may have to use an elaborate cooling system to carry away this heat so it won't melt the engine.

In a political sense, then, we "conserve" energy when we degrade as little of it as possible to heat. We can do this in two ways. First, we can simply limit the number of big energy transformations: use bicycles instead of cars, go without air conditioning in the summer, and so on. Second, we can use devices that transform most of the input energy into a usable form. When a device produces most of its output energy where we want it, we say it is *efficient*. If it produces a great deal of unusable energy compared to what it takes in, it is *inefficient*.

efficiency

Definition: The *efficiency* of a device is the ratio of the useful energy it produces to the total energy it takes in:

$$\text{Efficiency} = \frac{\text{useful energy output}}{\text{total energy input}} \qquad (9\text{-}4)$$

For instance, the motor in Fig. 9-6 produced 18 kJ of useful work output when it took in 20 kJ of electric energy. Its efficiency is therefore (18 kJ)/(20 kJ), or 0.90, or 90 percent.

Table 9-2 lists typical efficiencies for some common energy-converting devices. Notice that heat engines (engines that use a burning fuel or some other source of heat to produce work) have fairly low efficiencies. This is what we said earlier: it is difficult to change heat to a "higher" form of energy. Still, we see that some engines are more efficient than others. A diesel engine, for instance, produces less waste heat than a gasoline engine with the same output.

We should also notice that no device can have an efficiency greater than 100 percent. Such a device would have to put out more energy than it takes in, which is clearly impossible in light of the law of conservation of energy.

Example 9-5 Energy Conversion in a Generator.

A certain electric generator has an efficiency of 98.6 percent. How much electric energy does it produce when 85.0 kJ of work is done in driving it?

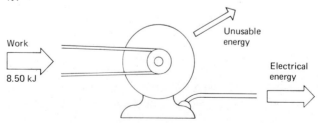

Work

8.50 kJ

Unusable energy

Electrical energy

TABLE 9-2   EFFICIENCIES OF SOME COMMON DEVICES*

Device	Efficiency, %
Bearings	
Journal	95–98
Roller	98
Ball	99
Drive mechanisms	
Spur gears, cast	93
Spur gears, machined	96
Worm gear	80–98
Belt and pulley	95–98
Chain and sprocket	94–97
Fluid engines	
Windmill blades, fans, propellers in air	up to 51
Water turbine	65–80
Heat engines (chemical energy to mechanical work)	
Reciprocating steam engine	12–16
Steam turbine	up to 45
Gasoline engine	up to 30
Gas turbine	25–44
Diesel engine	26–36
Electric devices	
Incandescent lamp	14
Fluorescent lamp	60
Transformer	99
Electric motor	60–96
Electric generator	up to 99
Chemical devices	
Dry-cell battery	90
Storage battery	72
Gas furnace	85
Oil furnace	65

* All values are approximate. The actual efficiency of a device depends strongly on the conditions of its use (temperature, pressure, operating speed, type of lubricant, etc.).

When we say that a device has an efficiency of 98.6 percent, we mean that 98.6 percent of the input energy goes where we want it. In this case, then, the electrical output is

Useful energy output = (efficiency)(total energy input)

$$= 98.6\% \ (85.0 \text{ kJ})$$

$$= (0.986)(85.0 \text{ kJ})$$

$$= 83.8 \text{ kJ}$$

How much energy goes into unusable forms? The rest of it does. This is 85.0 kJ − 83.8 kJ, or 1.2 kJ. We could also find this wasted energy by taking 1.4 percent of the original 85.0 kJ. ◄

Very often, we know how much useful energy we want to get out of a device. Then the question is, How much energy do we need to supply to it? We may answer such questions by rewriting Eq. (9-4):

$$\text{Total energy input} = \frac{\text{useful energy output}}{\text{efficiency}} \qquad (9\text{-}5)$$

Let's look at an example.

Example 9-6 Work Needed to Drive a Pump.

A certain pump has an efficiency of 86 percent. We want to use it to lift 870 L of water a total of 15 m. How much work is needed to operate the pump?

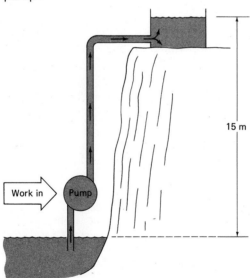

Since water has a density of 1.0 kg/L (Table 6-2), 870 L of water has a mass of 870 kg. The weight of this water is then 870 kg$_f$. Calculating the work done in lifting this water 15 m gives us

$$\text{Work} = (\text{force})(\text{distance})$$

$$= (870 \text{ kg}_f)(15 \text{ m}) \left(\frac{9.806\ 65 \text{ J}}{1 \text{ kg}_f \cdot \text{m}} \right)$$

$$= 130\ 000\ \text{J}$$

$$= 130\ \text{kJ}$$

This is the useful work *output* that we need.
    To find the work input, we use Eq. (9-5):

$$\text{Total energy input} = \frac{\text{useful energy output}}{\text{efficiency}}$$

$$= \frac{130\ \text{kJ}}{0.86}$$

$$= 150\ \text{kJ}$$

    Notice that this energy input is *greater* than the useful energy output. ◄

Exercises

13. A chain and sprocket with an efficiency of 96 percent is used to drive a motorcycle. What is the useful work output when the work input is 100 kJ?
     Answer: 96 kJ
14. A water turbine has an efficiency of 75.7 percent. What work input is necessary to produce a work output of 96.2 MJ?
     Answer: 127 MJ
15. A steam-turbine engine has an efficiency of 28 percent. How much heat energy is produced for each 1.0 MJ of useful work output?
     Answer: 2.6 MJ

9-6 Multiple Energy Conversions

We often encounter cases where the energy goes through a series of conversions before reaching its final useful form. Figure 9-7 shows a familiar example. Coal is dug from the ground, transported to a power plant, and pulverized to a fine dust. This dust is blown into a boiler where it is burned. Water circulating in pipes in the boiler absorbs some of this heat and changes it to steam. The steam is piped through a turbine engine whose drive shaft turns a dynamo. Electricity generated by the dynamo goes through a transformer and then to high-voltage transmission lines. After traveling some distance, this electric energy goes through a step-down transformer to the local power lines, then through another step-down transformer to a home. The overall effect is to use chemical potential energy to operate, say, a dishwasher.

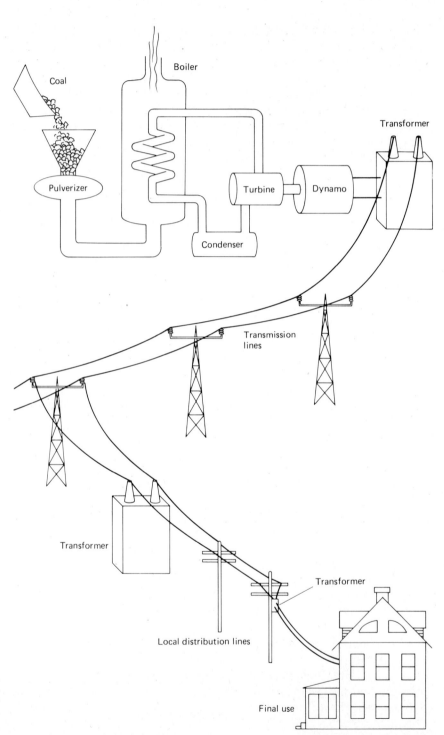

Fig. 9-7 Schematic of the series energy conversion that transforms chemical potential energy at a generating plant to electricity in the home.

In Fig. 9-8 we see the same process diagramed in terms of the energy conversions (beginning at the boiler). Of the chemical potential energy in the powdered coal, only 89 percent makes it to the turbine. Thus 89 percent is the efficiency of the boiler. The turbine converts 35 percent of this 89 percent to work, so now only 31 percent of the original energy is left to drive the dynamo. The dynamo changes 98 percent of this 31 percent to electric energy. And so on down the line. The total efficiency is the ratio of the final-use energy to the original chemical energy. This can be found by multiplying all the separate efficiencies:

Total efficiency = (0.89)(0.35)(0.98)(0.99)(0.985)(0.99)(0.99)(0.99)

= 0.29

Thus the total efficiency of this process is only 29 percent.

Suppose that the homeowner uses the final electric energy in an electric resistance heating system. As we said earlier, such a system converts 100 percent of this electric energy to heat. But is it really such a good idea? Probably not. It takes nearly 4 J of chemical energy to produce just 1 J of heat this way. If the coal were burned directly in the home at an efficiency of even 75 percent, 4 J of chemical energy would produce 3 J of useful heat. Because of all the waste along the way, electric resistance heating is an expensive way to heat a home. Electricity usually has much better uses.

Notice that when we have a series of energy conversions, the total efficiency is lower than the lowest efficiency in any single step. Thus if the efficiency of the entire process is to be improved, we should concentrate on the "weakest link."

We may also benefit by looking for a way to completely eliminate one step. Suppose that we have a gear train with five pairs of meshing gears. Each pair has an efficiency of 96 percent, so the total efficiency is

Efficiency of five-pair gear train = (0.96)(0.96)(0.96)(0.96)(0.96)

= 0.82, or 82%

If we can find a way to eliminate one pair of gears, the efficiency is

Efficiency of four-pair gear train = (0.96)(0.96)(0.96)(0.96)

= 0.85, or 85%

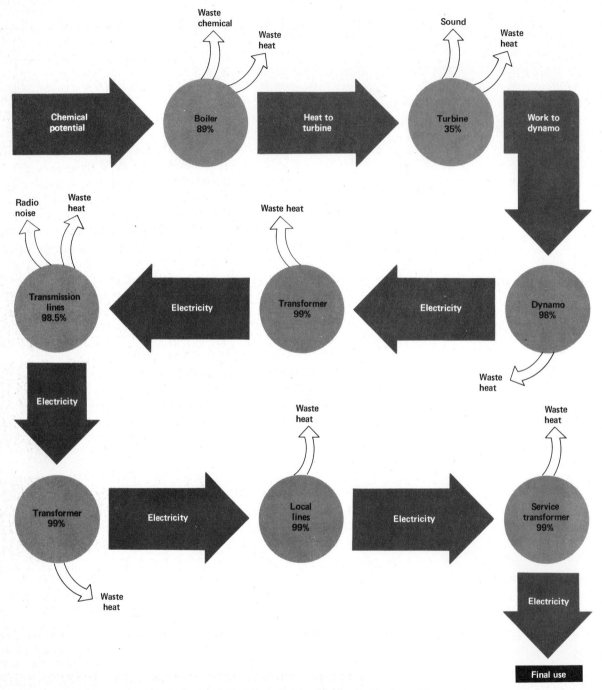

Fig. 9-8 Energy flow in the generation and transmission of electricity. Percentages are the approximate efficiencies of individual transformations.

This second gear train would require about 3 percent less input energy.

The weak link in generating electricity is the turbine engine. Its efficiency can be improved very little. It would do better if it operated at a higher temperature, but unfortunately high temperature steam breaks down into hydrogen and oxygen, particularly in the presence of steel. The steel parts absorb this hydrogen and become very brittle, which leads to premature failure of the engine. Since the solution to this problem lies in the use of exotic and very expensive materials, most large steam turbines are restricted to a peak temperature of only 566°C [1050°F]. We will have more to say about the problems of engine efficiency later.

Lately, there has been a great deal of interest in trying to generate electricity on a large scale without any engine at all. If the burning coal could produce electricity directly, this would replace the first three steps in Fig. 9-8 by just one step. In theory, this may be done with a magnetohydrodynamic (MHD) generator. The principle is shown in Fig. 9-9. The coal would be burned, and the resulting plasma would be purified and doped with a small amount of potassium to improve its electric conductivity. It would then expand through a funnel-shaped chamber held between the poles of a very large and very strong magnet. This sets up a flow of electricity between the two conducting sides of the funnel. (Remember, a plasma contains many free electrons.) Additional energy could be extracted from the hot exhaust by putting it through a conventional generating cycle. At present, there are two main problems with the MHD generator. One is in building the large and powerful magnets needed to make the process efficient. The second is the old problem of finding suitable materials for the corrosive atmosphere.

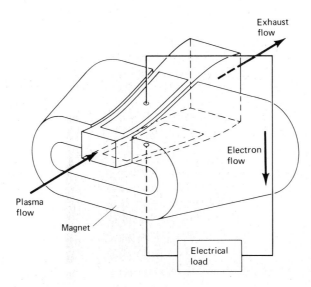

Fig. 9-9 Principle of the magnetohydrodynamic (MHD) generator. A conducting plasma is forced at high speed through a duct held between the poles of a very strong magnet. The magnetic force pushes electrons through the external circuit connecting the top and bottom of the duct.

Exercises

16. A water turbine has an efficiency of 76 percent. It is used to drive a generator with an efficiency of 95 percent. What is the total efficiency of the conversion of gravitational potential energy to electricity?
    Answer: 72 percent
17. To obtain the necessary speed reduction from a stationary engine, three separate belt-and-pulley systems are linked together. Each has an efficiency of 95 percent. (*a*) What is the total efficiency? (*b*) What is the total efficiency if the efficiency of each step could be increased to 96 percent? (*c*) What is the total efficiency if one step is completely eliminated?
    Answers: (*a*) 86 percent, (*b*) 88 percent, (*c*) 90 percent
18. Construction and research costs aside, how efficient would an MHD generator have to be to make it competitive with a conventional steam-driven generating plant? Use the efficiency values in Fig. 9-8.
    Answer: Better than 27 percent

9-7 Other Energy Units

Before the mid-1800s, no one knew that work, heat, electricity, sound, and light are all forms of energy. Consequently, they were all measured in different units. Add to this the fact that the old U.S. Customary System developed separately from the metric system, and we can understand why more than 30 different energy units are used today.

Of course, the SI unit of energy is the joule. Together with its SI multiples (kJ, MJ, GJ), it is finding widespread acceptance across the world. Even so, electric energy will probably continue to be measured in kilowatthours (kWh). Other energy units, like the calorie (cal), the British thermal unit (Btu), the kilogram-meter (kg$_f$·m), and the horsepower-hour (hp·h), are gradually being phased out (at least in printed works). For reference, Table 9-3 shows how some of these other energy units are related to the joule.

9-8 Power

Suppose that you need 8 m³ of concrete for a foundation footer. You could mix it by hand, a little at a time, and eventually you would get the job done. Or you could rent a small, gasoline-powered mixer and get the same job done quicker. More likely, though, you would order the concrete from a firm that mixes and delivers it in a heavy truck designed for the purpose. Whichever way you do it, mixing the concrete takes the same amount of work. But the mixer truck does this work very quickly by using power equipment.

TABLE 9-3 CONVERSION FACTORS FOR UNITS OF ENERGY

	J	cal	kWh	ft·lb	Btu	hp·h
1 joule =	1	0.239 01	$2.777\ 7 \times 10^{-7}$	0.737 56	$9.484\ 5 \times 10^{-4}$	$3.725\ 1 \times 10^{-6}$
1 calorie =	4.184	1	$1.162\ 2 \times 10^{-6}$	3.086 0	0.003 968 3	$1.558\ 6 \times 10^{-6}$
1 kilowatthour =	3.6×10^{6}	$8.604\ 2 \times 10^{5}$	1	$2.655\ 2 \times 10^{6}$	3414.4	1.341 0
1 foot-pound =	1.355 8	0.324 05	$3.766\ 2 \times 10^{7}$	1	0.001 285 9	$5.050\ 5 \times 10^{-7}$
1 British thermal unit =	1054.4	252.00	$2.928\ 8 \times 10^{-4}$	777.65	1	$3.927\ 5 \times 10^{-4}$
1 horsepower-hour =	$2.684\ 5 \times 10^{6}$	$6.416\ 2 \times 10^{5}$	0.745 70	$1.980\ 0 \times 10^{6}$	2546.1	1

1 erg = 10^{-7} J
1 electronvolt (eV) = $1.602\ 2 \times 10^{-19}$ J
1 kg_f·m = 9.806 65 J
1 liter-atmosphere = 101.33 J = 74.737 ft·lb
1 ton of TNT (nuclear equivalent) = 4.20×10^{9} J

The simplest equipment can do a great deal of work if we're in no hurry to get it done. But if we want to do work very fast, we need equipment with a large power rating.

Definition: *Power* is the rate at which energy is converted from one form to another: *power*

$$\text{Power} = \frac{\text{energy converted}}{\text{time}} \qquad (9\text{-}6)$$

Since the SI unit of energy is the joule (J), the SI unit of power is joules per second (J/s). This is a commonly used unit, and it has been given a name of its own, the *watt* (W):

$$1\ W = 1\frac{J}{s}$$

If a light bulb is rated at 40 W, this means that it will normally convert 40 J of electric energy to other forms each second.

Example 9-7 Power in a Hydraulic Log Splitter.

This device was described in Example 9-1. There we calculated that the hydraulic ram had to supply some 1600 J of work to split the log.

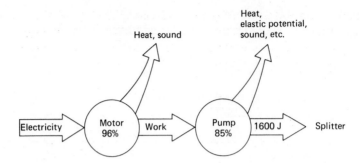

Suppose that we want the device to split a log in 5.0 s. The useful power is then

$$\text{Power} = \frac{1600 \text{ J}}{5.0 \text{ s}}$$

$$= 320 \text{ W}$$

This 320 W is the power output. But the pump has an efficiency of only around 85 percent. It is driven by an electric motor with an efficiency of maybe 96 percent. The overall efficiency is therefore (0.85)(0.96), or 82 percent. Therefore, *more* than 1600 J of electric energy has to be supplied to the motor. How much? Using Eq. (9-5), we get

$$\text{Total energy input} = \frac{\text{useful energy output}}{\text{efficiency}}$$

$$= \frac{1600 \text{ J}}{0.82}$$

$$= 1950 \text{ J}$$

Thus the actual power "consumption" of the motor is

$$\text{Power} = \frac{\text{energy converted}}{\text{time}}$$

$$= \frac{1950 \text{ J}}{5.0 \text{ s}}$$

$$= 390 \text{ W} \qquad\qquad \blacktriangleleft$$

The log splitter in Example 9-7 converted electric energy at a rate of 390 W, but it did work at a rate of only 320 W. If we calculate the ratio of the useful output power to the input power, we get (320 W)/(390 W) = 0.82. This last figure is exactly the same as the device's overall efficiency. So although we defined efficiency as a ratio of output *energy* to input *energy,* we get the same thing if we use power:

$$\text{Efficiency} = \frac{\text{useful power output}}{\text{total power input}} \tag{9-7}$$

For instance, a gear train may have an efficiency of 88 percent and a power input of 25 kilowatts (kW). If so, its power output is (0.88)(25 kW), or 22 kW. During *each second* of operation, this transmission accepts 25 kJ of input energy and converts it to 22 kJ of useful output energy.

In the old USCS units, we express work and other mechanical energy in foot-pounds. This gives us foot-pounds per second (ft·lb/s) as the unit of power. If a device converts energy at a rate of 550 ft·lb/s, its power rating is 1 hp. Heat energy is sometimes measured in British thermal units (Btu): 1 Btu heats 1 lb_m of water 1°F. Furnaces, air conditioners, and heat pumps are commonly given a power rating in units of British thermal units per hour (Btu/h): notice that this is the *rate* of energy conversion. Because there are so many energy units, a large number of power units are also in use. Table 9-4 lists the more important ones.

TABLE 9-4   *CONVERSION FACTORS FOR UNITS OF POWER*

	W	cal/s	Btu/h	ft·lb/s	hp
1 watt =	1	0.239 01	3.414 4	0.737 56	0.001 341 0
1 calorie per second =	4.184	1	14.286	3.086 0	0.005 610 8
1 Btu per hour =	0.292 88	0.069 999	1	0.216 01	$3.927\ 5 \times 10^{-4}$
1 foot-pound per second =	1.355 8	0.324 05	4.629 3	1	0.001 818 2
1 horsepower =	745.70	178.23	2546.1	550	1

1 cheval vapeur (metric horsepower) = 75 kg_f·m/s = 735.50 W
= 0.986 32 hp
1 kg_f·m/s = 9.806 65 W = 0.013 151 hp
1 ton (refrigeration) = rate of heat transfer that will freeze 1 ton of water in a 24-h day = 12 000 Btu/h = 3516.9 W
1 kW = 1.341 0 hp

Example 9-8 Electric and Gas Heat.

A certain gas furnace has a rated output of 80 000 Btu/h. This means that it will put out 80 000 Btu of useful heat energy in 1 h of continuous operation. Suppose that we want to get the same rate of heat production from electric baseboard resistance heaters. What is the total electric power requirement in kilowatts?

   At first glance, this may seem like a complicated problem. Actually, it amounts to no more than a unit conversion. From Table 9-4, we see that

$$1 \text{ W} = 3.414 \text{ } 4 \text{ } \frac{\text{Btu}}{\text{h}}$$

and by using an SI multiplier we can write

$$1 \text{ kW} = 1000 \text{ W}$$

Then we have

$$80\ 000 \text{ } \frac{\text{Btu}}{\text{h}} = \left(80\ 000 \text{ } \frac{\text{Btu}}{\text{h}}\right) \left(\frac{1 \text{ W}}{3.414 \text{ } 4 \text{ Btu/h}}\right) \left(\frac{1 \text{ kW}}{1000 \text{ W}}\right)$$

$$= 23.4 \text{ kW}$$

   This 23.4 kW is the electric *power* requirement. It amounts to a conversion of 23 400 J of electric energy to heat *each second*. The consumption of electricity at this rate may cost more than $1.00 per hour◄

Exercises

19. An oil furnace has a rated input of 130 000 Btu/h and a rated output of 105 000 Btu/h. (a) What is the furnace's efficiency? (b) What is the power loss resulting from heat and unburned chemicals in the chimney exhaust?
     Answers: (a) 81 percent, (b) 25 000 Btu/h

20. A certain electric motor has a power input of 8.22 kW. Its efficiency is 92 percent. What is the useful power output in (a) kilowatts, (b) horsepower?
     Answers: (a) 7.56 kW, (b) 10.1 hp

21. A certain crane can lift a weight of 4300 lb through a vertical distance of 21 ft in 6.0 s. What is the useful power output in (a) feet-pounds per second, (b) horsepower, (c) kilowatts?
     Answers: (a) 15 000 ft·lb/s, (b) 27 hp, (c) 20 kW

22. A certain diesel engine has a rated power output of 185 hp. It is to be used to drive an electric generator with an efficiency of 97 percent. What is the generator's electric power output in kilowatts?
Answer: 134 kW
23. A small factory uses 960 kW·h of electricity in a typical 8-h workday. What is the average power consumption?
Answer: 120 kW

We may rewrite Eq. (9-6) in terms of the energy converted in a process: **9-9 Power and Time**

$$\text{Energy converted} = (\text{power})(\text{time}) \qquad (9\text{-}7)$$

Thus if a 1-hp motor operates continuously for 1 h, the work done is (1 hp)(1 h), or 1 hp·h. This same work could be done by a 2-hp motor running for 0.5 h, or a ¼-hp motor running for 4 h. The horsepower is a unit of power, while the horsepower-hour is a unit of energy.

Similarly, an air conditioner might draw 1 kW of electric power. If it does this for 10 h, the electric energy converted is (1 kW)(10 h), or 10 kWh. The consumer is billed on this total energy used, not on the power. A 2-kW unit would use this same energy in only 5 h, and a 10-kW unit in only 1 h, but in all cases the electric bill is based on the 10 kWh that has been supplied.

We need to be very careful when using units like the horsepower-hour and the kilowatthour, because many people take them to be power units when, in fact, they are used to measure energy. The reader should take the time to carefully study Tables 9-3 and 9-4 to avoid any later confusion about the units for these different quantities.

Exercises

24. An engine produces a power output of 52.0 hp for 32.0 min. How much work is done in (a) horsepower-hours, (b) foot-pounds, (c) megajoules?
Answers: (a) 27.7 hp·h, (b) 5.49 × 10^7 ft·lb, (c) 74.4 MJ
25. A 100-W light bulb burns continuously for 1 h. How much heat does it produce in (a) kilojoules, (b) British thermal units, (c) calories? *Note*: All the energy supplied to the bulb eventually is transformed to heat.
Answers: (a) 360 kJ, (b) 340 Btu, (c) 86 000 cal

Summary   Every physical event—that is, everything we see happening around us—can be described as an energy transformation. Energy is never actually created or used up in such transformations; it merely changes form. There are many different forms of energy, but they all fall into three main categories: work, which is the action of a force moving through a distance; potential energy, which is work that has been stored so it can be recovered later; and "traveling energy," which moves from one place to another.

Every energy transformation produces some heat or thermal energy as a byproduct. Thus only a portion of the transformed energy can usually be used. The *efficiency* is the ratio of this usable energy to the total input energy. It is usually expressed as a percent.

The rate at which energy is transformed is called *power*. Two machines may do the same amount of work, but the one with the greater power output will do it quicker. Thus a car with a 150-hp engine can accelerate to 50 mph in less time than a similar car with a 100-hp engine.

Terms You Should Know

work	efficiency
joule	power
potential energy	watt
kinetic energy	perpetual motion
law of conservation of energy	

Problems

1. The hammer of a pile driver has a mass of 8430 kg. It is lifted 6.65 m in 4.92 s. Find the power requirement in (a) kilowatts, (b) metric horsepower, (c) horsepower.

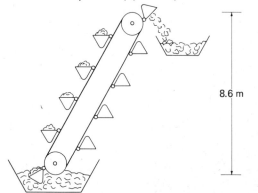

2. A conveyor belt lifts coal through a vertical height of 8.6 m. The conveyor's efficiency (work to gravitational potential energy) is 73 percent. (a) How much work must an engine do in making the device lift 10.0 t of coal? (b) If this is to be done in 10 min, how much power in kilowatts must the engine put out? (c) What should the engine's output power be in horsepower?

3. A certain pump is driven by a 2.0-hp electric motor. (This is the motor's output and the pump's input.) The pump lifts 24 000 L of water 8.4 m in $3\overline{0}$ min. Estimate the pump's efficiency.

4. A motor with a 25.0-hp output drives a large pump. The motor's efficiency is 95 percent,

Application: Perpetual-Motion Machines of the First Kind

Inventors have long dreamed of building a machine so efficient that, once started, it would produce work continuously with no fuel or other energy input. Such a device might power cars and ships, manufacturing and mining operations, refrigerators and heat pumps. And, of course, it would make the inventor very rich.

As early as 400 B.C., Egyptian galley slaves found their workload lightened by the invention of the sail. About 100 B.C., the water mill was invented in northern Greece. Although both inventions were tremendously successful, neither produced true *perpetual motion* since they depended on external and unreliable energy sources.

Then in 1235 A.D., a French architect hit upon an ingenious idea. Rather than depend on wind, water, or muscle to produce motion, he would mount seven heavy swinging hammers on the rim of a large wheel and axle. Once it was set into motion, there would always be more weight on the descending side of the wheel than on the ascending side, and so the motion would be perpetual. The inventor's enthusiasm was scarcely dampened by the fact that the contraption refused to work; he advocated its use for such things as sawing wood, grinding grain, and raising weights.

In the next several centuries, dozens of perpetual-motion machines were designed on this unbalanced-wheel principle. One of them, filled with 50-lb iron balls that rolled around, impressed King Charles I when it took a very long time to stop. But none of these devices could keep themselves going indefinitely, much less produce a surplus work output.

Robert Fludd, an English doctor, took a different approach. He noticed that the water wheel would be a perpetual-motion machine if it didn't have to depend on a natural source of running water. Why not use the wheel to drive a pump that lifted the fallen water back to the top of the millrace? The same water would be recycled again and again. The device was first built in about 1618, and it failed to work. Many variations on this closed-cycle mill were also tried, but they all quickly proved fruitless. Even so, as late as 1871, an American patent attorney complained that many inventors still submitted the idea to him each year.

There were also those who tried magnetism. In one version (1674), a powerful magnet was to draw an iron ball up a ramp. Just before it came to the magnet, the ball was to fall through a hole in the ramp and roll back to its starting point, where the process would begin again. The device failed for lack of a strong enough magnet. Two hundred years later, with the availability of stronger magnets, it failed again. A magnet strong enough to draw the ball up the ramp simply will not allow it to drop through a hole.

In 1810, a respected member of the English Parliament invented a perpetual-motion machine powered by a continuous chain of sponges. The sponges on one side of the chain soaked up water, while those on the other side were squeezed dry. When built, the device refused to keep itself moving.

The most successful perpetual-motion machine was built by John Keely around 1874. This amazing engine seemed to run on a charge of water that it never consumed. It could rip apart steel cables, bend iron bars, and fire bullets through 1-ft-thick planks. Keely rented a suite in New York City, where he demonstrated his engine and attracted scores of hungry investors. He then founded the Keely Motor Company, capitalized at $1 million (quite a bit in those days), and spent the next 14 years improving the invention and devising more impressive demonstrations. By the time he died in 1898, sources say the engine could generate 40 hp at 8000 rpm on less than a thimbleful of water and could keep doing it for 15 days with the same water. But the company never built a single engine for commercial sale, nor did it generate a dime of profit. When Keely died, his stockholders were horrified to find that his secret died with him. A subsequent search revealed that his totally fraudulent devices operated from cleverly concealed compressed-air lines.

and the pump's efficiency is 87 percent. (a) What is the electric power input in kilowatts? (b) What is the fluid power output in kilowatts? (c) Approximately how many British thermal units of heat are generated each hour?

5. Electricity is billed at a rate of $.052 per kilowatthour. A large electric motor has a power output of 8.75 hp and an efficiency of 96.0 percent. (a) What does it cost to operate this motor continuously at full load for one month (30.0 days)? (b) What would it cost if the efficiency were only 92.0 percent for this same power output?

6. A car has a mass of 1460 kg. It accelerates from rest to 80.0 kmph on a level road in 7.72 s. (a) What useful work has been done? (b) Approximately what is the average power needed to do this?

7. A truck encounters a total drag of 323 kg$_f$ when it is traveling at a speed of 80.0 kmph.

In the 1880s, electric motors and generators first became commercially available. Immediately, thousands of would-be inventors independently came up with the "perpetual generator." The idea was to use an electric motor to turn a generator, whose output was then used to power the motor. A day doesn't pass without someone new thinking of this great idea. Unfortunately, it is nothing more than a sophisticated version of Fludd's mill, and it doesn't work at all.

Why did all these devices fail? Because of the law of conservation of energy. You simply cannot get more energy out of a device than you put into it. When you subtract the energy transformed to unrecoverable heat, there is no way a machine can have enough energy left to even keep itself going. Certainly, then, there is no way it can generate a surplus.

What output power must the engine develop to keep the truck traveling at this speed on a level road? *Hint:* Calculate the work that must be done each second.

8. An overshot water wheel has an efficiency of 42 percent. Of the remaining 58 percent of the input energy, 9 percent is converted to heat and the other 49 percent is in the kinetic energy of the water leaving the wheel. To extract additional energy from this water, a second-stage turbine is used. Its efficiency is 55 percent. What is the efficiency of the entire system? *Hint:* Draw an energy-flow diagram; this is not the same as the generating plant example.

9. An elevator has a mass of 980 kg. It is driven by an electric motor with an actual output of 10.0 hp. Approximately how long will it take this motor to lift the elevator 92 m if the drive mechanism has an efficiency of 91 percent?

10 Friction

Introduction   We have already mentioned friction a number of times in the last few chapters. Friction between tires and a road surface allows us to control a car in a variety of maneuvers. Friction between the air and an airplane wing produces a drag that must be overcome by the thrust of the plane's engines. Friction is important in many other applications. Sometimes, as with the moving parts of a car engine, it needs to be kept very low. At other times, as with clutch plates and grinding wheels, a lot of friction is desirable. In this chapter, we look into the causes and effects of friction, as well as a number of practical applications.

10-1 Static and Sliding Friction   Try to push a refrigerator across the room. Chances are, it won't be easy to get it started. Friction is holding it back. Friction always tries to prevent one surface from sliding over another, never to help it along. Fortunately, friction is not always successful in preventing motion, and things do move. But it always takes more effort to move them than it would if there were no friction.

Figure 10-1 shows a highly magnified view of the interface between two objects. We see many little humps and valleys in both surfaces. Sometimes a hump on the top surface fits into a valley in the bottom surface. Sometimes a hump rests on a hump, and the pressure developed on this small area "welds" them together. Depending on the

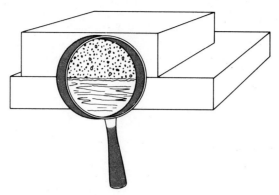

Fig. 10-1 Magnified view of the interface between two objects in contact.

actual surface materials, there may also be forces of adhesion. Consequently, it takes a certain amount of force to break one surface loose from the other and start it sliding. It also takes a continuing force to keep it sliding as the humps and valleys bump over one another.

Friction depends on a number of factors. One obvious factor is the type of material in contact. Thus there is very little friction when a piston slides in a lubricated cylinder, but quite a bit when a rubber tire slides on a concrete road surface. Another factor is the amount of force pressing the surfaces together: there is a little friction when a loose V-belt slips on a steel pulley, but quite a bit more when the belt is tightened. Friction also depends on temperature, although different substances react differently and there is no general rule. But contrary to popular belief, friction actually depends very little on the area of the surfaces in contact. In Fig. 10-2 it takes just as much force to start the brick moving no matter which side it is lying on. In fact, in some instances an increase in contact area actually *reduces* the friction, because the lower pressure doesn't weld any of the humps together. This is the case, for instance, with a wood crate sliding on a rough concrete floor.

Let's go back to pushing on the refrigerator. We find it takes a great deal of effort just to get the refrigerator to *start* to move. But once it's moving, it is not as hard to keep it moving. (If you don't believe this, try it.) This means that we have to talk about *two* kinds of friction: one when there is sliding motion and one when there is not.

Definition: *Static friction* is the force which resists the starting of one surface's movement across the other when the two surfaces are in contact.

static friction

Definition: *Sliding friction* is the force that acts to oppose the movement of one surface across the other after the motion is established.

sliding friction

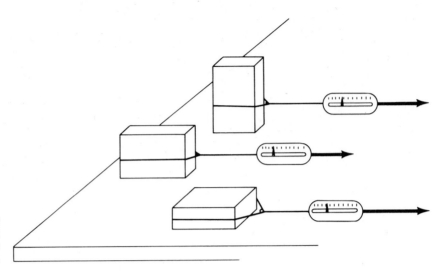

Fig. 10-2 Friction depends very little on the contact area. It takes the same force to pull the brick in all three cases.

The point is that static friction is always greater than sliding friction when all the other factors are the same. Try to drive a car out of a snowbank. As long as the tires don't slip very much, we have static friction and there is little problem. But get overanxious and spin the wheels, and there may be difficulties; the lower friction is probably not enough to move the car. (This is something to remember when driving up a snow-covered hill. There is more traction, or friction, if the drive wheels aren't slipping.)

If we examine the operation of a tire more closely, we find that the transition from static to sliding friction is not abrupt. If a car is coasting at 80 kmph, the tire tread is rotating about the rear axle at this same speed. Thus there is no slipping of the tread over the road at the contact surface, and we have static friction. Now apply the brakes. The tread is held back, and it rotates at a speed slightly less than the car's forward speed. This means the tread is creeping backward over the contact surface. The greater this "creep," the greater the friction—up to a point. When the tire is rotating at a speed some 15 percent less than the car's forward speed, the friction has reached a maximum. We say that there is 15 percent "slip" between the tire and the road, and that the tread is at the *breakaway* point. With greater amounts of slip, the tire quickly loses some of its grabbing ability and the friction decreases. Up to the breakaway point, we say that the tire is governed by static friction; afterward it is governed by sliding friction. The effect is shown in Fig. 10-3. There are no numbers on the vertical axis because the exact values depend on all the other factors we've mentioned.

Although we have discussed the transition from static to sliding friction in terms of car tires, the same general behavior is true of any pair

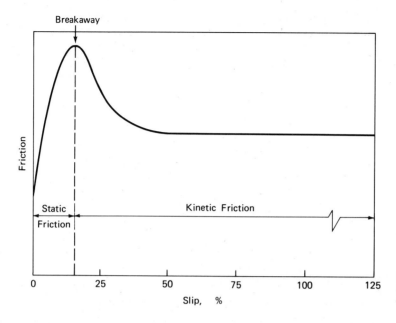

Fig. 10-3 Friction versus slip for automobile tires. Friction decreases after breakaway.

of surfaces. As we push one surface across another, the force resisting our efforts is greatest just after the motion has barely begun. It then quickly drops to a lower value as the speed is increased. With stiff materials, like metals or ceramics, this transition happens fairly quickly, and any noticeable slipping at all puts us past breakaway and into the sliding friction region.

We said that friction depends on the force pressing the surfaces together. Such a force is called a normal force, because it acts normal, or perpendicular, to the contact interface.

10-2 The Normal Force

> Definition: A *normal force* is any force acting perpendicular to two surfaces and pressing them together.

normal force

For an object resting on a horizontal surface, the normal force is just equal to the object's weight. For an object on a hill, the normal force is reduced: the steeper the incline, the lower the force pressing the surfaces together. Table 10-1 shows how the normal force varies with the inclination of the surface. These data are for any unit weight. For a 10-unit weight, the values should be multiplied by a factor of 10, for a 30-unit weight a factor of 30, and so on.

TABLE 10-1   NORMAL FORCE NEEDED TO SUPPORT A 1-UNIT WEIGHT AT DIFFERENT INCLINATIONS TO THE HORIZONTAL

Inclination	Normal force, unit
0°	1.000
15°	0.966
30°	0.866
45°	0.707
60°	0.500
75°	0.259
90°	0.000

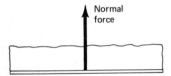

Example 10-1 Snow Load on a Roof.

Freshly fallen snow has a density of about 0.13 g/cm³ [8 lb$_m$/ft³], while wet snow and sleet may be as dense as 0.48 g/cm³ [30 lb$_m$/ft³]. Thus a 15-cm [6.0-in] snowfall can easily place a load of 3000 kg$_f$ [6600 lb] on a typical home roof. On a large roof, the load will be much greater.

Suppose that a 3000-kg$_f$ snow load accumulates on a flat roof (Fig. 10-4). Obviously, the roof must be capable of supporting 3000 kg$_f$ without bending out of shape. Now suppose that the same structure has a roof pitched at a 30° angle. This means that the rafters have to be longer to bridge the same span. But because the snow falls vertically, the same amount accumulates, and the total weight is still 3000 kg$_f$. Consulting Table 10-1, we see that the normal force needed to support each kilogram is now only 0.866 kg$_f$. Thus the total normal force is just (0.866) (3000 kg$_f$), or 2600 kg$_f$.

What if we make the roof steeper yet? At 45°, the normal force supporting 3000 kg$_f$ is around 2100 kg$_f$, and at 60° it is just half the weight, or 1500 kg$_f$. Thus there is an obvious advantage to steep-roofed buildings in Northern climates.

Now it may seem that we're getting something for nothing: the steeper roof doesn't have to be as strong to support the same load. Actually, it still takes a total supporting force of 3000 kg$_f$ to hold up 3000 kg$_f$ of snow. What happens is that the steeper roof provides some of this support as friction, which, in turn, is transmitted as a compression in the rafters down to the bearing surfaces on the side walls. With less *bending* force on the rafters, the steeper roof really doesn't have to be as strong. Of course, if the roof is *very* steep, the snow will slide right off. ◀

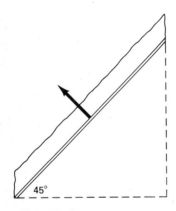

Fig. 10-4 Snow load on roofs. The weight of the snow is the same in all three cases, but the steeper roof has to supply less normal force.

In the last example, how did the roof manage to know how much normal force to supply? It certainly doesn't do a calculation as the snow falls. Somehow, it responds to the changing load and automatically provides the proper normal force.

What happens is this: Any object transmits its weight to the supporting surface. This force causes the surface to deform elastically. The elastic deformation gives rise to the normal force. The greater the load, the greater the deformation and the greater the normal force. Of course, if the load is too heavy, the supporting structure will fail and there is no longer a normal force. Everything comes crashing down.

So far, we have talked about normal forces that arise in reaction to an object's weight. However, a normal force arises in reaction to any force pressing one object into another. Thus a brake pad pressed into a brake disk and a chisel pressed against a grinding wheel each has a normal force at the interface.

Exercises

1. A 1600-kg automobile rests on a road inclined at 15°. What is the normal force?
   Answer: 1550 kg$_f$
2. A chute is built to slide used slate from an old roof down to a truck. The chute is inclined at 75°. If a typical piece of slate has a mass of 0.65 kg, what is the normal force on it in the chute?
   Answer: 0.17 kg$_f$

Both static and sliding friction depend on the normal force. In general, if we double the normal force, we just about double the force of friction. For sliding friction, we may write the relationship this way:

10-3 Friction Coefficients

$$\text{Force of sliding friction} = \left(\text{Coefficient of sliding friction}\right)(\text{normal force}) \qquad (10\text{-}1)$$

The *coefficient of sliding friction* is a number that depends on the materials in contact and on their condition. Representative values of this quantity are listed in Table 10-2.

Example 10-2 Force on a Skidding Vehicle.

A car driver locks the wheels to make a sudden stop, and the car begins to skid. What is the decelerating force?

Let's say that the car in Fig. 10-5 has a mass of 1620 kg and that the road is level. This means that the total normal force is 1620 kg$_f$. Since the tires are skidding on the road surface, we have sliding friction. From Table 10-2, we see that the coefficient of sliding friction for rubber on concrete is 1.02. Then Eq. (10-1) gives us

$$\begin{aligned}\text{Force of sliding friction} &= 1.02\,(1620\text{ kg}_f)\\ &= 1650\text{ kg}_f\end{aligned}$$

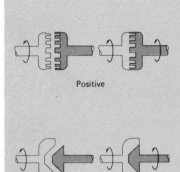

Positive

Conical

Contracting-band

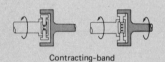

Radially expanding
(centrifugal)

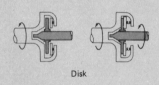

Disk

CLUTCHES

Application: The Clutch

The clutch is a device that permits the connection and disconnection of rotating shafts. Although clutches are sometimes found on stationary engines, their most common use is in land vehicles. With the vehicle at rest and the engine running, the crankshaft must be connected to the drive train to start the vehicle moving. But if this connection is made too abruptly, the large reaction force may stall the engine or damage the drive-train components. Therefore clutches must be designed to engage smoothly when the engine is turning, with a certain amount of slipping allowed as the drive train accelerates to the engine's speed. A positive clutch would quickly destroy itself if installed in a car.

Smooth engagement of two shafts rotating at different speeds can be done with a friction clutch. Friction clutches are of four general types: conical clutches, radially expanding clutches, contracting-band clutches, and disk clutches. Most cars with manual transmissions use a dry, single-disk clutch. In any friction clutch, the idea is to bring the driven surface in contact with the driving surface, then increase the normal force until static friction takes over. On a car, this is done by the action of pressure-plate springs that compress the clutch disk between a pressure plate and the engine flywheel. The clutch is disengaged by forcing a clutch release bearing (throw-out bearing) against release levers on the pressure plate, which then back the clutch disk away from the flywheel. The clutch disk has a splined hub that allows it to slide back and forth on the driven shaft, but still forces the disk and shaft to rotate as one unit.

During the period of slipping, any clutch converts mechanical work to heat. Therefore heavy-duty clutches are often surrounded by a fluid to carry away the heat. Dry clutches may have vents or even fans to circulate cooling air over the contact surfaces. Even so, the clutch will not stand up to lengthy periods of slipping—which is why student drivers are always cautioned not to "ride the clutch."

TABLE 10-2 COEFFICIENTS OF SLIDING FRICTION*

Surfaces in contact	Coefficient of sliding friction		
	Dry	Lightly lubricated	Well lubricated
Bronze and cast iron	0.21	0.16	0.077
Cast iron and cast iron	0.4	0.15	0.064
Cast iron and hardwood	0.49	0.19	0.075
Hardwood and hardwood	0.48	0.16	0.067
Hard steel and babbitt	0.35	0.16	0.06
Hard steel and hard steel	0.42	0.12	0.03
Plastic and steel	0.35	—	0.05
Rubber and concrete	1.02	0.9 (wet)	—
Soft steel and soft steel	0.57	0.19	0.09

* Values listed are for ordinary pressures and temperatures, but may still vary somewhat from sample to sample.

This is the total force acting to decelerate the car. The actual deceleration is 1.02g (numerically the same as the coefficient of sliding friction).

Notice, however, that this total 1650 kg$_f$ is unevenly distributed between the front and rear tires. Since most of the weight is in front, the normal force is greater on the front tires and so is the friction. In a panic stop like this, additional weight is transferred to the front wheels, making the friction in front even greater. Thus the front brakes usually wear out before the rear ones, and they also generate more heat.

But there is something more serious here than a question of brake wear. The large decelerating force in front makes the car very sensitive to any imbalance in the braking forces on the two front ties. Bumps or other surface variations in the road may make the car swerve or, in extreme cases, go into a spin. In a skid situation like this, motion of the car is determined mainly by the properties of the tire-road interface, and any attempt at steering has a reduced effect. ◀

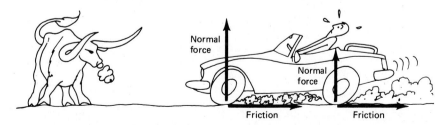

Fig. 10-5 Car in a panic stop. Because most of the weight is in front, most of the friction occurs at the front wheels.

As long as two surfaces are moving over each other, the force of friction has the value given by Eq. (10-1). But when there is no motion, the static friction can have a whole range of values. We see how this happens in Fig. 10-6. The sample block is initially at rest. If we don't pull on it, there is no friction (remember the first condition for equilibrium). If we pull with, say, 0.05 kg_f and the block doesn't move, then there is 0.05 kg_f of friction opposing our pull. If we pull with 0.20 kg_f and the block doesn't move, then there is 0.20 kg_f of friction holding it back. Eventually we reach the breakaway point, where the static friction is as large as it can get. Although we may now notice the block "creeping" very slowly, our applied force is still exactly balanced by the friction holding back the block. But if we exceed this value, the friction drops drastically and our applied force not only moves the block but *accelerates* it. If you are walking on a roof and you begin to slide, you will actually pick up speed on your way toward the edge. Just hope for a strong gutter.

Now although static friction can have a whole range of values, the value at the breakaway point is most important. Here the friction is as great as it can get. This value is related to the normal force in this way:

$$\text{Static friction at breakaway} = \left(\begin{array}{c} \text{coefficient of} \\ \text{static friction} \end{array} \right) (\text{normal force}) \qquad (10\text{-}2)$$

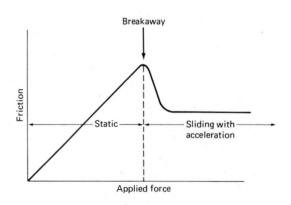

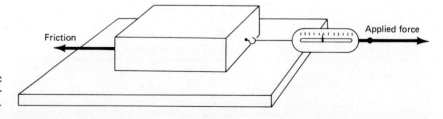

Fig. 10-6 The force of static friction can have a range of values up to the breakaway point.

*TABLE 10-3 COEFFICIENTS OF STATIC FRICTION, AS MEASURED AT
            BREAKAWAY**

Surfaces in contact	Coefficient of static friction	
	Dry	Well lubricated
Hard steel and babbitt	0.42	0.17
Hard steel and hard steel	0.78	0.11
Soft steel and soft steel	0.74	0.11
Soft steel and lead	0.95	0.5
Plastic and steel	0.5	0.1
Rubber and concrete	1.5	1.0 (wet)

* Values listed are for ordinary pressures and temperatures, but may still vary somewhat
from sample to sample.

The *coefficient of static friction* depends on the surfaces and their
condition. Representative values are listed in Table 10-3.

Example 10-3 Traction of a Locomotive.

A railroad locomotive has steel drive wheels resting on steel rails. The
traction it develops is limited by the breakaway static friction at the
wheel-to-rail interface. Once the wheels start to slip, their friction drops
to the lower sliding value.

A diesel locomotive has a typical mass of 110 000 kg [240 000 lb$_m$]. What is the maximum force it can exert in pulling a train?

The rails are never far from level (maximum grade is typically 3 percent), so the normal force is 110 000 kg$_f$. For a dry soft-steel wheel on a soft-steel rail, the coefficient of static friction from Table 10-3 is .074. Thus,

$$\text{Static friction at breakaway} = 0.74 \,(110\ 000\ \text{kg}_f)$$

$$= 81\ 000\ \text{kg}_f$$

Can this force really pull a train? Let's say that the locomotive is coupled to 20 cars with an average mass of 45 000 kg (99 000 lb$_m$). (For longer and heavier trains, there will usually be more than one locomotive.) The weight of each car is supported on axle bearings in which hard steel rests on well-lubricated babbitt metal. The total normal force supporting the 20 cars is (20) (45 000 kg$_f$), or 900 000 kg$_f$. The coefficient of static friction, from Table 10-3, is 0.17. Multiplying these gives a total static friction at breakaway of 150 000 kg$_f$ in all the bearings. Since there is only 81 000 kg$_f$ of traction, it may seem that this train can never start moving.

Or does it? If you've ever listened to a train start, you know that the couplings always have some "play," or lost motion. You hear the first car start with a clang, then another clang as it takes up the play and jerks the second car along, then another clang as the third car starts, and so on down the line. It takes 7650 kg$_f$ to start the first car (0.17 × 45 000 kg$_f$), so there is more than enough engine traction for this. But as soon as it starts, the force needed to keep the car moving drops to the sliding value of 2700 kg$_f$. Why sliding? Because the axle slides against the babbitt bearing. Friction at the rail causes free rolling of the wheel, but doesn't hold back the car.

To get all 20 cars moving, then, it takes only 59 000 kg$_f$ if they start *one at a time*. The 81 000 kg$_f$ of traction is more than enough. In fact, this force not only starts the train, but accelerates it at about 0.027g after the motion has begun.

This experiment was actually done at a railroad yard. Oak wedges were driven into each coupling in a train to take up the slack. The engine was then started, and it did nothing but spin its wheels. "Play" in the couplings performs the important task of allowing static friction to be overcome one car at a time. ◀

Very powerful locomotives are purposely built very heavy so friction can transmit the driving force. We frequently use this fact that an increase in normal force increases the friction. Some people prepare for winter driving by throwing a few bags of sand in the trunk, over the drive wheels. Using a sanding block, you press the sandpaper *into* the wood as you slide it. If a V-belt is slipping, you tighten it to increase the normal force, which increases the friction. The harder the brake pads are pressed into the disks, the greater the friction acting to stop a car's wheels. You may think of other examples.

By the same token, decreasing the normal force decreases the friction. We have already seen that the normal force decreases with increasing angle of inclination. Thus less traction is available to a car on a hill than on the level. And there is less friction holding a roofer on a steep roof than there is on a shallow-pitched roof.

Exercises

3. A piece of hard steel is held against another piece of hard steel with a normal force of 22 kg$_f$. (*a*) What force is needed to start the motion if the surfaces are dry? (*b*) What force is needed to maintain the motion if the surfaces are dry? (*c*) What forces are needed to start and maintain the motion if the surfaces are well lubricated?
    Answers: (*a*) 17 kg$_f$, (*b*) 9.2 kg$_f$, (*c*) 2.4 and 0.66 kg$_f$
4. A car decelerates on a dry, level road where the coefficient of static friction between the tires and the road is 1.5. The car has a mass of 1800 kg. (*a*) What is the maximum force available to decelerate the car? (*b*) What force decelerates the car if the wheels lock and the car skids?
    Answers: (*a*) 2700 kg$_f$, (*b*) 1800 kg$_f$

Although we have listed values for the friction coefficients, in practice friction is not quite this predictable. For one thing, the irregularities in a surface are bound to vary from point to point. This causes the coefficient of friction to change slightly as an object moves along. For another thing, there is always a certain amount of dirt or dust contaminating a surface. On very smooth surfaces, this increases the friction. But on rough surfaces, this contamination may actually reduce friction by filling in some of the valleys. This is the principle behind the use of graphite as a lubricant. Its own cohesion is so low that the graphite breaks up and fills in the surface irregularities.

10-4 Effect of Temperature and Surface Condition

There is also the problem of temperature. Sliding friction generates heat, which raises the temperature of the surfaces. An increase in temperature may reduce the friction, particularly when lubricants are present. Even dry brake pads are less effective when they get hot. But sometimes just the opposite happens. The use of "gumball" tires on drag racers is based on their tendency to get "gummy" at high temperatures.

Application: Road Friction and Vehicle Handling

A driver controls the car by using the gas pedal, brake, and steering wheel, but the actual control forces are transmitted as friction between the tire patches and the road surface. When breakaway occurs, the tires skid on the pavement rather than rolling. Friction is then reduced to its sliding value, and the car is probably "out of control." Thus the static friction at breakaway limits a car's ability to corner, brake, and accelerate in general.

It happens that the limiting breakaway force is the same whether the tire begins to slip forward, backward, or sideways. Suppose that a car executes a tight turn at the highest possible speed. Static friction (acting laterally to the tire patch) is near the breakaway limit. Now the driver applies the brakes without straightening the wheel. Since the additional rearward friction combines with the sideways friction, the total control force must now be greater than the breakaway limit. (See the diagram.) Of course, this is impossible. So the tires break loose from the road, and the car begins to skid off the outside of the turn. To recapture control, the driver must release the brakes and/or decrease the lateral force by straightening the front wheels.

Sudden braking in corners often proves disastrous to the inexperienced driver, particularly when road conditions are poor. Since the resulting skid is a condition of oversteer, the steps needed to correct it are exactly opposite to normal instinct: most drivers will actually press the brake harder and try to steer a tighter turn. But this just keeps the tires sliding. Proper control can be regained only if the wheels are allowed to begin rolling again in their direction of travel.

Speeding up while cornering is much less risky, at least when the drive wheels are in the rear. Now the rear tires break away before the front ones, and this sends the car into an oversteer condition. Natural instinct is then to let up on the gas and straighten the wheels slightly—which very quickly brings the car back under control. In fact, racing drivers are frequently seen executing this maneuver as they try to gain the edge in a turn.

So with a certain amount of driving skill, speeding up in a turn (within reasonable limits) is not dangerous. But braking in a turn can easily lead to total loss of control. Does this mean it is impossible to safely slow down in a turn? No. It just means that any braking should be very gentle and that most of the braking should be done *before* entering the turn. When it is necessary to decelerate in the turn itself, the best bet is to downshift. Then if a skid develops, it will be confined to the rear wheels and the resulting oversteer is easily corrected.

The same idea applies when driving a car up a snow-covered grade. Since friction is greatest near breakaway, it does no good at all to madly spin the drive wheels on the way up. The good driver will listen carefully to the sound of the engine (or watch the tachometer) and ease off the gas at the first sign of excessive slipping. If this doesn't get the car up the grade, spinning the wheels certainly won't.

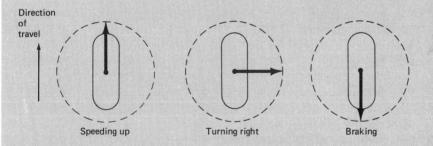

Speeding up   Turning right   Braking

(A) Regardless of the direction of the acceleration, the control force on a tire is limited to the breakaway value of static friction.

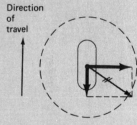

(B) If a vehicle is already near the breakaway limit in a turn, the addition of a braking force may exceed the breakaway limit—causing sliding and a loss of control.

The temperature increase is achieved by spinning the drive wheels at the start.

Some materials exhibit more complicated effects. Tungsten carbide, for instance, is a very hard material commonly used on saw blades, drill bits, and other high-speed machine tools. The coefficient of static friction between two pieces of this material at room temperature is about 0.47. As the surfaces are heated, this value decreases—for awhile. But at high enough temperatures, the coefficient of static friction increases again. The effect is shown in Fig. 10-7.

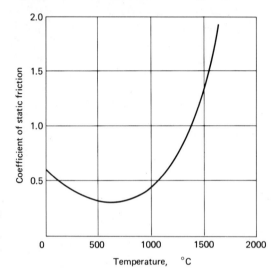

Fig. 10-7 Coefficient of static friction for tungsten carbide on tungsten carbide. Some materials exhibit complicated temperature effect.

10-5 Rolling Friction

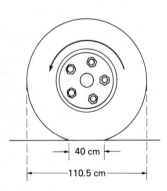

Fig. 10-8 Deformation of a heavy truck tire at full load and full inflation. The deformation gives rise to rolling friction.

Figure 10-8 shows a load-bearing tire on a flat surface. Since the tire deforms under the load, the contact region is a "patch" rather than a point. On a heavy truck tire 110 cm [43.5 in] in diameter, the vertical deflection under load can be as much as 4 cm. This gives a patch around 40 cm [16 in] long. As the tire rolls, this flattening presents a resistance known as *rolling friction*. Since the tread must flex and unflex more rapidly at higher speeds, rolling friction increases with speed.

A steel wheel on a steel rail develops very little rolling friction because the deformation is so slight. A pneumatic tire develops more rolling friction because it flattens more. Typically, a vehicle on pneumatic tires develops 10 times as much rolling friction as a similar steel-wheeled vehicle on rails.

Rolling friction is generally much lower than sliding friction between the same surfaces. Thus wheels and rollers are a very good way of moving things around. If, however, we try to drive on a flat front tire on an icy road in subzero weather, it may be easier for the tire to slide along than to roll.

There are many applications where a large amount of friction is desirable: brakes, clutches, grinding wheels, highway surfaces, vises, conveyor belts, and bulldozer tracks, to name a few. In many other applications involving moving parts friction is undesirable. Internal engine parts, transmissions, axles, and pump parts fall into this category.

A *bearing* is a device that accomplishes two purposes. First, it confines the motion to a certain path. Thus, an engine valve must slide back and forth exactly the same way; if the path changes even a bit, the valve won't seat properly. Second, a bearing must keep friction below a harmful level. Too much friction causes the surfaces to wear quickly, and may also generate enough heat to create other problems like warping and seizure resulting from expansion.

Bearings can be classified according to the kind of motion they allow or prevent (Fig. 10-9). A cylinder wall acts as a *slide bearing*, allowing a piston to slide up and down without rocking. A valve guide performs the same function, as does the slide-bolt cylinder on a rifle. A *thrust bearing* must be used to prevent a drill or propeller shaft from being forced back and drilling through itself as it rotates. Any rotating shaft must also have a *main bearing* (or perhaps several) to keep its axis of rotation from wandering.

A *journal bearing* is a simple sleeve that fits around a portion of a rotating shaft (called the journal). The sleeve may be split into two halves for easy replacement, and it may contain holes and grooves for the passage of a lubricant. The efficiency of a journal bearing is determined by sliding friction. Journal bearings are the most common bearings for low-speed applications like clock movements. They may also be used in high-speed applications where continuous forced lubrication is provided for. The main crankshaft bearings in an engine are usually of this type.

Ball bearings and *roller bearings* are used in places where very low friction is a requirement and forced lubrication is difficult or impossible. The wheel bearings on bicycles and motor vehicles are of this type. Ball bearings are also used as thrust bearings, and occasionally even as slide bearings. These types of bearing construction are shown in fig. 10-10.

10-6 Bearings

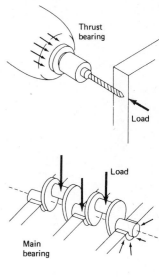

Fig. 10-9 Bearings can be classified according to the kind of motion they allow or prevent.

10-7 Lubrication

Sliding friction can usually be reduced by a suitable lubricant. The lubricant fills the little valleys in the mating parts and allows one surface to float over the other without material contact. Thus wear is also reduced. Generally, the lubricant should have the lowest viscosity that still provides a supporting film. If the viscosity is too high, the lubricant may actually increase the friction between the parts (as when petroleum jelly is placed between two pieces of glass).

Fig. 10-10 Journal, roller, and ball bearings.

Of course, the problem is complicated by the fact that friction generates heat and an oil's viscosity decreases with temperature. Thus the proper lubricant must be chosen very carefully according to a machine's operating conditions. It's always a good idea to follow the manufacturer's recommendations rather than experiment.

Figure 10-11 shows the supporting oil film in a journal bearing. At rest, the journal makes actual contact with the bearing surface. In starting, this contact point moves as the journal tries to roll uphill. Eventually, the oil film is established and friction is reduced. But journal bearings tend to wear much more rapidly when there is frequent starting and stopping.

Furthermore, the bearing pressure has a large effect on the friction. Figure 10-12 shows some typical results for lightly lubricated journal bearings. As the bearing pressure is increased, the oil film is less effective and the coefficient of sliding friction increases. Eventually, there is metal-to-metal contact. With a great enough pressure, adhesion between the metals begins to take over, and the bearing eventually "seizes."

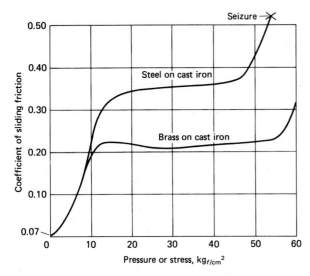

Fig. 10-12 Effect of pressure on the coefficient of sliding friction in a journal bearing.

Fig. 10-11 Supporting oil film in a journal bearing. Clearance has been exaggerated.

Friction is actually a very complicated force. The friction between two objects is affected by their surface roughness, the forces of adhesion and cohesion between the materials, their elasticity, the presence of surface contaminants, the presence of lubricants, the temperature, sometimes the area of contact, and always the normal force. All but the last of these are accounted for in a number called the coefficient of friction. Still, we need to have two of these coefficients: the static value at breakaway and the sliding value. And if any of the important factors change, these friction coefficients also change.

The effects of friction need to constantly be kept in mind in the trades. After all, every object we encounter is in contact with something else. As a result, there isn't a single thing that friction doesn't act on. We depend on friction to move and stop our land vehicles. Airplanes and boats have to fight against it. It wears out our machine tools and engines. It holds together our houses (friction is what keeps the nails from coming loose). And so on. In the next chapter, we look more closely at the role of friction in machines.

Summary

Terms You Should Know

static friction	coefficient of sliding friction
sliding friction	coefficient of static friction
breakaway	rolling friction
normal force	slide bearing

thrust bearing   roller bearing
main bearing   ball bearing
journal bearing

Problems

1. A railroad locomotive has a mass of 110 t, and it pulls 15 cars that average 45 t each. The wheels are soft steel, and so are the rails. The engine as well as each car has brakes. (*a*) What is the maximum stopping force that can be generated on a dry, level track? (*b*) What is the train's maximum deceleration in *g*'s?

2. A certain four-wheel-drive vehicle has a mass of 2160 kg. It climbs a hill with a 30° grade. The coefficient of static friction is an average 0.831. (*a*) What is the normal force supporting the vehicle on the hill? (*b*) What friction force can the tires generate in climbing the grade? (*c*) What friction force can the tires generate on a level road with the same coefficient of friction?

3. A certain car has a mass of 1500 kg. It makes a turn on a level road which requires a centripetal acceleration of 0.55*g*. The coefficient of static friction between the tires and the road surface is 0.75. (*a*) What maximum braking force can be generated in the turn before the car goes into a skid? (*b*) What maximum braking force could be generated if the car were traveling straight rather than turning?

(*Hint*: Make a scale drawing of the forces at the tire patch.)

4. A hard-steel axle rotates in a lightly lubricated babbitt bearing. The bearing places a vertical load of 3500 N on the axle. (*a*) What is the force of sliding friction between the axle and the bearing? (*b*) If the axle has a diameter of 2.8 cm, how much work in joules must be done in overcoming friction when the axle rotates one complete revolution? (*c*) If the axle rotates 20 times each second, how much power in watts is needed to overcome friction?

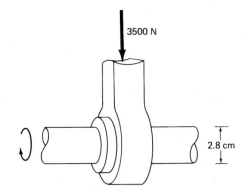

3500 N

2.8 cm

11

Simple Machines

We have seen that there are many devices for converting energy from one form to another. A simple machine is a device whose energy input and output are both in the form of mechanical work and whose parts are all rigid solids.

Why would we want to convert a work input to a work output? There may be several reasons. For one thing, we may want to apply a force at one place and have it do work somewhere else. Using pulleys, for instance, we can lift a scaffold to a roof by pulling on a rope from the ground. Second, we may have only a small force available to produce the work input when we need a large force in the output. This is the case with a car jack. By "pumping" the jack handle, we can lift a car that we'd hardly budge otherwise. Of course, we have to do a great deal of pumping to lift the car just a small amount.

Simple machines are usually classified in six categories: levers, pulleys, wheel and axles, inclined planes, screws, and wedges. Examples of these are shown in Fig. 11-1. More complicated machines, like lathes or surface grinders, are combinations of these six basic types of machine.

Introduction

Fig. 11-1 The six types of simple machine.

11-1 Mechanical Advantage

There are two important forces in any simple machine: effort and load. *Effort* is the force applied to the machine. *Load* is the force overcome by the machine in doing useful work. When we use a nutcracker, the effort is supplied by our hand squeezing the handle; the load is the elastic force in the nut being cracked.

Now the effort and the load are usually not equal. In fact, most simple machines are used in situations where the load is greater than the effort. You could never pull a 16d nail from a board with your fingernails. But by applying some effort to a crowbar, you can easily overcome the forces holding the nail in the wood. These forces are the load on the crowbar.

A machine's ability to move a load is described by its mechanical advantage.

Definition: A machine's *mechanical advantage* is the ratio of the load to the effort. It is often abbreviated as MA.

mechanical advantage

$$MA = \frac{load}{effort} \qquad (11\text{-}1)$$

Thus if a machine lifts a load of 8 kg$_f$ with an effort of 2 kg$_f$, its mechanical advantage is 4. Or if a machine has a mechanical advantage of 6, this means that each 1 kg$_f$ of effort overcomes a load of 6 kg$_f$. Notice that there is no unit for mechanical advantage, since it is a ratio of two forces. We need to be careful, however, to express the load and effort in the same units.

Exercises

1. A certain machine needs an effort of 18 kg$_f$ to move a load of 87 kg$_f$. What is its mechanical advantage?
   Answer: MA = 4.8
2. A load of 2.2 tons is to be lifted. A man can develop a 130-lb effort with his arms. What mechanical advantage is needed to allow the man to lift the load?
   Answer: MA = 34
3. A certain machine has a mechanical advantage of 5.0. What load can it move with an effort of 48 kg$_f$?
   Answer: 240 kg$_f$
4. If a machine has a mechanical advantage of 12.7, what effort is needed to overcome a load of 525 kg$_f$?
   Answer: 41.3 kg$_f$

We said earlier that no device can have an efficiency greater than 100%. Of course, this must hold true for simple machines. Just as before, the efficiency is

11-2 Efficiency of Simple Machines

$$\text{Efficiency} = \frac{\text{useful work output}}{\text{work input}} \qquad (11\text{-}2)$$

We may also rewrite this formula as

$$\text{Useful work output} = (\text{efficiency})(\text{work input}) \qquad (11\text{-}3)$$

Fig. 11-2 A screw jack. The mechanical advantage is high, but the efficiency is low.

Because the efficiency is always less than 100 percent, it follows that a machine's useful work output is always less than its work input.

It is quite possible for a machine to have a large mechanical advantage but a low efficiency. In other words, a small effort may handle a large load, yet the effort may have to move through a very large distance to get the load to move just a little. This is the case with the screw jack in Fig. 11-2. The work done in twisting the screw is much larger than the work done in lifting the load.

Simple machines would all have efficiencies near 100 percent if it weren't for sliding friction and rolling friction. When friction is greatest, as in the wedge or screw, the efficiency may be only 10 percent or less. For levers and wheel and axles, where friction is low, the efficiency may approach 99 percent. There is also some loss of efficiency as a result of elastic distortion of the machine under load. In most cases, this is a very minor effect.

11-3 Levers

A lever is simply a rigid bar which can pivot at some point along its length. The pivot point is called the *fulcrum.* The effort and the load are applied at points other than the fulcrum. There are three possible arrangements, as we see in Fig. 11-3. They are called the first-class, second-class, and third-class lever.

In the *first-class lever,* the effort and load are on opposite sides of the fulcrum. The effort balances the load on the same principle as the

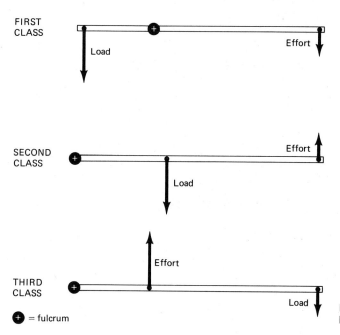

Fig. 11-3 The three classes of lever.

seesaw. Both the equal-arm balance and the unequal-arm balance of Chap. 8 are first-class levers. In Fig. 11-4 we see some others. Notice that the tongs are two first-class levers with a common fulcrum. The walking beam we saw on the atmospheric engine is also a first-class lever.

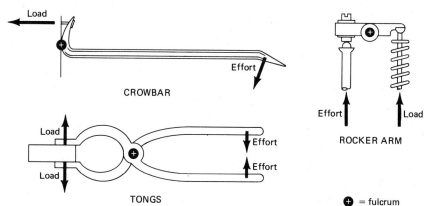

Fig. 11-4 Some first-class levers.

In the *second-class lever,* the load is between the effort and the fulcrum. A nutcracker is two second-class levers with a common fulcrum. Figure 11-5 shows some other examples. The oar and oarlock is a second-class lever because the boat is the load being moved. Ideally,

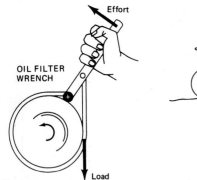

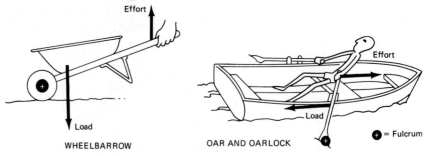

Fig. 11-5 Some second-class levers.

the oar blade should not move the water as much as the oarlock moves the boat. But if you use the oar to skim the water and splash some other boaters, the water becomes the load and the oar is now a first-class lever.

When the effort is between the load and the fulcrum, we have a *third-class lever*. Tweezers are two third-class levers with a common fulcrum. We see some other examples in Fig. 11-6. Third-class levers are not as common as the first- and second-class levers.

The distance between the effort and the fulcrum is called the *effort arm*, while the distance between load and fulcrum is the *load arm*. For all three types of lever, the mechanical advantage can be found from

$$\text{MA of lever} = (\text{efficiency}) \left(\frac{\text{effective effort arm}}{\text{effective load arm}} \right) \qquad (11\text{-}4)$$

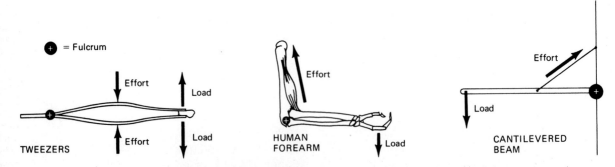

Fig. 11-6 Some third-class levers.

What do we mean by an *effective* effort arm? It is simply the perpendicular distance between the fulcrum and the line of action of the effort. This is the same as the effective torque arm developed by the effort. Similarly, the effective load arm is the perpendicular distance between the fulcrum and the line of action of the load. Although we already discussed this idea in Chap. 8, it is shown again in Fig. 11-7. Let's see how Eq. (11-4) is used.

Example 11-1 Mechanical Advantage of the Claw Hammer.

As the hammer pulls out the nail, it rocks on its curved claw. This is a first-class lever whose fulcrum moves as it does its job.

The effective effort arm is generally around $3\overline{0}$ cm [12 in]. This varies only a small amount as the fulcrum moves. The effective load arm, on the other hand, varies quite a bit. With the nail head close to the surface, the effective load arm is only around 1 cm [0.4 in]. With the hammer upright, this has increased to around 5.7 cm [2.2 in]. Now the hammerhead is touching the surface, and a spacing block is needed to pull the nail further (assuming it is not already out).

Because the effective load arm varies, the mechanical advantage of the claw also varies. Since there is little sliding or rolling friction, the efficiency is close to 100 percent. Then using Eq. (11-4), the mechanical advantage at the start is

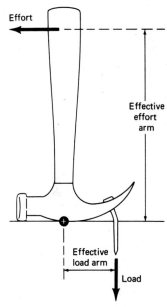

Fig. 11-7 Effective effort and load arms on a claw hammer.

$$MA = \frac{\text{effective effort arm}}{\text{effective load arm}}$$

$$= \frac{3\overline{0} \text{ cm}}{1 \text{ cm}}$$

$$= 3\overline{0}$$

This means that a 1-kg$_f$ effort will overcome a 30-kg$_f$ force holding the nail in the wood. Or a 20-kg$_f$ effort will produce a 600-kg$_f$ pulling force.

When the nail starts to move, the claw rocks and the MA decreases. It finally reaches a low value of

$$MA = \frac{3\overline{0} \text{ cm}}{5.7 \text{ cm}}$$

$$= 5.3$$

Now a 1-kg$_f$ effort develops only a 5.3-kg$_f$ pulling force, and a 20-kg$_f$ effort develops 106 kg$_f$. Since the most difficult part is getting the nail-pull started, this drop in mechanical advantage usually presents no problem. Actually, the claw was intentionally designed to give the greatest MA at the start of the pull. ◀

Exercises

5. A lever has an effective effort arm of 89 cm and an effective load arm of 3.3 cm. What is its mechanical advantage if the efficiency is (*a*) nearly 100 percent, (*b*) 97 percent, (*c*) 93 percent?
Answers: (*a*) MA = 27, (*b*) MA = 26, (*c*) MA = 25
6. What load can be lifted by the lever shown in the diagram? Assume that the efficiency is nearly 100 percent.
Answer: 1400 kg$_f$

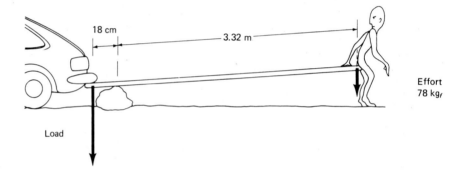

18 cm

3.32 m

Effort
78 kg$_f$

Load

7. Solve exercise 6 if the fulcrum is moved 6.0 cm to the left.
Answer: 2200 kg$_f$
8. Solve Exercise 6 if the fulcrum is moved 6.0 cm to the right.
Answer: 1100 kg$_f$
9. A second-class lever is needed with an MA of 7.0. The efficiency is near 100 percent. The load arm has to be 15.7 cm in length. (*a*) How far from the fulcrum should the effort be applied? (*b*) What load will be moved with an effort of 44 kg$_f$?
Answers: (*a*) 110 cm, (*b*) 310 kg$_f$

11-4 Speed Advantage   Is it possible for a lever to have an MA less than 1? Yes. Looking at Eq. (11-4), we see that this happens whenever the effective effort arm is shorter than the effective load arm. Thus, some first-class levers and all third-class levers have MAs less than 1.

But what would such a lever be good for? Since it requires an effort bigger than the load, it might seem that we could connect the effort directly to the load and completely forget the lever.

Actually, there are two reasons why we might settle for an MA of less than 1. In some stationary systems, such as the cantilever, we need to support a load some distance from the effort, and we are willing to use a lot of effort to do this. And in some moving machines, such as a typewriter action, we are more concerned with speed than with mechanical advantage.

A low mechanical advantage leads to a large speed advantage (assuming that the efficiency is not too low). In other words, the load moves faster than the effort when the MA is less than 1. This is true not only of levers, but of all machines.

The human arm and wrist are a series of third-class levers, each with an MA less than 1. It is very difficult to hold a weight at arm's length because the MA is too low. But if you throw a baseball, these same levers might give it a speed of, say, 100 kmph. This is much faster than the rate at which the muscles contract.

11-5 Pulleys

A single pulley changes the direction of a force without changing its size. In Fig. 11-8, the 25-kg$_f$ load is lifted by an effort only slightly larger than 25 kg$_f$.

Since the rope or cable generally stretches very little, the pulley's efficiency is limited mainly by friction in its bearing. As we have seen (Table 9-2), the efficiency of a ball bearing is typically 99 percent. A roller bearing's efficiency might be 98 percent, while for a journal bearing the efficiency is only slightly lower. These figures, then, are also the efficiencies of pulleys using these bearings. As a result, a ball-bearing pulley has an MA of 0.99, and a roller-bearing pulley has an MA of 0.98. A journal-bearing pulley, if lubricated, generally has an MA between 0.95 and 0.98.

11-6 Pulley Systems

Although a single pulley doesn't provide a great mechanical advantage, a system of several pulleys does. In Fig. 11-9 we see a two-pulley system called a *block and tackle*. The upper pulley is fastened to a stationary support, while the lower pulley moves with the load. Notice that the load is now supported by two parallel sections of cable. Thus a 25-kg$_f$ cable tension will support a load of about 50 kg$_f$. This gives an MA of 2.0.

Now actually the MA is a little less than this. The effort must also lift the weight of the lower pulley and cable along with the load. In most cases, this effect is small. But in addition, the effort must overcome the friction in both pulley bearings. If each bearing has an efficiency of 98 percent, the total efficiency is (0.98)(0.98), or 96 percent. Thus the MA is really about 96 percent of 2.0, or 1.9.

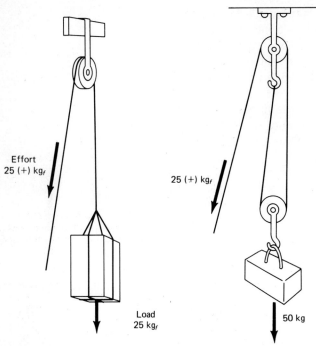

Effort
25 (+) kg$_f$

25 (+) kg$_f$

Load
25 kg$_f$

50 kg

Fig. 11-8 A single pulley has a
mechanical advantage equal to
the efficiency of the bearing. In
most cases, this is close to 1.

Fig. 11-9 A two-pulley block
and tackle. This system is also
called a gun tackle.

We can also make a block and tackle with three, four, five, or even
more pulleys. The mechanical advantage is increased each time there is
an additional strand supporting the load. Figure 11-10 shows the
arrangements using three and four supporting strands.

The mechanical advantage of a block and tackle can be found from
the following formula:

$$\text{MA of block and tackle} = (\text{total efficiency})(\text{number of} \qquad (11\text{-}5)$$
$$\text{supporting strands})$$

Example 11-2 shows the use of this formula.

Example 11-2 Mechanical Advantage of a Six-Strand Block and Tackle.

The device is shown in Fig. 11-11. There are two blocks: a fixed upper
one and a movable lower one. The arrangement of cables (strands) is

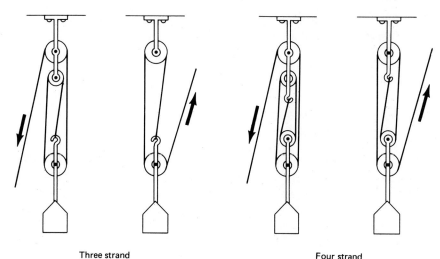

Three strand

Four strand

Fig. 11-10 Three- and four-strand block and tackles. Only the strands supporting the load are counted.

called the tackle. In practice, each block will have three pulleys mounted side by side. This allows the greatest amount of travel before the blocks collide. To analyze the system, it is easier to sketch the pulleys one above the other. The two systems are mechanically equivalent.

Let's say that each pulley has an efficiency of 98 percent. The total efficiency of six pulleys is then (0.98)(0.98)(0.98)(0.98)(0.98)(0.98), or 0.89, or 89 percent.

We can count the six strands supporting the load. Notice that the cable where the effort is applied does not support the load directly. Then using Eq. (11-5), we have

$$MA = \text{(total efficiency)(number of supporting strands)}$$

$$= 0.89\,(6)$$

$$= 5.3$$

This means, of course, that each 1 kg_f of effort lifts a load of 5.3 kg_f. At this rate, an 82-kg [180 lb_m] person can lift a load of 430 kg_f [950 lb] without his or her feet leaving the ground. ◄

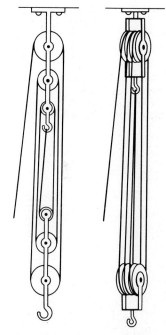

Fig. 11-11 A six-strand block and tackle. To get maximum clearance between the blocks, each set of three pulleys may be arranged side by side. To analyze the system, it is easier to sketch the pulleys one above the other.

Exercises

10. Find the MA of the block and tackle in Fig. 11-11 if each pulley bearing has an efficiency of only 95 percent.
    Answer: MA = 4.4

11. Sketch a five-strand block and tackle with the effort applied from the ground. Calculate the MA if each pulley has an efficiency of 97 percent.
    Answer: There are three pulleys on the upper block and two pulleys on the lower one; the MA is 4.3.
12. What is the highest possible MA for a block and tackle with (a) two supporting strands, (b) three supporting strands, (c) four supporting strands, (d) five supporting strands?
    Answers: (a) 2.0, (b) 3.0, (c) 4.0, (d) 5.0
13. A certain block and tackle has four pulleys and four supporting strands. Each pulley has an efficiency of 96 percent. The system is used to lift a 190-kg$_f$ load through a height of 1.6 m. (a) What is the total efficiency? (b) What is the mechanical advantage? (c) What is the useful work output? (d) What is the work input?
    Answers: (a) 85 percent, (b) 3.4, (c) 300 kg$_f$·m, or 3.0 kJ, (d) 3.5 kJ
14. A block and tackle is needed with an MA of at least 4.2. The pulleys have an efficiency of 97 percent. How many supporting strands are needed?
    Answer: 5

11-7 The Wheel and Axle

When a wheel rolls freely on a bearing, it functions as a pulley and its mechanical advantage is the same as the bearing's efficiency. This is the case with a paint roller or the idler wheel on a contracting-band clutch. But when a wheel is rigidly connected to an axle so that the two turn together, we have quite another thing. The wheel and axle can be used to generate a large mechanical advantage; or, used in reverse, it will give a large speed advantage. Figure 11-12 shows some example of this simple machine.

The mechanical advantage may be found from

$$\text{MA of wheel and axle} = (\text{bearing efficiency})\left(\frac{\text{wheel radius}}{\text{axle radius}}\right) \quad (11\text{-}6)$$

Often it is more convenient to use diameters rather than radii. Then the equation is

$$\text{MA of wheel and axle} = (\text{bearing efficiency})\left(\frac{\text{wheel diameter}}{\text{axle diameter}}\right) \quad (11\text{-}7)$$

Let's look at a numerical example.

Example 11-3 The Windlass.

The device is shown in Fig. 11-13. The effort is applied perpendicular to the crank handle; since this handle moves in a circle, it is effectively a wheel. The drum is the axle.

Let's say that the diameter of the cranking circle is 62 cm, while the drum's diameter is 18 cm. The bearing efficiency is 98 percent. Then Eq. (11-7) gives

$$MA = (\text{bearing efficiency}) \left(\frac{\text{wheel diameter}}{\text{axle diameter}} \right)$$

$$= 0.98 \left(\frac{62 \text{ cm}}{18 \text{ cm}} \right)$$

$$= 3.4$$

Again, this means that each 1 kg_f of effort lifts a load of 3.4 kg_f, or each 1 lb of effort lifts a load of 3.4 lb.

Notice that the crank handle moves much faster than the load, because it travels the circumference of a larger circle in the same time. How much faster? Just the ratio of the two circumferences. This is the same as the ratio of the diameters, or about 3.4. To lift the load at a rate of, say, 5.0 cm/s, the crank handle must move at (3.4)(5.0 cm/s), or 17 cm/s.

Screwdriver

Steering wheel

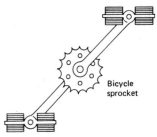

Bicycle sprocket

Fig. 11-12 Examples of the wheel and axle.

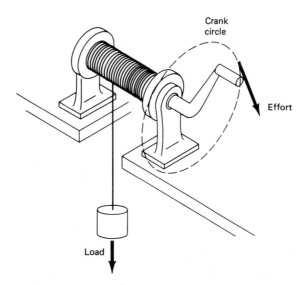

Crank circle

Effort

Load

Fig. 11-13 A windlass is a wheel and axle. The effort is applied at the crank circle, which is effectively a wheel. The cable wraps around the drum, which is the axle.

What happens when the effort is removed? Obviously, the load drops, unwinding the cable and spinning the handle. And, again, the handle spins 3.3 times as fast as the load moves. This has led to serious industrial accidents, even though the device is usually equipped with a ratchet mechanism to keep this from happening. Every precaution should be taken when working near a windlass (or any rotating machine, for that matter). ◄

The *torque* applied to a wheel is transmitted by the axle with very little loss. The same is true if the torque is applied to the axle and transmitted by the wheel. The relationship is

$$\text{Torque out} = (\text{bearing efficiency})(\text{torque in}) \qquad (11\text{-}8)$$

Because bearing efficiencies are usually so high, the output and input torques are usually considered to be equal. Thus a wheel and axle can be a useful device for transmitting torque from one place to another. Examples are the screwdriver and the automotive drive shaft.

Example 11-4 Torque and Acceleration in an Automobile.

A 1720-kg car is to accelerate from rest to 90.0 kmph in 11.1 s. The road is level. Using the methods of Chap. 7, we find that the thrust needed is 395 kg$_f$ [871 lb]. But what torque is needed to produce this acceleration?

Let's look at the rear wheel and axle in Fig. 11-14. The wheels have a tread diameter of 71.2 cm [28.0 in]. Their radius is therefore 35.6

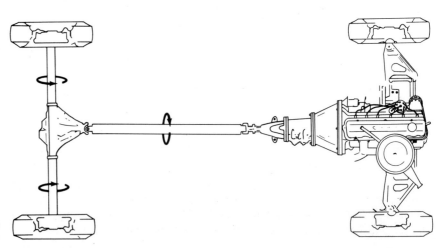

Fig. 11-14 Torque conversion in an automobile.

cm, and this is the effective torque arm between the road-contact patch and the axle hub. The torque, from Eq. (8-1), is

$$\text{Torque} = (\text{force})(\text{effective torque arm})$$

$$= 395 \text{ kg}_f \ (0.356 \text{ m})$$

$$= 141 \text{ kg}_f \cdot \text{m}$$

In old USCS units, this is 1020 lb·ft.

To produce this much torque at the drive wheels, 141 kg$_f$·m of torque must be applied to the axle. But a conventional car engine does not develop this much torque. Under the conditions of this example, 33.5 kg$_f$·m [242 lb·ft] is a more typical value for the peak engine torque.

Where, then, does the additional torque come from? It comes from the transmission and rear end, or from the transaxle if the car has its engine over the drive wheels. These drive-train components increase the torque at the expense of speed. Thus the rear axle may turn at only one-fifth the speed of the engine while it transmits 5 times the engine torque to the drive wheels. ◀

Exercises

15. A wheel and axle has a bearing efficiency of 96 percent, a wheel diameter of 38.2 cm, and an axle diameter of 2.7 cm. (a) What is the mechanical advantage? (b) What is the force at the axle if the force at the wheel is 18.5 kg$_f$? (c) What is the torque input at the wheel? (d) What is the torque output at the axle?
    Answers: (a) MA = 14, (b) 250 kg$_f$, (c) 3.53 kg$_f$·m, (d) 3.4 kg$_f$·m

16. A wheel has a radius of 8.21 cm while its axle has a radius of 1.92 cm. The bearing efficiency is 97.8 percent. (a) What is the mechanical advantage? (b) What force is needed at the wheel to develop 115 kg$_f$ at the axle? (c) What is the torque input at the wheel? (d) What is the torque output at the axle?
    Answers: (a) MA = 4.18, (b) 27.5 kg$_f$, (c) 10.5 kg$_f$·m, (d) 10.3 kg$_f$·m

11.8 Simple Belt Drives

We have seen that a single wheel and axle can multiply force, but it does little to change the torque. At the same time, it has no effect on rotational speed: each revolution of the wheel produces just one revolution of the axle. A *system* of wheels and axles, on the other hand, can be used to change both speed and rate of rotation. The simplest, and cheapest, way to do this is to connect the wheels with a continuous friction belt.

Figure 11-15 shows a simple belt drive. The wheels are usually called "pulleys," even though they aren't. A true pulley rotates freely on a bearing; wheels and axles are rigidly connected and rotate together, supported by the axle bearing. So this is really a system of two wheels and axles. We will follow standard practice, however, and call them pulleys. The pulleys are commonly grooved to accept a V-belt, but flat belts and smooth pulleys are found in some applications.

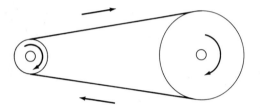 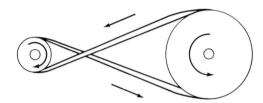

Fig. 11-15 The belt drive. The pulleys normally rotate in the same direction. If the belt is twisted and crossed, they will rotate in opposite direction.

The pulley connected to the power source (motor, hand crank, etc.) is called the *driver,* or *driver pulley.* The other pulley is the *driven pulley.* Ordinarily, both pulleys rotate in the same direction. If the belt is given a half-twist and crossed, the pulleys will rotate in opposite directions. This is not a good idea in high-speed applications, because heat generated by the rapid flexing and unflexing can contribute to premature belt failure.

The ratio of the two pulley diameters is called the drive ratio:

drive ratio

Definition: The *drive ratio* of a belt drive is the ratio of the diameter of the driven pulley to the diameter of the driver:

$$\text{Drive ratio} = \frac{\text{driven pulley diameter}}{\text{driver diameter}} \quad (11\text{-}9)$$

If a belt drive has a 5-cm pulley driving a 10-cm pulley, the drive ratio is 2. This number describes the fact that the driver must rotate twice to get the driven pulley to rotate once. Sometimes people will also say that this drive ratio is "2 to 1." Like mechanical advantage, drive ratio has no unit.

The drive ratio tells us how the belt drive changes the torque:

$$\text{Output torque} = (\text{efficiency})(\text{input torque})(\text{drive ratio}) \quad (11\text{-}10)$$

As we saw in Table 9-2, the efficiency of a belt-and-pulley system is 95 to 98 percent. So if the drive ratio is 2.0, the output torque is nearly twice the input torque.

Now a belt drive changes the speed of rotation as well as the torque. Rather than calling it speed of rotation, though, we usually talk about the frequency.

Definition: The *frequency* of any rotating device is the number of complete revolutions per unit time. It is usually expressed in units of revolutions per minute (rpm) or revolutions per second (rps).

frequency

Electric motors usually operate at a frequency of 1750 rpm. Car engines idle at a frequency of 500 to 800 rpm. If an engine turns at 1800 rpm, this means that its flywheel makes 1800 revolutions each minute. This corresponds to 30 rps.

We have seen that when a little pulley drives a bigger pulley, the torque is increased. But at the same time, the big (driven) pulley rotates slower than the little (driver) pulley. Torque can be increased only by decreasing the frequency. The formula is

$$\text{Output frequency} = \frac{\text{input frequency}}{\text{drive ratio}} \qquad (11\text{-}11)$$

This formula assumes that the belt doesn't slip. We may also write it as

$$\text{Input frequency} = (\text{drive ratio})(\text{output frequency}) \qquad (11\text{-}12)$$

Let's look at an example.

Example 11-5 The Variable-Speed Belt Drive.

With some power tools, it's a great advantage to be able to change the operating speed or frequency. When wood or metal is being drilled on a drill press, a high rotational frequency is usually desirable. But if plastic is being drilled, the bit must turn very slowly to keep the work from melting as a result of friction. Similarly, the rotational frequency of a lathe chuck should be set according to the metal being worked. In general, softer metals should be cut at higher speeds than harder metals.

Figure 11-16 shows a pulley system that gives three different frequencies from one motor. The two stepped pulleys are mounted facing in opposite directions. Of course, the motor must be stopped to change the drive ratio.

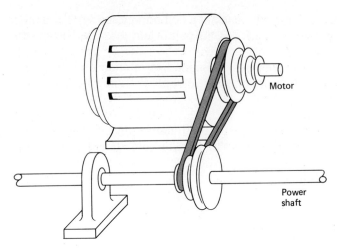

Fig. 11-16 A variable-speed belt drive using stepped pulleys.

Let's say that the pulleys are identical and that the three steps have diameters of 2.00, 3.00, and 4.00 in. (The corresponding metric dimensions are 5.08, 7.62, and 10.16 cm.) There are three possible drive ratios. If the belt connects the 2.00-in step on the driver pulley with the 4.00-in step on the driven pulley, the drive ratio is 4.00/2.00, or 2.00. When the belt connects the 3.00-in steps of both pulleys, the drive ratio is 3.00/3.00, or 1.00. The third possibility is having the 4.00-in step drive the 2.00-in step, giving a drive ratio of 2.00/4.00, or 0.500.

Suppose that the motor has a 1.00-hp output at a frequency of 1750 rpm. Such a motor will develop a torque of 3.00 lb·ft [0.415 kg$_f$·m]. With a drive ratio of 0.500, the output frequency from Eq. (11-11) is

$$\text{Output frequency} = \frac{\text{input frequency}}{\text{drive ratio}}$$

$$= \frac{1750 \text{ rpm}}{0.5}$$

$$= 3500 \text{ rpm}$$

At the same time, the output torque from Eq. (11-10) is

$$\text{Output torque} = (\text{efficiency})(\text{input torque})(\text{drive ratio})$$

$$= (0.96)(3.00 \text{ lb·ft})(0.500)$$

$$= 1.44 \text{ lb·ft}$$

where we have assumed an efficiency of 96 percent. The speed is stepped up, but there is a loss of torque. It will be easy to stall the machine. Notice that the output power is 0.96 hp (96 percent of 1.00 hp).

When the drive ratio is 1.00, the output frequency is 1750 rpm (the same as the motor). The output torque is (0.96)(3.00 lb·ft)(1.00), or 2.88 lb·ft. The output power is still 0.96 hp.

When the drive ratio is 2.00, the output frequency is 875 rpm. This is a speed reduction of 50 percent. The output torque, however, increases to 5.76 lb·ft [as calculated from Eq. (11-10)]. The output power is still 0.96 hp. ◄

Exercises

17. Calculate the drive ratios for the following simple belt drives.

	Driver diameter, cm	Driven diameter, cm
(a)	4.0	8.0
(b)	4.1	17.2
(c)	12.3	3.9
(d)	5.5	5.5

    Answers: (a) 2.0, (b) 4.2, (c) 0.32, (d) 1.0
18. A 1750-rpm motor must turn a grinder at 820 rpm. (a) What drive ratio is needed? (b) What driver diameter is needed if the driven pulley diameter is 10.2 cm?
    Answers: (a) 2.1, (b) 4.8 cm
19. A belt drive has a drive ratio of 3.0 and an input frequency of 3600 rpm. What is the output frequency?
    Answer: 1200 rpm
20. A belt drive has a drive ratio of 0.25 and an input frequency of 30 rpm. What is the output frequency?
    Answer: 120 rpm
21. A belt drive has an output frequency of 1200 rpm. If the drive ratio is 1.75, what is the input frequency?
    Answer: 2100 rpm
22. In Exercise 18, the efficiency is 95 percent and the input torque is 1.2 kg$_f$·m. What is the output torque?
    Answer: 2.39 kg$_f$·m
23. A belt drive has a drive ratio of 2.75, an input torque of 6.65 kg$_f$·m and an efficiency of 97 percent. What is the output torque?
    Answer: 17.7 kg$_f$·m

11-9 Multiple Belt Drives

Figure 11-16 shows the most common use for stepped pulleys. But they may also be used in a multiple belt drive, as shown in Fig. 11-17. Here a small pulley (A) drives a large one (B)—producing both a frequency reduction and a torque increase. This first large pulley has a smaller pulley on the same shaft, which drives another larger pulley (C) and produces a second frequency reduction. The process is repeated to get a third frequency reduction at D. Suppose that each small pulley has a diameter of 4.0 cm and the large ones measure 12.0 cm across. Then the first drive ratio is 3.0, which means that the frequency is cut to one-third. The second belt works on a drive ratio of 3.0 again, which means that pulley C is rotating at one-third the frequency of B and one-ninth the frequency

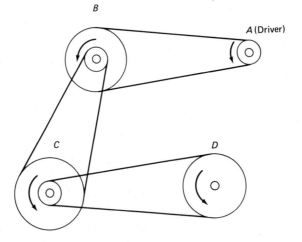

Fig. 11-17 A multiple belt drive.

of A. Then D rotates at one-third the frequency of C, or 1/27 the frequency of A. Thus the overall drive ratio is 27. If A rotates at 1750 rpm, D rotates at just 65 rpm.

We can generalize this:

Overall drive ratio = (drive ratio 1)(drive ratio 2)(drive ratio 3)   (11-13)

As for the torque, we can write

$$\frac{\text{Output}}{\text{torque}} = (\text{overall efficiency})(\text{input torque})(\text{overall drive ratio})   \quad (11\text{-}14)$$

Example 11-6 Two-Step Reduction Drive.

The belt system shown is driven by a motor whose frequency is 1750 rpm and whose torque is 1.2 $kg_f \cdot m$. Pulley A has a diameter of 3.0 cm. Pulley B has two steps with diameters of 3.0 and 15.0 cm. Pulley C measures 12.0 cm across.

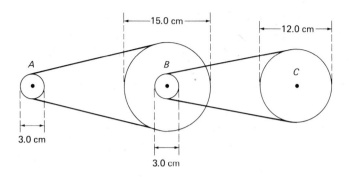

From A to B, the drive ratio is (15.0 cm)/(3.0 cm), or 5.0. From B to C, the drive ratio is (12.0 cm)/(3.0 cm), or 4.0. Then from Eq. (11-14), the overall drive ratio is

$$\text{Overall drive ratio} = (\text{drive ratio 1})(\text{drive ratio 2})$$

$$= 5.0\,(4.0)$$

$$= 20$$

What is the frequency of rotation at C? Using Eq. (11-11), we get

$$\text{Output frequency} = \frac{\text{input frequency}}{\text{drive ratio}}$$

$$= \frac{1750\ \text{rpm}}{20}$$

$$= 88\ \text{rpm}$$

If the efficiency of each belt drive is 96 percent, the overall efficiency is (0.96)(0.96), or 0.92, or 92 percent. The output torque, from Eq. (11-14), is then

$$\text{Output torque} = \left(\begin{array}{c}\text{overall}\\\text{efficiency}\end{array}\right)\left(\begin{array}{c}\text{input}\\\text{torque}\end{array}\right)\left(\begin{array}{c}\text{overall}\\\text{drive ratio}\end{array}\right)$$

$$= (0.92)(1.2\ \text{kg}_f\cdot\text{m})(20)$$
$$= 22\ \text{kg}_f\cdot\text{m}$$

This is a fairly large torque. The belts will have to be very tight to keep them from slipping. ◀

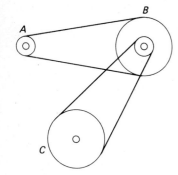

Exercises

24. Pulley A in the diagram rotates clockwise at 1750 rpm. Its diameter is 5.3 cm. Pulley B's outer diameter is 14.7 cm while the diameter of its small step is 4.8 cm. Pulley C's diameter is 15.2 cm. The efficiency of each drive is 97 percent. (a) What is the overall drive ratio? (b) What is the overall efficiency? (c) In what direction does pulley C rotate? (d) What is the frequency of pulley C?
    Answers: (a) 8.8, (b) 94 percent, (c) clockwise, (d) 2̄00 rpm

25. The small pulleys in the diagram each have a diameter of 5.0 in. The large pulleys measure 12.5 in across. Pulley A has an input of 16.2 $kg_f \cdot m$ at 15 rpm. The efficiency of each belt is 96 percent. (a) What is the overall drive ratio? (b) What is the overall efficiency? (c) If pulley A rotates clockwise, in what direction does pulley D rotate? (d) What is the frequency of pulley D? (e) What is the torque at the shaft of pulley D?
    Answers: (a) 0.064, (b) 88 percent, (c) counterclockwise, (d) 230 rpm, (e) 1.0 $kg_f \cdot m$

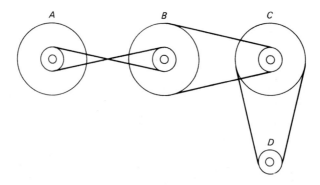

11-10 Gears

Like the belt drive, gears are basically wheels and axles. The gear teeth transmit a driving force from one gear to another. When a little gear drives a big one, the frequency is reduced and the torque is increased. When a big gear drives a little one, the frequency increases but the torque is reduced.

Gear systems are governed by the same principles as belt-and-pulley systems. The major difference is that meshing gears rotate in opposite directions.

The drive ratio of a pair of gears is the ratio of the driven diameter to the driver diameter, just as in the belt drive. Unfortunately, it is difficult to measure the effective diameter of a gear, because the teeth overlap as they mesh. It is easier to work out the drive ratio by counting the teeth:

$$\text{Drive ratio for gears} = \frac{\text{number of teeth on driven gear}}{\text{number of teeth on driver}} \quad (11\text{-}15)$$

Thus if a 20-tooth gear drives a 45-tooth gear, the drive ratio is 45/20, or 2.25.

Gears may be cast or even stamped from sheet metal, but for high speeds or large loads they must be machined. This makes gears very expensive compared to the belt drive. To keep friction as low as possible, gear teeth must be shaped so they roll on one another as they mesh (rather than sliding). Even so, conventional spur gears (Fig. 11-18) "chatter" at high speeds. This problem is avoided by using helical gears. With helical gears, a portion of several pairs of teeth is engaged at the same time, so the transition from tooth to tooth is smooth.

Once we know the drive ratio for a pair of meshed gears, we can use the belt-drive formulas to find the other quantities. For instance, the torque is found from Eq. (11-10):

Output torque = (efficiency)(input torque)(drive ratio)

and Eq. (11-11) gives the frequency:

Fig. 11-18 Spur and helical gears.

$$\text{Output frequency} = \frac{\text{input frequency}}{\text{drive ratio}}$$

Let's look at an example.

Example 11-7 The Bevel-Gear Drive.

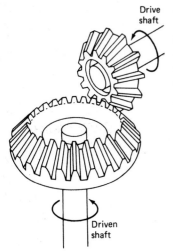

Drive shaft

Driven shaft

Fig. 11-19 A bevel-gear drive.

Figure 11-19 shows a pair of meshed bevel gears. Such a system is used in the automotive rear end, as well as other applications where the driven shaft must point in a direction different from the drive shaft.

In this case, the driver has 12 teeth and the driven gear has 24 teeth. Let's say that the input torque is 48.5 kg$_f$·m [351 lb·ft] at a frequency of 1800 rpm, and the efficiency is 97.8 percent. Let's find the output torque and the output frequency.

The first step is to calculate the drive ratio. From Eq. (11-15), this is

$$\text{Drive ratio} = \frac{\text{number of teeth on driven gear}}{\text{number of teeth on driver}}$$

$$= \frac{24}{12}$$

$$= 2.00$$

Then Eq. (11-10) gives us the torque:

$$\text{Output torque} = (\text{efficiency})(\text{input torque})(\text{drive ratio})$$

$$= (0.978)(48.5 \text{ kg}_f\text{·m})(2.00)$$

$$= 94.9 \text{ kg}_f\text{·m}$$

To get the output frequency, we use Eq. (11-11):

$$\text{Output frequency} = \frac{\text{input frequency}}{\text{drive ratio}}$$

$$= \frac{1800 \text{ rpm}}{2.00}$$

$$= 900 \text{ rpm}$$

Notice the direction of rotation of the output shaft in Fig. 11-19.◄

Exercises

26. A 36-tooth gear rotates at 750 rpm. It drives a 90-tooth gear. (*a*) What is the drive ratio? (*b*) What is the frequency of the driven gear?
     Answers: (*a*) 2.5, (*b*) 300 rpm
27. A 56-tooth gear rotates at 750 rpm. It drives an 18-tooth gear. (*a*) What is the drive ratio? (*b*) What is the frequency of the driven gear?
     Answers: (*a*) 0.32, (*b*) 2300 rpm
28. A pair of gears has a drive ratio of 3.0 and an efficiency of 98 percent. The torque on the driver is 18 $kg_f \cdot m$. What is the torque output?
     Answer: 53 $kg_f \cdot m$

11-11 Gear Trains

If two gears are fastened to the same shaft, they have to rotate at the same frequency. In a train of many gears, as in Fig. 11-17, the overall drive ratio is based only on the pairs of gears that mesh. If we calculate the drive ratio of each meshing pair, we can multiply all these drive ratios to get the overall drive ratio. The principle is the same as with the multiple belt drive.

Example 11-8 A Gear Reduction Drive.

The small gears in Fig. 11-20 have 11 teeth each, while the large ones have 18 teeth each. The driver on top rotates at 950 rpm. Let's find the frequency of rotation of the last gear on the bottom.

There are three pairs of meshing gears. Each pair has a drive ratio of 18/11, or 1.636. The overall drive ratio is then

$$\text{Overall drive ratio} = (1.636)(1.636)(1.636)$$

$$= 4.38$$

Using Eq. (11-11), we get an output frequency of

$$\text{Output frequency} = \frac{\text{input frequency}}{\text{drive ratio}}$$

$$= \frac{950 \text{ rpm}}{4.38}$$

$$= 217 \text{ rpm}$$

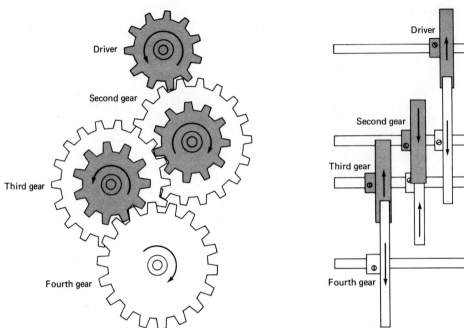

Fig. 11-20 A gear train.

Notice that each gear rotates in the opposite direction from the gear it meshes with. Thus the third gear in the train rotates in the same direction as the driver, and the last driven gear rotates opposite to the driver. ◄

Most of our examples have dealt with frequency reduction rather than frequency increase. But if the power is applied to the last gear in a reduction drive, the transmission will work in reverse to produce a high rotational frequency at the other end. If the gear on the bottom in Fig. 11-20 is driven at 950 rpm, the small gear on the top rotates 4.38 times as fast, or at nearly 4200 rpm. This is why we should be careful when downshifting a vehicle, particularly on a downhill slope. A car's overall drive ratio in low gear (transmission plus rear end) may be around 11. At a speed of 90 kmph [56 mph], the car's wheels are rotating at about 780 rpm. If the transmission is in low gear at this speed, the wheels drive the engine at a frequency of (11)(780 rpm), or 8600 rpm. This is beyond the red line of most engines, and serious damage can result.

Exercises

29.   The small gears in the diagram each have 18 teeth while the large gears have 45 teeth. Gear A rotates at 3600 rpm. Calculate the rotational frequencies of the other four gear shafts.

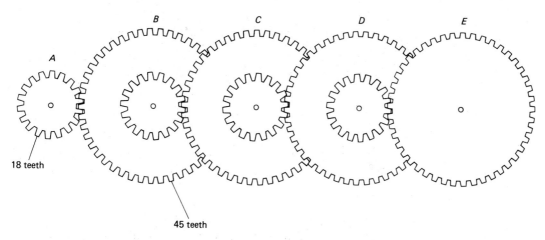

18 teeth

45 teeth

Answers: B, 1440 rpm; C, 576 rpm; D, 230 rpm; E, 92 rpm
30. If gear A in Exercise 29 rotates counterclockwise, in what direction does gear D rotate? In what direction does gear E rotate?
Answers: D, clockwise; E, counterclockwise

11-12 The Inclined Plane

This is the fourth of our simple machines. It is easier to push a piano up a ramp, for instance, than to lift it directly. The ramp provides a mechanical advantage.

Of course, we all know that a ramp is less effective when it is steeper. We see this effect in Fig. 11-21, where the MA is graphed against the angle of inclination. Only for fairly shallow angles is the MA much bigger than 1. We also see that the coefficient of friction has a large effect on the MA. When friction is high, the MA of an inclined plane can actually dip below 1. In such a case, it would be easier to lift a load directly than to push it up a ramp.

In Fig. 11-22, we see how the inclination affects the inclined plane's efficiency. The plane is most efficient when it is steep (and its MA is low). Remember that the efficiency is a measure of how much work we get out for a given amount of work input. So the inclined plane is not a very good machine to use when we want a high MA along with high efficiency.

At takeoff, an airplane wing is angled upward slightly so that it rides up on the air. In this case, the wing functions as an inclined plane. Another important application is the cam (Fig. 11-23). This is basically a continuous inclined plane that forces the cam follower to move along its surface.

Used in reverse, an inclined plane allows a load to drop without having it gain too much speed. Here the low efficiency is actually desirable. Manufactured parts can be sent down a shallow chute to get

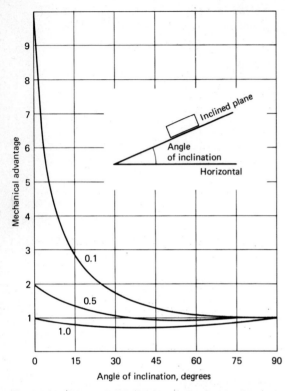

Fig. 11-21 The mechanical advantage of an inclined plane depends on the angle of inclination from the horizontal and on the coefficient of friction. Curves shown are for friction coefficients of 0.1, 0.5, and 1.0.

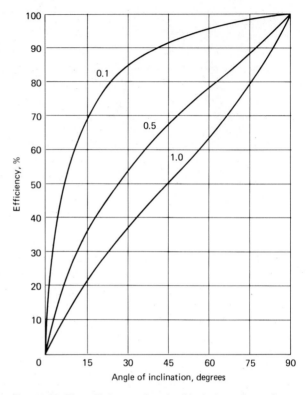

Fig. 11-22 The efficiency of an inclined plane depends on the angle of inclination and on the coefficient of friction. Curves shown are for friction coefficients of 0.1, 0.5, and 1.0.

from one machine to another. In railroad yards, individual freight cars are often allowed to roll down a "hump" and are then switched to the proper tracks to make up the trains. And in a dump truck, the bed is inclined until the load begins to slide off.

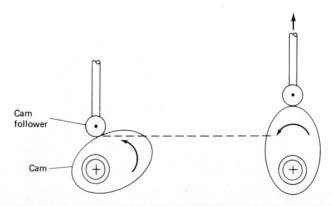

Fig. 11-23 The cam is a continuous inclined plane.

Exercises

31. A hardwood crate is pushed up a dry hardwood ramp inclined at 15°. (*a*) What is the MA? (*b*) What is the efficiency? (*c*) What is the efficiency if the angle is increased to 30°?
    Answers: (*a*) 1.4, (*b*) 35 percent, (*c*) 55 percent
32. A wheeled ore hopper has journal bearings with a coefficient of friction of 0.10. The hopper has a mass of 550 kg. It is drawn up a ramp inclined at 30°. (*a*) What is the MA? (*b*) What effort is needed to move the load? (*c*) What is the efficiency? (*d*) What work input is needed to raise the hopper 15 m?
    Answers: (*a*) 1.7, (*b*) 320 kg$_f$, (*c*) 85 percent, (*d*) 7000kg$_f$·m, or 69 kJ

Although a screw may be thought of as an inclined plane wrapped around a cylinder, here we will consider it as a separate type of simple machine. Normally, a screw has two parts, called male and female. The male part, which itself may be called a screw, is usually useless without a female counterpart. The female part may be a nut that can be turned, or it may simply be a threaded hole. In the case of a wood screw or a self-tapping machine screw, the male part cuts its own thread in the female.

11-13 The Screw

We are all familiar with the use of the screw as a fastener. When the screw in Fig. 11-24 is tightened, there is a small amount of friction between the male and female threads. When the nut is drawn up tight, the threads deform elastically, and this causes a considerable increase in the friction. This increased friction holds together the male and female threads. But if the screw is overtightened, the threads may deform plastically or even fail completely. Since a screw will not hold if its

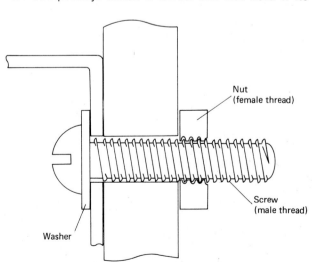

Fig. 11-24 The screw as a fastener.

Application: Furniture Repair

Most wood furniture is glued together, possibly in addition to being nailed or screwed. With glued joints the bond has a large surface area, and so it tends to be very strong.

But in Northern climates, the dry indoor air in the winter encourages moisture to evaporate from the wood. This causes the wood to shrink slightly, and over a long period of use this often leads to loosened glue joints. Many people try to repair loose furniture with screws, nails, or metal brackets, but this is never a long-term solution. A chair back, for instance, has a high mechanical advantage as a lever and quickly works loose any makeshift fasteners.

The diagram shows the use of the screw and wedge principles in regluing a chair. The chair is disassembled, and the pieces that extend through the bottom are slotted. The joints are cleaned, reglued, and put back together; then small wooden wedges are driven into the slots from the bottom. The legs and stretchers are held tight by making simple turnbuckles from ropes and sticks. The rope is tightened on the same principle as a screw, by twisting the stick. Of course, the sticks have to be propped in place so the ropes don't unwind. Although the sketch shows only one rope turnbuckle on the chair, as many as are needed can be used at the same time.

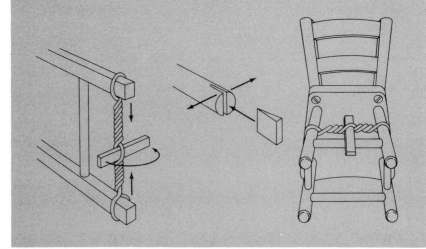

threads are stripped, care must be taken in applying that last twist. A good mechanic gains a "feel" for the amount of elastic distortion that she or he has imposed on the threads. Still, in applications like engine head bolts, it's much safer to consult the manufacturer's recommendations and use a torque wrench. Cylinder head bolts are typically tightened to a torque of about 11 kg$_f$·m [80 lb·ft].

Screws are used not only as fasteners. Screw jacks, vises, and screw-operated presses are machines that use the screw to generate large mechanical advantages.

When a screw is twisted through one complete revolution, it advances one thread. This distance is called the screw's pitch.

> Definition: The *pitch* of a screw thread is the distance between two adjacent threads.   *pitch*

Figure 11-25 shows what we mean by this definition. If the screw has, say, five threads per centimeter, then the threads must be 0.2 cm apart. This 0.2 cm is the pitch.

A screw's mechanical advantage depends on the pitch and the radius of the turning circle where the effort is applied.

$$\text{MA of screw} = \frac{(\text{efficiency})(2\pi)(\text{radius of turning circle})}{\text{pitch}} \qquad (11\text{-}16)$$

Of course, the efficiency depends on the amount of friction. With lubrication, the efficiency is usually greater than 50 percent and may be as high as 98 percent.

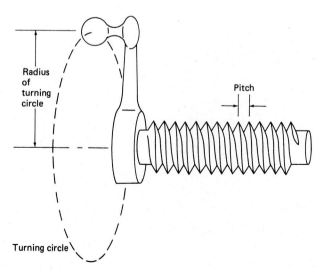

Fig. 11-25 Screw-thread nomenclature.

Example 11-9 The Jackscrew.

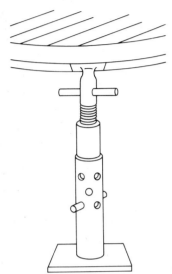

Figure 11-26 shows a jack post used to correct sagging floor joists. The screw has five threads per centimeter, and the turning radius is 20 cm. With light lubrication, the efficiency is 60 percent. What is the MA?

The pitch is 1/4 cm, or 0.25 cm. Then using Eq. (11-16), we have

$$MA = \frac{(\text{efficiency})(2\pi)(\text{radius of turning circle})}{\text{pitch}}$$

$$= \frac{(0.60)(2\pi)(20 \text{ cm})}{0.25 \text{ cm}}$$

$$= 3\overline{0}0$$

Notice that this is a much larger MA than in any other machine we've discussed. This jackscrew can lift 1 t with an effort of only 3.3 kg$_f$! ◄

Fig. 11-26 A jack post and jackscrew.

One other important application of the screw is the worm gear (Fig. 11-27). The worm is the driver; it consists of a screw turning against a thrust bearing. The screw threads engage the teeth of a gear. The gear advances one tooth each time the worm makes one revolution. This gives a tremendous frequency reduction:

Drive ratio of worm gear = number of teeth on driven gear   (11-17)

Fig. 11-27 The worm gear.

A drive ratio of 40 or more is easily obtained from a relatively small worm gear. Thus a 1750-rpm motor can be reduced to around 40 rpm in a single step. At the same time, because its efficiency is so high when well lubricated, a worm gear's torque advantage is nearly the same as its drive ratio. If the motor's torque is 1.2 $kg_f \cdot m$, the torque at the shaft of the driven gear is nearly 40 times as great, or about 48 $kg_f \cdot m$.

One other thing about screws and worm gears: the effort and load usually cannot be interchanged as they can with other simple machines. If the effort is applied to the gear, the worm is forced against the thrust bearing, and it binds rather than turning. The device is effective only when the effort twists the screw thread.

Exercises

33. A screw has 6.2 threads per centimeter. It is turned by a wrench whose effort arm is 32 cm. The efficiency is 62 percent. (*a*) What is the MA? (*b*) What load can be moved with an effort of 9.5 kg_f?
     Answers: (*a*) 770, (*b*) 7300 kg_f
34. A worm gear is used to crank the turret of a windmill so its sails face the wind. The turret has 360 gear teeth around its circumference. The pitch is 2.5 cm. (*a*) What is the drive ratio? (*b*) How many revolutions must the worm make to drive the turret through one complete revolution? (*c*) If the input torque is 0.52 $kg_f \cdot m$, what is the output torque?
     Answers: (*a*) 1/360, or 0.0028, (*b*) 360, (*c*) about 180 $kg_f \cdot m$

This is the last of our six simple machines. Knives, chisels, axblades, saws, snowplows, and shears are all wedges. Wedges are generally used for cutting or breaking apart things. A pitched roof acts as a wedge in deflecting rain and snow, and the bow of a ship is a wedge that deflects the water out around the hull (Fig. 11-28).

11-14 The Wedge

Wedges always encounter a large amount of friction in their operation. For this reason, it is difficult to write a formula for their mechanical advantage. In general, we can say that the sharper the edge, the greater the MA. Of course, a very sharp-edged wedge is easily damaged.

A nail is also a wedge. As it is driven into a board, it forces apart the wood fibers. The elastic forces holding the wood together act on the nail and give rise to the friction that keeps the nail in the wood. But if the wood is hard and the nail is too sharp, the nail acts more like a chisel and splits the board. Obviously, this should be avoided. When the wood is subject to splitting, carpenters will dull each nail by standing it on its head and whacking the point with a hammer. Since a dull nail has a smaller mechanical advantage, it has less tendency to split the wood. Instead of spreading apart the wood fibers, it just forces them downward and breaks them.

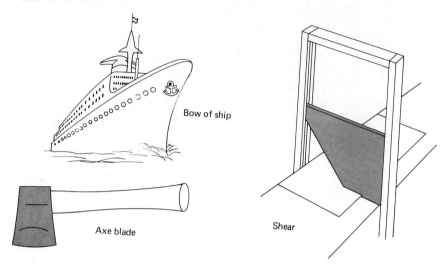

Fig. 11-28 Examples of the wedge.

Summary Although there are only six categories of simple machines, these may be combined to make more complicated machines. We have seen, for instance, that wheels and axles may be combined in various arrangements to get belt drives and multiple-gear transmissions. A ratchet is a third-class lever combined with a short inclined plane, and a glass cutter is a pulley and a wedge. A more complicated mechanism, like a printing press, contains many levers, wheels and axles, pulleys, and so on.

We said that all simple machines are made of solid, fairly rigid parts. It turns out that it is also possible to gain a mechanical advantage (or even a speed advantage) by using fluids. We see how this is done in the next chapter.

Terms You Should Know

mechanical advantage   frequency
drive ratio   pitch

Problems

1. What simple machines are used in the following devices?
   (a) paint roller
   (b) screwdriver
   (c) shovel
   (d) paper stapler
   (e) turnbuckle
   (f) pipe wrench
2. A certain lever-operated jack has an effective effort arm of 58.6 cm. The effective load arm is 1.62 cm. To lift a load of 1.00 t, a measured effort of 28.3 kg$_f$ is required. What is the efficiency of the jack?
3. Estimate the mechanical advantage of the hand punch shown in the diagram.

4. A six-strand block and tackle is used to lift a 432-kg$_f$ load through a vertical distance of 8.52 m. The effort is supplied from below. The efficiency of each pulley hearing is 97.6 percent. (a) What is the total efficiency? (b) What is the MA? (c) What effort is needed? (d) What is the total work input? (e) How far does the effort move? (In other words, how much cable is pulled out of the system in lifting the load?)
5. Pulley A in the diagram is the driver. It rotates at 2230 rpm. (a) What is the drive ratio? (b) What is the frequency of pulley B? (c) What is the linear speed of the belt in meters per second?

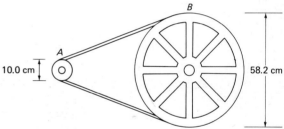

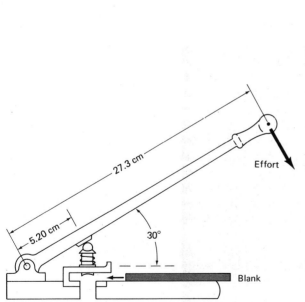

6. A geared transmission is needed with an overall drive ratio of 27.0. To keep the device from being too big, this must be done in several stages. (a) If two identical stages are used, what should the drive ratio of each stage be? (b) If three identical stages are used, what should the drive ratio of each stage be? (c) What if four identical stages are used?

12

Hydraulic and Pneumatic Principles

Introduction   Fluids can be made to do many useful things, from operating machinery to floating coal barges. If the fluid is at rest, or else moving very slowly, it is governed by the principles of *hydraulics*. Because gases are more easily compressed than liquids, they present some special problems. The behavior of gases, especially air, is studied in a branch of hydraulics known as *pneumatics*.

In this chapter, we look into some of the basics of hydraulics and pneumatics, as well as some of their important applications.

12-1 Buoyancy   We have seen that a fluid's pressure increases with depth. Suppose that we submerge a solid cylinder in a liquid (Fig. 12-1). The liquid exerts a pressure on the cylinder's top that tries to push the cylinder down. At the same time, the fluid pressure on the bottom of the cylinder tries to push it up. Because the bottom is deeper, the pressure there is greater, and the net fluid force acts in the upward direction. This fluid force is called a buoyant force.

buoyant force   Definition: A *buoyant force* is the upward fluid force acting on any object placed in a stationary fluid.

Does the object have to be a cylinder? No, there is a buoyant force regardless of the object's shape. Does the object have to be light in

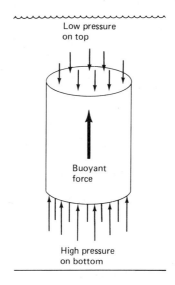

Fig. 12-1 A solid submerged in a liquid experiences an upward buoyant force, because the pressure is greater on the solid's bottom than on its top. The pressure on the sides balances out and is not shown.

weight? No, there is a buoyant force even on heavy objects, although it may not be as noticeable as with light objects. Does the fluid have to be a liquid? No, a buoyant force is produced by gases as well.

Many people assume that there is a buoyant force only on floating objects. But since the buoyant force arises from the increased fluid pressure at greater depths, we see that this force must actually act on everything. The only exception is an object in a vacuum. Right now, buoyant forces are actually acting on both you and this book; they result from the slight variation in atmospheric pressure with elevation.

Then why do some things float while others sink? The reason is that the buoyant force is not the only force acting on an object. Things have weight, too, and this weight tries to pull them toward the earth. If an object's weight is greater than the buoyant force, the object sinks. If its weight is less than the buoyant force, the object rises and floats at a level where the upward and downward forces are in equilibrium.

While working on the problem of Hiero's crown (Chap. 5), Archimedes formulated the principle of hydraulics that still bears his name:

12-2 Archimedes' Principle

Archimedes' principle: The buoyant force on any object is the same as the weight of the fluid it displaces.

Archimedes' principle

Suppose, for instance, that we place a 1000-cm³ aluminum block in a bucket of water (Fig. 12-2). Obviously, the block must *displace*, or push aside, 1000 cm³ of water. Water has a density of 1.0 g/cm³, so 1000 cm³ of water weighs about 1.0 kg$_f$. According to Archimedes' principle,

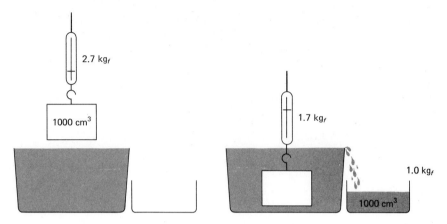

Fig. 12-2 Archimedes' principle applied to a 1000-cm³ block of aluminum placed in a bucket of water.

then, the buoyant force on the block is 1.0 kg$_f$ (the same as the weight of the displaced fluid).

Let's go a step further. From Table 5-1 we see that the density of aluminum is 2.7 g/cm³. Our 1000-cm³ block therefore weighs 2700 g$_f$, or 2.7 kg$_f$. But when the block is underwater, the 1.0-kg$_f$ buoyant force acts to oppose this weight. The *apparent weight* of the block underwater is only 2.7 kg$_f$ − 1.0 kg$_f$, or 1.7 kg$_f$. It is always easier to lift a submerged object than it is to lift the same object in the air.

Example 12-1 Raising a Sunken Boat.

The sunken boat in Fig. 12-3 has a mass of 14 200 kg. It is to be refloated by pumping air into plastic bags placed in its hull by divers. What

Fig. 12-3 A sunken ship may be raised by pumping air into rubber or plastic bags placed in its hull.

volume of water needs to be displaced by the air before the boat begins to float to the surface?

This is a straightforward application of Archimedes' principle. Since the boat weighs about 14 200 kg$_f$, we need about 14 200 kg$_f$ of buoyant force to float it. Thus about 14 200 kg of water must be displaced by the air bags. Since fresh water has a density of 1.00 kg/L, the bags must expand to a volume of about 14 200 L.*

If the boat has sunk in saltwater, the requirement is a bit lower. A 14 200-kg$_f$ buoyant force is still needed, but from Table 6-2 we see that the density of seawater is 1.025 kg/L. The volume of seawater having a mass of 14 200 kg is, from Eq. (5-3).

$$\text{Volume} = \frac{\text{mass}}{\text{density}}$$

$$= \frac{14\ 200\ \text{kg}}{1.025\ \text{kg/L}}$$

$$= 13\ 900\ \text{L}$$

After the air bags have expanded to this volume, the boat begins to float toward the surface.

There are big advantages to using a compressor and air bags, rather than cranes and winches, to raise the boat. For one thing, the mechanical equipment has to be positioned fairly precisely over the sunken boat, whereas with a compressor the salvage boat can drift around a bit as long as the air hose still reaches. Furthermore, the air bags give the boat enough flotation to be towed to land before repairs are attempted or it is dismantled for scrap. ◀

Example 12-2 Flotation of an Oil Drum.

Empty oil drums are often used to provide flotation for docks, floating bridges, and other temporary floating platforms (Fig. 12-4). The standard drums have a volume of 55 gal and may weigh 45 lb. How much weight can such a drum support when floating in fresh water?

The biggest weight the drum can support is the weight that submerges it completely. When submerged, the drum displaces 55 gal of water (its own volume). Since 1 gal = 0.133 7 ft³ this is the same as

---

*There is an approximation in this solution: the buoyant force on the sunken hull itself has been neglected. In other words, the *apparent weight* of the boat is somewhat less than 14 200 kg$_f$ when it is resting on the bottom. The answer of 14 200 L of air is therefore just a bit on the high side. As a practical matter, the complications of a more detailed solution would not make the result any more useful.

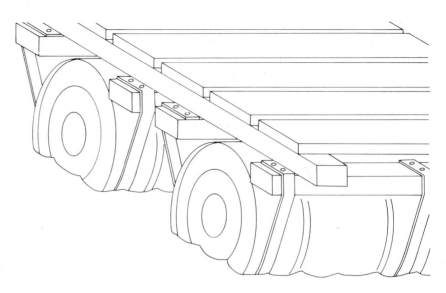

Fig. 12-4 Empty 55-gal oil drums are often used for flotation.

7.35 ft³ of water. The mass of this much water is given by Eq (5-2).

$$\text{Mass} = (\text{density})\ (\text{volume})$$

$$= 62.3\ \frac{\text{lb}_m}{\text{ft}^3}\ (7.35\ \text{ft}^3)$$

$$= 458\ \text{lb}_m$$

The weight of the displaced water is therefore about 458 lb.

According to Archimedes' principle, the buoyant force on the drum is the same as the weight of the water it displaces. Thus the buoyant force on a submerged oil drum is about 458 lb. Since the drum itself weighs 45 lb, the additional weight it can support without sinking is 458 lb − 45 lb, or about 413 lb. This, then, is our answer: an empty, sealed oil drum floating in fresh water can support about 413 lb without sinking.

If a platform is built on six floating oil drums, it can support nearly 2500 lb before the structure gets wet. By using this result, then, a raft can quickly be designed to support any given load. ◄

In the last two examples, both the air bags and the oil drums were filled with air. We didn't worry about the weight of the air because we know it doesn't amount to very much when compared to the weight of the displaced water. But what if the drums were filled with gasoline? They would float, of course, but they couldn't support much additional weight. Now the weight of the gasoline itself must be added to the weight of each drum. After this total weight is subtracted from the buoyant force, there won't be a great deal left over to float a platform.

Let's summarize what Archimedes' principle says about things that float and things that sink.

Things that float: The buoyant force balances the total weight of the object. The object displaces a *weight* of fluid equal to its own *weight*.

things that float

Things that sink: The buoyant force is less than the weight of the object. The object displaces a *volume* of fluid equal to its own *volume*, and the weight of this volume of fluid equals the buoyant force.

things that sink

Exercises

1. A certain barge weighs 152 t and can displace 1630 m³ of water before it is swamped. How many metric tons of cargo will bring the water level to the deck?
    Answer: 1480 t
2. A certain ship weighs 12 300 t, and it carries a cargo weighing 97 700 t. (*a*) What volume of water does the ship displace if the water is pure? (*b*) What volume of water does the ship displace if it floats in the ocean? (*c*) If the ship travels from the St. Lawrence River (fresh water) into the ocean, what volume of ocean water should it take on as ballast to keep it floating at the same level?
    Answers: (*a*) 110 000 m³, (*b*) 107 000 m³, (*c*) about 3000 m³
3. A certain tanker ship can safely carry 260 000 t of kerosene. How many metric tons of fuel oil can it carry and still float at the same level?
    Answer: 260 000 t
4. A certain tanker ship can safely carry 165 000 m³ of kerosene. How many cubic meters of fuel oil can it carry and still float at the same level?
    Answer: 145 000 m³
5. A 55-gal oil drum is filled with kerosene. The drum itself weighs 45 lb. (*a*) Does the drum float or sink in fresh water? (*b*) If the drum is filled with fuel oil, does it float or sink in fresh water? (*c*) If the drum is filled with fuel oil, will it float or sink in saltwater?
    Answers: (*a*) floats, (*b*) sinks, (*c*) sinks, but barely
6. Sand is often dredged from river bottoms. One cubic foot of dry sand weighs about 94 lb. What is the apparent weight of 1 ft³ of sand underwater?
    Answer: 32 lb
7. A pine beam measures 1.00 ft wide by 0.500 ft thick by 8.00 ft long. (*a*) How much weight will the beam support if it is floating in water?

(*b*) How much weight would a similar beam support if it were made of oak? (*Hint*: Refer to Table 5-1.)
Answers: (*a*) 156 lb, (*b*) 72 lb

12-3 Center of Buoyancy; Stability

We have seen that every object has a point where its weight appears to be concentrated; we called this point the *center of gravity*. There is also a point where the buoyant force can be considered to be concentrated. Appropriately enough, this point is called the center of buoyancy.

center of buoyancy

> Definition: An object's *center of buoyancy* is the point where the buoyant force acting on the object can be considered to be concentrated. This point is the same as the center of gravity of the displaced fluid.

When an object is totally submerged, its center of buoyancy is usually the same as its center of gravity. But for floating objects, the two points may be quite different.

Figure 12-6 shows a front view of a floating deep-hulled ship. In the first case, the ship is very light and its center of gravity (c.g.) is above

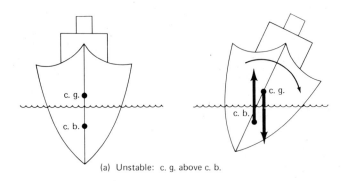

(a) Unstable: c. g. above c. b.

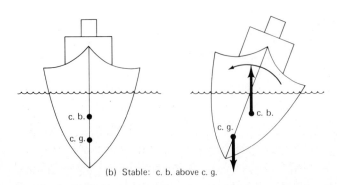

Fig. 12-5 The stability of a deep-hulled vessel depends on the relative positions of the c.g. and the c.b.

(b) Stable: c. b. above c. g.

its center of buoyancy (c.b.). When the ship is rocked slightly (say by a wave hitting it broadside), we see that a torque is generated which acts to capsize the ship. Thus an arrangement of c.g. above c.b. leads to instability in deep-hulled vessels and is something to be avoided.

In the second diagram of Fig. 12-5, the ship's c.g. is below its c.b. Now a disturbing force generates a torque that tends to stand the ship back upright. To gain stability in deep-hulled ships, then, the c.g. needs to be kept very low. This is why such ships are often ballasted.

With the shallow-draft vessels like rafts and barges, the c.g. can be higher than the c.b. without leading to stability problems. We see why in Fig. 12-6. This is the front view of a wide barge that has a high c.g. When the barge is tipped slightly, the c.b. moves to the side where most of the water is displaced. Thus the torque generated acts to keep the barge upright. This same principle contributes to the stability of twin-hulled boats (catamarans) and boats with outriggers.

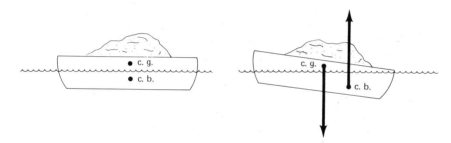

Fig. 12-6 A shallow-draft vessel can have a high c.g. and still be stable if its width is large compared to its draft. When the barge is rocked, the c.b. moves toward the side where most of the water is displaced.

Exercises

8. Would a tanker be more stable if its hold were filled with crude oil or kerosene?
   Answer: Crude oil
9. Will a ship be more stable floating in fresh water or saltwater?
   Answer: Fresh water
10. A piece of steel is fastened to a piece of aluminum as shown. Both pieces are 10.0 cm thick. The assembly is placed underwater. (*a*) Where is the c.b.? (*b*) Where is the c.g.? (*c*) If the assembly is allowed to sink to the bottom, which side will probably land faceup?
    Answers: (*a*) The c.b. is in the geometrical center, or at the midpoint of the contact area between the steel and the aluminum. (*b*) The c.g. is in the steel section—actually about 2.4 cm above the interface. (*c*) The aluminum one

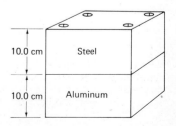

12-4 Specific Gravity

One consequence of Archimedes' principle is that a substance floats if its density is lower that of the fluid in which it's placed. And a substance sinks if its density is higher than that of the fluid. Since the fluid of interest is usually water, a special name has been given to the ratio of a substance's density to the density of water.

specific gravity

Definition: A substance's *specific gravity* is its density divided by the density of water:

$$\text{Specific gravity} = \frac{\text{density of substance}}{\text{density of water}} \qquad (12\text{-}1)$$

Specific gravity has no units, since it is density divided by density. Or, in other words, specific gravity has the same numerical value whether we express the density in old USCS units or metric units. But actually no calculation is required. Since the density of water is 1.00 g/cm³, it follows that the specific gravity of any substance is the same as the numerical value of its density in grams per cubic centimeter. For example, the density of kerosene (from Table 6-2) is 0.82 g/cm³. Thus the specific gravity of kerosene is 0.82.

Anything with a specific gravity greater than 1 sinks in water (assuming it doesn't mix or dissolve). And anything with a specific gravity less than 1 floats on water (with the same assumption). Furthermore, things with the lowest specific gravities float the highest. The specific gravity is actually the same as the fractional volume that is underwater. Ice, for instance, has a specific gravity of 0.917. So a block of ice floating in water has 91.7 percent of its volume submerged; only 8.3 percent of its volume projects above the surface. But for an iceberg in the ocean, 89.5 percent is submerged and 10.5 percent is visible above the surface. Why the difference? Because seawater is denser than fresh water. Specific gravity is based on the density of pure water only.

A specific-gravity measurement quickly determines the effectiveness of a car engine's antifreeze coolant. Pure ethylene glycol has a specific gravity of 1.109, while water has a specific gravity of 1.000. So if the coolant's specific gravity is close to 1.00, it is mostly water; if it is close to 1.10, it is mostly antifreeze. The measurement is made with a device known as a *hydrometer* (see the application section).

The same idea can be used to check the charge in a lead-acid storage battery. When the battery is fully charged, the electrolyte is around 38 percent sulfuric acid, and the specific gravity is about 1.28. When the battery is discharged, the solution is only around 26 percent acid and the specific gravity is about 1.18. Again, the specific-gravity measurement is made with a specially designed hydrometer.

Exercises

11. Find three substances that float in liquid mercury and three that sink.
    Answers: Brick, wood, and lead float; gold, tungsten, and uranium sink. There are many others in both categories.
12. Find the specific gravity of (*a*) sand, (*b*) aluminum, (*c*) fuel oil, (*d*) seawater.
    Answers: (*a*) 1.5, (*b*) 2.699, (*c*) 0.93, (*d*) 1.025
13. Douglas fir logs are to be floated downstream from a logging site to a sawmill. The logs have diameters ranging up to 110 cm. What (approximately) should the minimum depth of the stream be?
    Answer: About 55 cm
14. A lightweight plastic bottle is filled with denatured alcohol and sealed. If the bottle is floated in fresh water, approximately what percentage of the volume is submerged?
    Answer: About 79 percent

We have seen that fluid pressure is responsible for buoyancy. This happens in both liquids and gases. Let's now look at some other applications of pressure in gases, particularly air. A good starting point is Eq. (6-2), which tells us how much force is generated by a pressure acting on some surface:

12-5 Pneumatic Pressure

$$\text{Fluid force} = (\text{pressure})\,(\text{area})$$

Thus the same force can be generated by a large pressure acting on a small area as by a small pressure acting on a large area.

The car in Fig. 12-7 has its weight supported by four pneumatic (air-filled) tires. If we look closely, we see that the tires are a little bit flat on the bottom. This does not necessarily mean that they are under-inflated. It happens because there has to be some contact *area* on which the air pressure can act to support the car's weight. This is called the *contact patch* (of the tire). If we increase the tire pressure, this area decreases (the tire gets rounder). If we decrease the tire pressure, this contact area increases (the tire gets flatter). In any case, the same force is generated—just enough to support the car.

You can measure a tire's contact area fairly accurately if you are willing to use some muscle. Park the car on a level surface and jack one wheel off the ground (taking all the usual precautions). Then take a thin, smooth board or, better, a steel plate, and place it under the tire. On top of this put a piece of clean paper; and slowly lower the wheel until it is fully loaded. Now jack the wheel off the ground a second time and remove the paper. It will bear a print of the tire patch. The tire patch can

Application: The Hydrometer

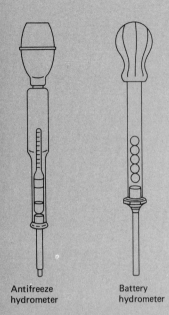

Antifreeze
hydrometer

Battery
hydrometer

A hydrometer is a device for measuring a substance's specific gravity. Although most hydrometers are used with liquids, some versions of the device are designed to determine the specific gravity of solids (dirt samples, for instance).

In the most common design, the liquid is drawn into a glass vial by squeezing and releasing a rubber bulb. Inside this vial is a smaller vial, sealed full of air and weighted at the bottom so that it floats upright. This smaller vial floats high when the test liquid has a high specific gravity, and it floats low in the liquid when the specific gravity is low. The specific gravity may be read directly from a scale on this inner vial.

Since the specific gravity of an antifreeze solution depends on the proportions of water and ethylene glycol, a specific-gravity measurement can be used to predict the mixture's freezing point. Table 12-1 lists the freezing points of some mixtures of ethylene glycol and water.

The specific gravity of a storage battery's electrolyte is an indication of the charge on the battery. In this case, the numerical value of the specific gravity is not as important as whether it is high or low. Therefore, hydrometers designed for battery testing usually have three to five floating balls instead of the graduated inner vial. If all the balls float, the electrolyte's specific gravity is greater than 1.28, and the battery is fully charged. If just a few balls float and the rest sink, the battery is partly discharged. If all the balls sink, the specific gravity is less than 1.15, and the battery is dead.

TABLE 12-1 SPECIFIC GRAVITIES AND FREEZING POINTS OF ETHYLENE GLYCOL ANTIFREEZE SOLUTIONS

Specific gravity	Percent of ethylene glycol, by volume	Freezing point, °C
1.000	0	0
1.013	9.2	−3.6
1.026	18.3	−7.9
1.040	28.0	−14.0
1.053	37.8	−22.3
1.067	47.8	−33.8
1.079	58.1	−49.3
1.109	100	−17.4

Fig. 12-7 A tire's air pressure acting on the contact area with the road generates the force that supports the vehicle. The size of the "patch" can be measured from a paper print.

now be measured as accurately as you want it. Of course, the tire must be dirty for this to work, and it may help to use printer's ink or lampblack on the tread before lowering it.

Example 12-4 Contact Patch on a Car Tire.

A certain car weighs 1980 kg_f [4370 lb]. Its four tires are inflated to a gauge pressure of 2.20 kg_f/cm^2 [31.3 lb/in²]. What is the contact area between the tires and the road?

The pneumatic pressure develops a force of 2.20 kg_f on each 1 cm². To support 1980 kg_f, then, requires a total contact area of

$$\text{Area} = \frac{1980 \text{ kg}_f}{2.20 \text{ kg}_f/cm^2}$$

$$= 90\overline{0} \text{ cm}^2$$

This area is distributed over four tires. If the load distribution happens to be equal (there may be a lot of luggage in the trunk, for instance), each tire must have a flat spot (patch) with one-fourth of this area, or 225 cm² [34.9 in²].

Suppose that the tire is 12.7 cm [5.00 in] wide. Then the patch is 17.7 cm [6.97 in] long. This may seem like a big flat spot, but then this is a fairly heavy car. The car will probably have large-diameter tires, so this patch doesn't present too much rolling resistance.

Of course, rolling resistance can also be reduced by increasing the inflation pressure. This reduces the total contact area and therefore shortens the patch so the tire doesn't distort as much as it rolls. But each tire has its limits, and overinflation doesn't do the tire any good either.

There is another alternative, as we see in Fig. 12-8. If very wide tires are used, the same contact area is shorter and wider. A tire 15.2 cm [5.98 in] wide has a patch just 14.8 cm [5.83 in] long under the previous conditions. This is one of the advantages of wide tires. By distorting less in the direction they roll, wide tires present less rolling resistance. They run cooler, and this contributes to a longer tread life. ◄

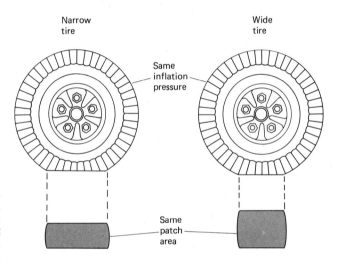

Narrow tire   Same inflation pressure   Wide tire

Same patch area

Fig. 12-8 Wide tires reduce rolling friction. A narrow tire must be more out-of-round to produce the same contact area as a wider tire.

We have all noticed that tractor trailers have larger tires than ordinary vehicles. They also have more of them (up to 16 supporting the trailer part alone). In addition, the inflation pressures are very high— ranging up to 8.1 kg$_f$/cm^2 [about 115 psig]. Notice that this is close to 9 atm, absolute. Of course, there is no mystery to all this. To support very heavy loads, high pressures and large contact areas are called for. The contact area on a tire can get only so large before it is too flat to roll. Using large tires helps some, but using a large number of them helps even more. Notice the large number of large tires needed to support the house in Fig. 12-9.

Did you ever get a nail stuck in a tire? If so, you probably pulled it out easily with a pair of pliers. But why didn't the tire's air pressure pop it

Fig. 12-9 Moving a house. The air pressure in the tires (about 100 psig) must act on a large area to support this load. This calls for a large number of large tires.

right out? If the air pressure can support a car, why shouldn't it force out a nail? You certainly can't support a car (or even one-fourth of it) yourself, yet you can easily remove the nail yourself.

The answer to this kind of question is at the heart of what we're talking about. The nail has a very small cross-sectional area. If its diameter is 0.25 cm, its area is 0.049 cm². At a pressure of 2.2 kg$_f$/cm², the pneumatic force pushing out on the nail is only 0.11 kg$_f$ (less than ¼ lb). This same pressure manages to support the car only because of the larger area on which the pressure acts in that case.

Exercises

15. A pneumatically operated piston has a face area of 34.5 cm². How much force, in kilogram-force, is developed on the piston when the air pressure is (a) 13.3 kg$_f$/cm², (b) 12.9 atm, (c) 1.305 MPa?
    Answers: 459 kg$_f$ in each case

16. A trailer has four tires, each inflated to a pressure of 3.32 kg$_f$/cm². The contact patch of each tire cannot exceed 275 cm². Find the maximum weight the tires can support in (a) kilogram-force, (b) newtons, (c) pounds.
    Answers: (a) 3650 kg$_f$, (b) 35 800 N, (c) 8050 lb

17. A certain car has a mass of 1270 kg. Its four tires are each inflated to a pressure of 2.04 kg$_f$/cm² [20$\bar{0}$ kPa, or 29.0 psi]. What is the total contact area between the tires and the road in (a) square centimeters, (b) square inches?
    Answers: (a) 623 cm², (b) 96.5 in²

12-6 Boyle's Law

We said several times that gases are easily compressed, while liquids are not. Boyle's law tells us how a gas's pressure increases as it is compressed into a smaller volume.

Boyle's law

> *Boyle's law:* If a gas's temperature is not allowed to change, its absolute pressure is inversely proportional to its volume.

Here's what this means: If we compress a gas into half its original volume, and we do it slowly enough that its temperature doesn't change much, then the gas's absolute pressure doubles. If it starts out at 1.03 kg_f/cm^2, absolute (1 atm), its pressure increases to 2.06 kg_f/cm^2, absolute (2 atm). Or if we compress a gas into one-tenth its original volume, we find that its absolute pressure increases by a factor of 10.

Boyle's law also works the other way. If we allow a gas to expand to, say, 5 times its volume, then its absolute pressure drops to one-fifth its original value. Again, the temperature must be kept constant for this to be accurate. Figure 12-10 shows how Boyle's law applies to a gas that occupies a 1-m^3 volume at atmospheric pressure.

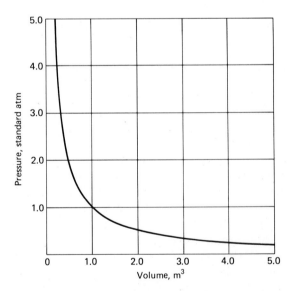

Fig. 12-10 Boyle's law applied to a gas whose volume is 1.0 m^3 at atmospheric pressure. The gas's temperature must not change if this is to be accurate.

Example 12-5 Compressed Air for a Salvage Operation.

In Example 12-1 we calculated the amount of water that had to be displaced by air bags to raise a certain sunken boat. Let's say that the boat is in 12 m [50 ft] of water. What volume of air needs to be pumped through the hose and at what pressure?

We already figured that 13 900 L of air is needed, but this is at the depth of the boat where the air will be compressed by the water pressure. At the surface, much more than this has to enter the pump.

The pressure under 12 m of water can be calculated from Eq. (6-3).

$$\text{Gauge pressure} = (\text{depth}) (\text{weight density})$$

$$= 12 \text{ m } (1025 \text{ kg}_f/\text{m}^3)$$

$$= 12 \text{ 300 kg}_f/\text{m}^2$$

where we have taken the density of seawater from Table 6-2. Then using the fact that 1 kg$_f$/cm^2 is the same as 10 000 kg$_f$/m^2 (Table 5-2), we can write this as

$$\text{Gauge pressure} = 12 \text{ 300 } \frac{\text{kg}_f}{\text{m}^2} \left(\frac{1 \text{ kg}_f/\text{cm}^2}{10 \text{ 000 kg}_f/\text{m}^2} \right)$$

$$= 1.23 \text{ kg}_f/\text{cm}^2$$

Since 1 atm = 1.03 kg$_f$/cm^2 (Table 6-4), the absolute pressure of the air in the bag must be

$$\text{Absolute pressure} = 1.23 \text{ kg}_f/\text{cm}^2 + 1.03 \text{ kg}_f/\text{cm}^2$$

$$= 2.26 \text{ kg}_f/\text{cm}^2$$

This, then, is the minimum pressure that must be supplied by the pump to get the air bag to expand in the hull.

Now let's answer the question of volume. Under the water, where the absolute pressure is 2.26 kg$_f$/cm^2, the air fills a volume of 13 900 L. At the surface, the air pressure is just 1.03 kg$_f$/cm^2, or 0.456 as much. According to Boyle's law, the volume at the surface is then 1/0.456 times the compressed volume:

$$\text{Volume} = \frac{1}{0.456} \text{ (13 900 L)}$$

$$= 30 \text{ 500 L}$$

We see, then, that to get 13 900 L of air at 2.26 kg$_f$/cm^2 pressure, we need to pump 30 500 L of air out of the atmosphere.

Exercises

18. Compressed air in a certain tank is at an absolute pressure of 33.2 kg_f/cm^2. The tank's volume is 78.5 m^3. What is this air's volume if its pressure is reduced to atmospheric?
    Answer: 2520 m^3

19. A large balloon is filled with 350 m^3 of helium at atmospheric pressure at sea level. What is the volume of the helium if the balloon rises to an altitude of about 4000 km, where the atmospheric pressure is 0.60 atm? Assume that the temperature does not change much.
    Answer: 580 m^3

20. Natural gas at a storage pressure of 55.2 psig is to be compressed into one-third its original volume. What pressure is needed to do this?
    Answer: 210 psia, or 195 psig

12-7 Hydraulic Systems

In many instances a force has to be transmitted over a distance. In an airplane, for instance, motion of the control stick and rudder pedals must operate the movable control surfaces some distance away. And in a car, a push on the brake pedal has to force the brake linings against the drums (or the pads against the disks) at the wheels. What is the best way to transmit forces like these?

A system of cables and pulleys might do the job, but this gets very complicated when allowance is made for the car's wheels being steered. And in the plane, the control linkages would have to be designed to make many bends and thread through narrow passageways. Cables, or other mechanical linkages like levers, also present problems of lubrication, corrosion, and dirt contamination. Fluid forces, on the other hand, are easily transmitted through flexible pipes. In a car, a hydraulic line can be routed around obstacles like mufflers and drive shafts; and if it is flexible, it will not interfere with steering.

Figure 12-11 shows an automobile brake system with drum-type brakes. The *hydraulic system* is some plumbing that connects a *master* (or *control*) *cylinder* to the four *slave cylinders* at the wheels. Each cylinder contains a movable piston, and the entire system is filled with a light oil. Pushing on the brake pedal causes the piston to push on the fluid in the master cylinder. The fluid then transmits the increased pressure to the wheel cylinders, where it moves the pistons that force the brake shoes against the rotating drums. The force is transmitted a large distance in four different directions, but the fluid itself moves very little.

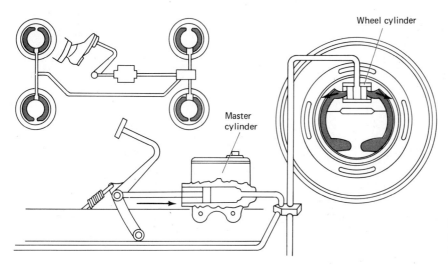

Wheel cylinder

Master
cylinder

Fig. 12-11 Principle of the
automotive hydraulic braking
system.

Most modern vehicles have dual braking systems. The front and
rear brakes are connected to separate hydraulic lines and separate
halves of a dual master cylinder. Since a leak in one system does not
affect the other, one set of brakes will always work. In some cars, each
half of the dual braking system operates the brakes on three of the four
wheels. This calls for dual slave cylinders on two of the wheels.

If the car has disk brakes, the hydraulic system is basically the
same. The only difference is that each brake has two connected wheel
cylinders that force the pistons toward each other rather than forcing
them apart. The pistons now press two brake pads against opposite
sides of a disk that rotates between the pads.

Many hydraulic systems are based on the idea that a fluid will
transmit a force without the fluid itself moving very much. When we use
the word "hydraulic," we usually mean that the fluid is a liquid. Various
light oils are usually preferred because they are self-lubricating and do
not promote corrosion.

Since liquids are practically incompressible, hydraulic systems
can be adjusted to very fine tolerances. Every 1 cm³ of hydraulic fluid
forced out of the master cylinder forces 1 cm³ of fluid into the slave
cylinders (0.25 cm³ apiece if there are four of them). But trouble starts if
some air leaks into the system. Now 1 cm³ of fluid may be forced from the
master cylinder; but because of the compression of air in the lines, only a
fraction of 1 cm³ travels into the slave cylinders. As a result, the slave
pistons do not move far enough to do their job. Thus all hydraulic
systems are built with some provision for "bleeding" out trapped air after
the system is filled.

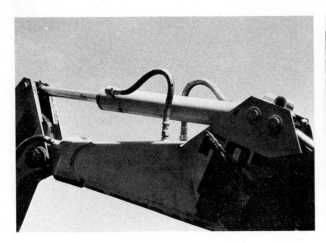 

Fig. 12-12 Pump-actuated hydraulic machines are very common.

In many heavy-duty hydraulic machines, the master cylinder is replaced by a pump. By closing a valve, the operator can control the fluid pressure developed in the slave cylinder. Figure 12-12 shows two examples: a bulldozer lift and a loading boom on a truck. Each one uses several different slave cylinders. Can you pick them out?

12-8 Pneumatic Systems

Although air is compressible, it can still be used to transmit a force through a pipe or air line. But since the system cannot depend on the transfer of a fixed volume of air, pedal-operated pistons cannot be used. Instead, *pneumatic systems* are powered by air pumps or atmospheric pressure or both.

Figure 12-13 shows a pneumatic hammer. Compressed air is supplied by a pump (compressor). The operator's handle grip has a control lever which regulates the air admitted to the hammer. When air flows into the hammer's outer cylinder, the air forces the piston upward. This compresses the air trapped above the piston and eventually pops open the rocker valve. Now compressed air enters the inner cylinder directly, quickly driving the piston downward and delivering a blow to the tool. When the piston uncovers the exhaust port, the pressure in the inner cylinder drops and the incoming compressed air forces the rocker valve shut. Now the compressed air enters the outer cylinder again, and the process is repeated. The exhaust air is released to the atmosphere. Although the air hammer is relatively inefficient, it offers the advantage of easy setup since there are no hydraulic lines to connect, fill, and bleed. And by isolating the compressor from the tool, very high pressures can be used without making the hammer too heavy for one person to use. A direct-drive power hammer would need a very large motor to make it

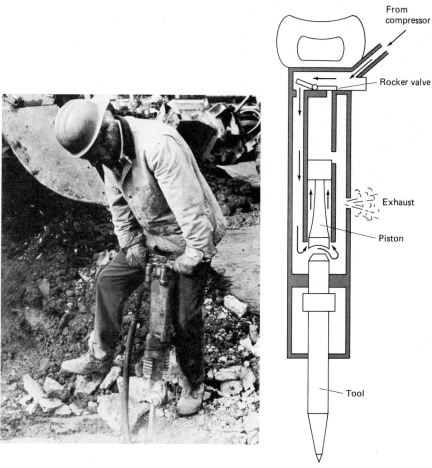

From compressor

Rocker valve

Exhaust

Piston

Tool

Fig. 12-13 Principle of the pneumatic hammer (air hammer).

capable of breaking concrete, and then the shock of the hammering would quickly damage most motor bearings.

Compressed air is also used for the air brakes on trains (Fig. 12-14). Again, the advantage is that air lines can easily be connected and disconnected. In a train, the system is normally kept at a high pressure, which keeps the brakes disengaged. Compressed air is also stored in a tank at each brake. When the high-pressure source is cut off, the compressed air in storage acts on the braking pistons to stop the train. Thus the train automatically stops if the main air line is broken or the compressor fails.

Although most hydraulic systems are used to solve the problem of transmitting a force from one place to another, it is also possible to generate a mechanical advantage in this way. In other words, if we design the

12-9 Mechanical Advantage in Hydraulic Systems

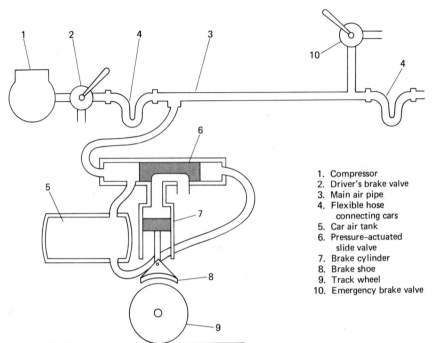

Fig. 12-14 A train's air-brake
system. When the pressure
drops in the main air pipe, compressed air in the car's air tank
actuates the slide valve and enters the brake cylinder. Here it
expands against a piston that
forces the brake shoe against
the wheel.

1. Compressor
2. Driver's brake valve
3. Main air pipe
4. Flexible hose
   connecting cars
5. Car air tank
6. Pressure–actuated
   slide valve
7. Brake cylinder
8. Brake shoe
9. Track wheel
10. Emergency brake valve

system right, a small force at the master cylinder provides a large force
at the slave cylinder.

Figure 12-15 shows a simple hydraulic system with a small master
cylinder and a large slave cylinder. An effort on the master piston
increases the fluid's pressure. This pressure acts on the entire inside of
the system, including the underside of the slave piston. Suppose that the
pressure increase is 1.0 kg_f/cm^2. Then a force of 1.0 kg_f is transmitted to
each 1 cm^2 of the slave piston. Since the slave piston has a bigger area
than the master piston, it follows that the force on the slave piston is
greater than the effort which generated the pressure.

The mechanical advantage of a hydraulic system is governed by
the relative areas of the master and slave pistons. Since the piston area
is the same as the inside area of the cylinder, we can write the
relationship this way:

$$\text{MA of hydraulic system} = (\text{efficiency})\left(\frac{\text{area of slave cylinder}}{\text{area of master cylinder}}\right) \qquad (12\text{-}2)$$

If the liquid has a low viscosity and the pistons don't move too fast,
the efficiency is close to 100 percent. But if there is some air in the

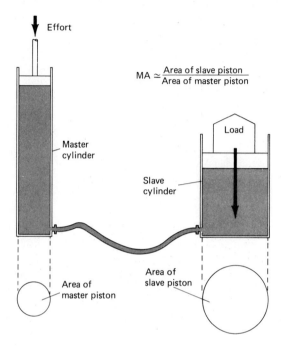

$$MA \simeq \frac{\text{Area of slave piston}}{\text{Area of master piston}}$$

Effort

Master cylinder

Load

Slave cylinder

Area of slave piston

Area of master piston

Fig. 12-15 A hydraulic system gives a mechanical advantage greater than 1 if the slave cylinder has a larger cross-sectional area than the master cylinder.

system, much of the motion of the master piston is wasted in compressing the air, and the efficiency drops drastically.

Suppose that the master cylinder has a cross-sectional area of 7.0 cm² and the slave cylinder has an area of 56 cm². This gives the system a mechanical advantage of nearly 8.0. An effort of 30 kg$_f$ can then lift a load of nearly 240 kg$_f$. Notice that the cylinder *areas* determine the MA, not the cylinder diameters.

Example 12-6 Mechanical Advantage of a Hydraulic Jack.

The jack is shown in Fig. 12-16. The slave cylinder has an inside diameter of 6.0 cm while the master cylinder's inside diameter is 2.0 cm. Let's estimate the MA of the system.

Since the master cylinder is so small, one stroke of the master piston transfers just a small amount of fluid into the slave cylinder, and the load is lifted just a little. To lift the load farther, the master piston must be stroked up and down a number of times. This action pumps oil from a reservoir through a pair of check valves that function on the same principle as the lift pump in Chap. 6. To lower the load, a release valve is screwed open, and the fluid is transferred directly from the slave cylinder back to the reservoir.

But all these details shouldn't obscure the principle. The MA is still found from Eq. (12-2).

Application: Pressure Gauges

Although the barometer is useful for measuring atmospheric pressure, it is simply not rugged enough for many applications. There are other liquid-filled instruments (manometers and inverted bells, for instance) that have the same drawback.

In industrial applications, pressure gauges must be tough, reliable, and easy to use. There are three general types in common use:

1. *The Bourdon-tube gauge.* If the pressure is increased in a section of coiled hose, the hose tries to uncoil. This happens because the outside of the coil has more area subjected to the fluid pressure. In the Bourdon-tube gauge, a short, curved tube is sealed at one end. When the inside pressure increases, the tube straightens out. When the pressure decreases, the tube's elasticity returns it to its original shape. The movable end of the tube is linked mechanically to a point on a scale. Automotive compression gauges are usually of this type.
2. *The bellows gauge.* This gauge has a piston that moves against a spring when the pressure is increased. To avoid the lubrication and sealing problems of a piston moving in a cylinder, expansion bellows are used. Again, the movable piston is linked mechanically to a pointer on a scale. Bellows gauges tend to be more accurate but somewhat less rugged than Bourdon-tube gauges. Bellows gauges can measure pressures both above and below atmospheric pressure.
3. *Pressure transducers.* Certain crystals can generate small electric voltages when they are placed under extreme pressures. Reading the voltage with a voltmeter then gives a measure of the pressure. The advantage of pressure transducers is that the actual measurement can be made some distance from the transducer itself.

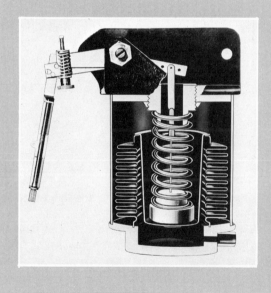

Elements of Applied Physics

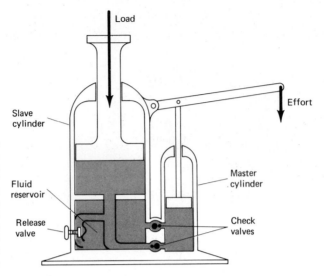

Fig. 12-16 A hydraulic jack. The small piston pumps the fluid into the large cylinder. The check valves allow the master cylinder to be refilled from a reservoir on each upstroke. The release valve allows the load to be lowered, at the same time forcing fluid from the slave cylinder directly back into the reservoir.

Since the master cylinder's diameter is 2.0 cm, its radius is 1.0 cm and its area is 3.14 cm². The slave cylinder's 6.0-cm diameter gives it a radius of 3.0 cm and an area of 28.3 cm². The efficiency is nearly 100 percent. Then, using Eq. (12-2), we have

$$\text{MA} = (\text{efficiency}) \left(\frac{\text{area of slave cylinder}}{\text{area of master cylinder}} \right)$$

$$= 1 \left(\frac{28.3 \text{ cm}^2}{3.14 \text{ cm}^2} \right)$$

$$= 9.0$$

Notice that one diameter is just 3 times the other, but since the ratio of the *areas* is 9, the MA is 9.

The jack in Fig. 12-16 gains an additional MA by coupling the master piston to a lever. If this lever has an MA of 4, the total MA is (4)(9), or 36. Thus a 10-kg$_f$ effort lifts a 360-kg$_f$ load, or a 50 kg$_f$ effort lifts a load of 1800 kg$_f$. ◀

Exercises

21. Find the MA of a hydraulic system if the master cylinder has a cross-sectional area of 7.07 cm², the slave cylinder has a cross-sectional area of 44.2 cm², and the efficiency is nearly 100 percent.
Answer: MA = 6.25

22. Find the MA of a hydraulic system if the master cylinder has a diameter of 4.22 cm and the slave cylinder has a diameter of 12.86 cm. The efficiency is 95.1 percent.
    Answer: MA = 8.83
23. A certain master cylinder has an area of 10.0 cm². What should the area of the slave cylinder be to produce an MA of (*a*) 2.0, (*b*) 5.0, (*c*) 1$\bar{0}$? Assume 100 percent efficiency.
    Answers: (*a*) 2$\bar{0}$ cm², (*b*) 5$\bar{0}$ cm², (*c*) 1$\bar{0}$0 cm²
24. A certain master cylinder has a diameter of 4.00 cm. The efficiency is nearly 100 percent. What should the diameter of the slave cylinder be to produce an MA of (*a*) 4.0, (*b*) 9.0, (*c*) 16, (*d*) 10.0?
    Answers: 8.0 cm, (*b*) 12 cm, (*c*) 16 cm, (*d*) 12.6 cm

12-10 Hybrid Systems

When a system uses both air and a liquid to accomplish some task, it is called a *hybrid system.*

Perhaps the simplest hybrid device is the *siphon*, shown in Fig. 12-17. Atmospheric pressure on the surface of the liquid forces it through the pipe, even if the pipe runs uphill for awhile. Of course, the pipe cannot go too high, because the back pressure of the liquid in the pipe must be less than the atmospheric pressure acting on the exposed surface. Also, the pipe's open end must be below the liquid level outside the tank. And even if both these conditions are met, the siphon still will not start itself. To start the flow, the pipe must first be filled with the liquid. This may be done by drawing a vacuum on the open end.

Siphons have long been used at wineries to draw the wine from vats without stirring up the sediment. But siphon action can also sometimes lead to problems. In home drains, for instance, emptying a

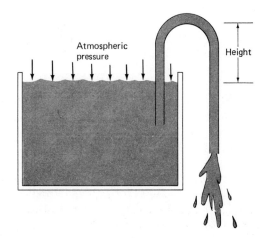

Fig. 12-17 The siphon. The maximum height of the siphon tube is limited to the height of a column of the liquid that can be supported by atmospheric pressure.

bathtub may siphon water from the traps on the sinks, toilet, etc., as well as from the tub trap itself. The result is a "gurgling" noise in traps that may be some distance from the fixture being drained. To prevent this, most plumbing codes call for reventing all drains to keep them at atmospheric pressure (Fig. 12-18). This is equivalent to putting a hole near the top of the siphon. For the same reason, the drain hoses from automatic washers may have to be equipped with antisiphon valves. Such valves allow air to enter the drain hoses, but do not allow water to come out.

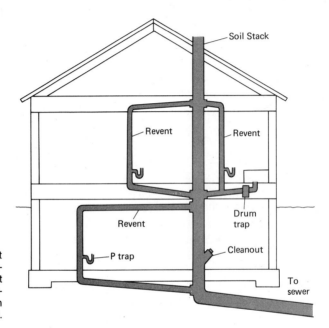

Fig. 12-18 Drain-waste-vent (DWV) lines in a home. Reventing the drains maintains them at atmospheric pressure and prevents the siphon action that can empty the traps.

The hydraulic lift (Fig. 12-19) is another example of a hybrid system. Here compressed air is pumped into an oil reservoir that also acts as a master cylinder. This air pressure forces oil into the slave cylinder, where it raises the lift piston.

Exercises

25. Calculate the maximum height (in centimeters) between the liquid surface and the top of the siphon tube if the liquid is (a) water, (b) kerosene, (c) mercury. Assume standard atmospheric pressure. *Answers:* (a) 1030 cm, (b) 1260 cm, (c) 76 cm
26. Suppose that the slave cylinder in Fig. 12-18 has a diameter of 32.3 cm. The ramp and slave piston have a total mass of 460 kg.

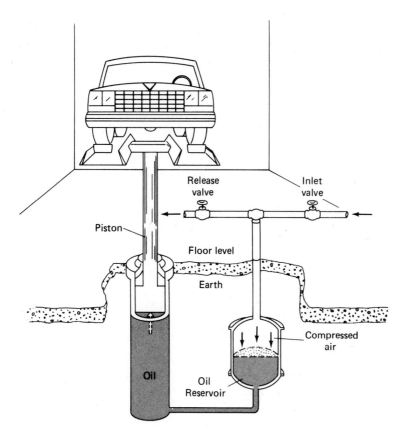

Fig. 12-19 The "hydraulic" lift is actually a hybrid system, using both compressed air and oil to transmit the lifting force.

Assuming 100 percent efficiency in the hydraulic portion, what air pressure is needed to lift a car whose mass is 1450 kg?
Answer: 2.3kg$_f$/cm² , or 230 kPa

Summary

A stationary fluid automatically develops a force on any object placed in it. This force is called a buoyant force, and it results from the fluid's increased pressure at greater depths. The buoyant force on an object can be predicted with the help of Archimedes' principle.

Confined fluids may also be used to transmit a force from one place to another. When this is done with a liquid, we have a hydraulic system. When it is done with a gas (usually air), we have a pneumatic system. In a hybrid system, both air and a liquid are used. Hydraulic and pneumatic systems can also be used to develop a mechanical advantage. Hydraulic devices generally have high efficiencies. Efficiencies of pneumatic and hybrid devices are lower, because of the compressibility of air (Boyle's law).

Except for the pneumatic hammer, all our examples in this chapter have assumed that the fluid is at rest or else moving very slowly. In the next chapter, we look into some of the effects of rapidly moving fluids.

Terms You Should Know

buoyant force	hydraulic system
Archimedes' principle	pneumatic system
center of buoyancy	master cylinder
specific gravity	slave cylinder
hydrometer	hybrid system
contact patch (of tire)	siphon
Boyle's law	

Problems

1. A certain barge is in the shape of a rectangular solid with a bottom area of 1500 ft² and a height of 8.0 ft (see the diagram). When empty, the barge weighs 195,000 lb. (a) What is the height of the waterline if the empty barge floats in fresh water? (b) How much weight can be added to the barge before it is swamped?

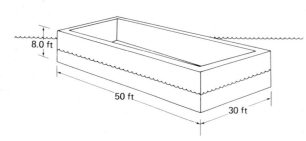

2. A certain compressor is capable of supplying air at a pressure of 10.2 kg$_f$/cm², gauge. (a) What piston area is needed to develop a force of 125 kg$_f$ from this pressure? (b) What is the piston's diameter?

3. The holding tank in the illustration has 75 percent of its volume filled with water and the other 25 percent with air at a pressure of 115 psig. Assume that no additional air gets into the tank as the water is drawn out. (a) What is the approximate pressure at the faucet? (b) What is the faucet pressure when one-third of the water has flowed out? (c) What is the faucet pressure when the tank is nearly empty?

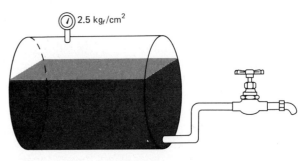

4. Calculate the air pressure needed to transfer the gasoline from the lower tank to the upper tank. Express this pressure in (*a*) kilogram-force per square centimeter, (*b*) kilopascals, (*c*) standard atmospheres. Use 0.67 g/cm³ as the density of gasoline.

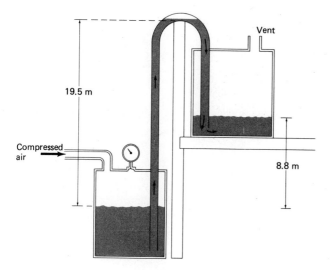

13
Fluids in Motion

Introduction In Chaps. 6 and 12, we saw how enclosed fluids may be used to transmit forces from one place to another. We classified all gases and liquids as fluids, even though gases are much more compressible. We studied the relationship between the pressure and the volume of an enclosed gas, and we made a distinction between gauge pressure and absolute pressure. We also looked at a number of applications of these principles, ranging from construction caissons to siphons to pneumatic tires.

In all this material, our fluids were either at rest or moving very slowly. In this chapter, we go a bit further and talk about how fluids behave when they are moving. We also look at some of the effects generated when objects move through a fluid at high speed.

13-1 Fluid Velocity We can very easily measure how fast a river is flowing by tossing in a stick and timing how long it takes the stick to float to a certain distance. Then, in the usual way, the river's speed is found from

$$\text{Speed} = \frac{\text{distance}}{\text{time}} \qquad (13\text{-}1)$$

This speed in a particular direction is called the *fluid velocity* of the river.

If we want to measure the speed of the wind, it becomes a little more difficult. We can try letting a balloon drift with the wind and timing it, but a more direct approach is to use an *anemometer*, shown in Fig. 13-1. The cone-shaped cups catch the wind and move with it, converting the linear motion of the wind into rotational motion of the shaft. The device is usually mounted on the roof of a building with the shaft running indoors to a gauge which reads the frequency of rotation. This frequency is directly proportional to the windspeed, so the instrument's scale may be calibrated directly in units of miles per hour, kilometers per hour, or knots. The wind speed measured by this device is also the fluid velocity of the moving air.

> Definition: *Fluid velocity* is the distance a fluid moves per unit time in a particular direction. Its magnitude is the same as the speed of an object moving with the fluid.

fluid velocity

The SI unit for fluid velocity is meters per second, abbreviated m/s. Of course, we may also use other units: kilometers per hour, miles per hour, feet per second, or knots, for instance.

Fluid velocity plays a part in many applications. For instance, an oxyacetylene flame burns at a rate of about 7.6 m/s. This is how fast the

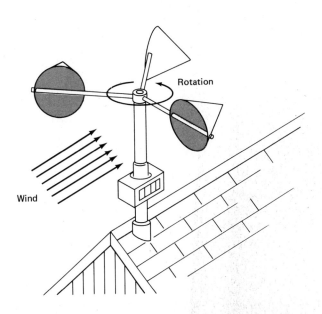

Fig. 13-1 An anemometer is used to measure windspeed.

flame would propagate through a container filled with the gas mixture. To successfully operate an oxyacetylene torch, then, the fluid velocity at the tip orifice must be equal to this burning velocity.

Let's look at some numerical examples involving fluid velocity.

Example 13-1 Finding the Fluid Velocity of a Stream.

Two bridges are 82.5 m [271 ft] apart. A log floats this distance in a measured time of 1 min 41 s. What is the fluid velocity of the stream?

Equation (13-1) tells us to divide the distance by the time. But we can't divide directly by 1 min 41 s. We can calculate with either minutes or seconds, but not both. Let's convert the time measurement to seconds. This gives

$$1 \text{ min } 41 \text{ s} = 60 \text{ s} + 41 \text{ s}$$

$$= 101 \text{ s}$$

Then we have

$$\text{Speed} = \frac{\text{distance}}{\text{time}}$$

$$= \frac{82.5 \text{ m}}{101 \text{ s}}$$

$$= 0.817 \text{ m/s}$$

This is the speed of the floating log. Since the log is moving with the current, it is also the fluid velocity of the stream.

We should point out that this result is accurate only at the stream's surface where the log is floating. Near the bottom and near the shores, the stream's fluid velocity is quite a bit lower. ◄

Example 13-2 Calculating Transport Time from Fluid Velocity.

A certain oil pipeline is 38.2 km long. Crude oil is pumped into one end at a fluid velocity of 25 cm/s. How long does it take before the oil reaches the other end?

This problem is very similar to some that we solved in Chap. 3. The only difference is that here we have a moving fluid rather than a car or a conveyor belt. We still use Eq. (13-1), but we rearrange it to solve for time:

$$\text{Speed} = \frac{\text{distance}}{\text{time}}$$

$$\text{Speed} \cdot \text{time} = \text{distance}$$ (multiply both sides by *time*)

$$\text{Time} = \frac{\text{distance}}{\text{speed}}$$ (divide both sides by *speed*)

Then, putting in the numbers,

$$\text{Time} = \frac{38.2 \text{ km}}{25 \text{ cm/s}}$$

Now we can't ignore the fact that the distance units are different. We have to make them the same if this calculation is to make any sense. Following the rules we developed in Chap. 1, we need to introduce the following conversion factors:

$$100 \text{ cm} = 1 \text{ m}$$

$$1000 \text{ m} = 1 \text{ km}$$

Then abbreviating the units in the standard way, we have

$$\text{Time} = \left(\frac{38.2 \text{ km}}{25 \text{ cm/s}} \right) \left(\frac{1000 \text{ m}}{1 \text{ km}} \right) \left(\frac{100 \text{ cm}}{1 \text{ m}} \right)$$

$$= 153\,000 \text{ s}$$

Since this is a lot of seconds, it makes sense to convert to a more convenient-sized unit—days (d), for instance. Then

$$\text{Time} = (153\,000 \text{ s}) \left(\frac{1 \text{ min}}{60 \text{ s}} \right) \left(\frac{1 \text{ h}}{60 \text{ min}} \right) \left(\frac{1 \text{ d}}{24 \text{ h}} \right)$$

$$= 1.77 \text{ d}$$

This is not as unreasonable an answer as it may seem. It actually took 42 d for the first crude oil to travel the length of the Alaskan pipeline (1300 km). ◀

Exercises

1. A barge floats 5.20 km down a river in 2 h 17 min. Nothing is pushing or pulling it. What is the stream's fluid velocity at the surface in (*a*) kilometers per hour, (*b*) miles per hour, (*c*) meters per second?

Answers: (a) 2.28 kmph, (b) 1.42 mph, (c) 0.633 m/s
2. Find the fluid velocity if oil travels through a 26-km pipeline in 31 h.
    Answer: 0.84 kmph, or 0.23 m/s
3. Smoke is carried away from an industrial smokestack by a 22-kmph wind. How long does it take the smoke to drift 9.5 km?
    Answer: 0.43 h, or 26 min
4. Oil in a pipeline is being pumped at a fluid velocity of 0.45 m/s. Once each hour, a plastic plug is placed in the line at the pumping station to separate the moving oil into segments. This limits the amount of oil lost if the pipeline ruptures. What is the distance between the plugs?
    Answer: 1600 m, or 1.6 km, or 1.0 mi

13-2 Fluid Flow Rate   If we are concerned with how fast we can fill a tank with fuel oil, or how long it takes to empty a cistern, then knowing the fluid velocity is of little help. What we really need to know is a quantity called the *flow rate.* It is defined as follows:

fluid flow rate   Definition: *Fluid flow rate* is the volume of fluid moving past a point per unit time:

$$\text{Flow rate} = \frac{\text{volume}}{\text{time}} \qquad (13\text{-}2)$$

Based on this definition, the SI unit for flow rate is cubic meters per second, abbreviated m³/s. As shown in Fig. 13-2, water flowing out of a pipe at 1 m³/s would fill a box 1 m on a side in just 1 s. This, then, is a fairly large unit. Other common units for flow rate are listed in Table 13-1, along with the conversion factors to and from the SI unit.

Example 13-3 Calculating Flow Rate.

The cylindrical tank shown is to be filled in 30.0 min. What flow rate is needed?
        Equation (13-2) calls for the fluid volume. From Chap. 3 we may calculate the volume of a cylinder by

$$\text{Volume} = \pi \, (\text{radius})^2(\text{height})$$

The radius here is 1.40 m (half the diameter), and the height is 4.20 m.

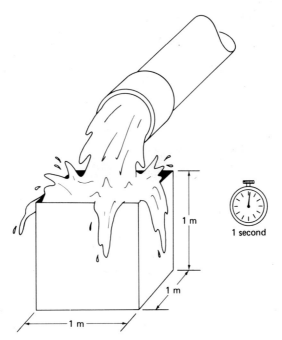

1 m

1 m

1 m

1 second

Fig. 13-2 Measuring the fluid velocity of a stream. The method is based on separate measurements of distance and time.

TABLE 13-1   *COMMON UNITS FOR FLUID FLOW RATE AND THEIR CONVERSION FACTORS*

	$\dfrac{m^3}{s}$	$\dfrac{m^3}{min}$	$\dfrac{L}{min}$	$\dfrac{gal^*}{min}$	$\dfrac{ft^3}{s}$	$\dfrac{ft^3}{min}$
1 cubic meter per second =	1	60	60 000	15 850	35.315	2 118.9
1 cubic meter per minute =	0.016 667	1	1 000	264.17	0.588 58	35.315
1 liter per minute =	0.000 016 67	0.001	1	0.264 17	0.000 588 6	0.035 315
1 gallon* per minute =	0.000 063 09	0.003 785 4	3.785 4	1	0.002 228 0	0.133 68
1 cubic foot per second =	0.028 317	1.699 0	1 699.0	488.83	1	60
1 cubic foot per minute =	0.000 471 95	0.028 317	28.317	7.480 5	0.016 667	1

* U.S. liquid.

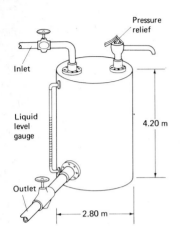

Inlet

Pressure relief

Liquid level gauge

4.20 m

Outlet

2.80 m

Then

$$\text{Volume} = \pi(1.40\ \text{m})^2(4.20\ \text{m})$$

$$= 25.9\ \text{m}^3$$

We may now use Eq. (13-2) to get the flow rate:

$$\text{Flow rate} = \frac{\text{volume}}{\text{time}}$$

$$= \frac{25.9\ \text{m}^3}{30.0\ \text{min}}$$

$$= 0.862\ \text{m}^3/\text{min}$$

This result, in turn, may be expressed in other acceptable units by using the conversion factors in Table 13-1. The reader can verify that it is equivalent to 228 gal/min, 864 L/min, or 30.5 ft³/min. ◀

Exercises

5. A certain airplane engine consumes 68.5 L of fuel each hour. What is the flow rate in the fuel line in liters per minute?
    Answer: 1.14 L/min
6. What is the flow rate if a pipe delivers 35.0 L of water in 2.21 min?
    Answer: 15.8 L/min, or 0.264 L/s
7. A tank holds 96 m³ of crude oil. What flow rate is needed if the tank is to be pumped dry in 7.0 h?
    Answer: 14 m³/h, or 3.8 L/s
8. A rectangular swimming pool is 36 ft long and 18 ft wide and has an average depth of 4.5 ft. If water can be supplied at a flow rate of 19.5 gal/min, how long does it take to fill the pool?
    Answer: Nearly 19 h

13-3 Pipes and Conduits   Obviously, a fluid's flow rate increases if its fluid velocity increases. But there is another factor governing flow rate, and that is the cross section of the pipe or conduit through which the fluid flows. If the fluid velocity is the same in a big pipe and a small pipe, the big pipe will have a greater volume of flow. We may write the relationship as follows:

$$\text{Flow rate} = (\text{cross-sectional area})(\text{fluid velocity}) \qquad (13\text{-}3)$$

Notice that the *area* of the pipe rather than its diameter appears in Eq. (13-3). Suppose that we have a pipe that cannot deliver the flow we need. We decide to replace it by a bigger pipe. If we double the pipe's diameter, its area increases by a factor of 4, and so the new pipe can deliver 4 times the flow at the same fluid velocity. The cross-sectional areas of standard pipe sizes are given in Fig. 13-3. Also listed are the relative flow rates at the same fluid velocity. For instance, a 3/4-in pipe will handle 2.25 times the flow of a 1/2-in pipe, and a 3-in pipe will carry 9 times the flow of a 1-in pipe. In most homes, the main waterline has a 3/4-in inside diameter, while spur lines to specific rooms are 1/2-in pipe. The waterline feeding a specific fixture may be even smaller; 3/8 in is common. Of course, someday these sizes will be replaced in the United States with metric sizes.

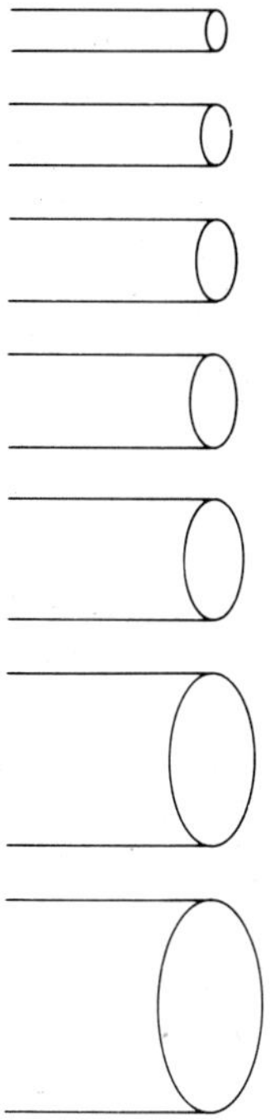

Inside diameter, inches	Area, square inches	Relative flow rate
1/2	0.196	1
3/4	0.442	2.25
1	0.785	4.0
1¼	1.23	6.28
1½	1.77	9.0
2	3.14	16.0
3	7.07	36.0

Fig. 13-3 Standard copper and plastic pipe sizes, their cross-sectional areas, and their relative flow-rate capacities at the same fluid velocity. Eventually these sizes will be replaced by metric sizes.

Suppose that a fluid flows from a large pipe into a small pipe, as shown in Fig. 13-4. Assuming that the fluid cannot be compressed, whatever volume flows into one end must flow out of the other end. In other words, the *flow rate* must be the same in both pipes. The only way this can happen is if the fluid velocity increases in the smaller pipe. In

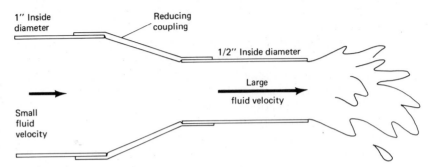

1″ Inside diameter

Reducing coupling

1/2″ Inside diameter

Large fluid velocity

Small fluid velocity

Fig. 13-4 The continuity principle. A decrease in cross-sectional area causes an increase in fluid velocity. An increase in cross-sectional area would have the opposite effect.

this case, the fluid velocity in the 1/2-in pipe is 4 times its velocity in the 1-in pipe. This phenomenon is called the *continuity principle.*

continuity principle

Continuity principle: As the cross-sectional area of a pipe or conduit decreases, the fluid velocity increases in proportion.*

The continuity principle can easily be demonstrated by putting your thumb over the stream of water coming from a spigot or hose (Fig. 13-5). By constricting the flow into a smaller cross section, you increase the fluid velocity. The resulting stream may squirt a considerable distance. The common hose nozzle operates on this same principle.

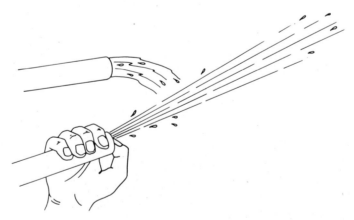

Fig. 13-5 Demonstrating the continuity principle. Constricting the cross section of the flow increases the fluid velocity.

The effect is also noticeable in many rivers. Everyone knows that rapids develop when the downhill slope, or *gradient,* of the river is large.

* This is exactly true only if the fluid cannot be compressed. But even for compressible fluids (air, for instance), it is still approximately true as long as the fluid velocity is not close to the speed of sound in the fluid.

But rapids also develop at places where the river gets very narrow. When a large volume of water is channeled into a narrow section, the river has no choice but to speed up.

The same thing happens with air currents. A moderate wind blowing into a modern city of skyscrapers picks up speed as it is forced into the channels between the buildings. Cases have been reported where a 20-kmph [about 12-mph] wind increases to over 100 kmph [62 mph] as it whips between tall buildings.

Exercises

9. Water flows into a 3/4-in-diameter pipe at a flow rate of 6.5 gal/min. What is the flow rate if the pipe narrows to a 1/2-in diameter?
   Answer: 6.5 gal/min
10. Water flows into a 3/4-in-diameter pipe at a fluid velocity of 5.4 ft/s. What is the fluid velocity if the pipe narrows to a 1/2-in diameter?
   Answer: 12 ft/s
11. A certain stream has an average depth of 2.2 m and a width of 37 m. It is flowing at 4.2 kmph. If it is dredged to double the average depth without increasing the width, what happens to the fluid velocity?
   Answer: Decreases to about 2.1 kmph

Of course, fluids do not move all by themselves. They need something to push them. This may be a pump, or the force of gravity, the uneven heating of the earth's surface, or a chemical explosion. Yet no matter which of these agents is at work, we can say that the fluid is moving in response to *pressure differences*. The direction of the fluid motion is from regions of high pressure to regions of low pressure.

Figure 13-6 shows the cross section of a steam-turbine engine. Steam is heated in a boiler until its pressure rises to over 10 MPa [100 kg_f/cm^2, or 1500 psi]. This steam is introduced into one end of the turbine cavity whose other end is at a pressure somewhat below atmospheric. Because of this great pressure difference, the live steam flows through the turbine at high speed, transferring some of its motion to the turbine blades it encounters along the way.

Air currents follow this same principle, moving from regions of high barometric pressure to regions of low barometric pressure. Because of the rotation of the earth, they also spiral as shown in Fig. 13-7. The sky tends to be clear when the barometer is high and overcast when it is low. If the barometer is extremely low, air currents rush in quickly and a storm is likely. Of course, weather prediction is much more complicated than

13-4 Pressure and Flow

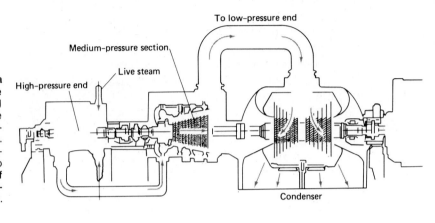

Fig. 13-6 Cross section of a steam-turbine engine. The working fluid is superheated steam, and it flows from the high-pressure end to the low-pressure end (the condensor). Notice that the direction of expansion alternates from right to left. This minimizes the stress of the thrust bearings on the turbine shaft.

Fig. 13-7 Air masses tend to move away from high-pressure regions and toward low pressure regions. The spiraling is due to the rotation of the earth.

this. The point is that air, being a fluid, follows the general pattern of moving from high-pressure regions to low-pressure regions.

You may remember seeing news photographs of the damage caused by a tornado. Houses look as if they exploded. In fact, that is exactly what happened. The air pressure in and near the funnel (Fig. 13-8) is so low that the normal atmospheric pressure in the house can blow it apart. If a tornado alert is sounded, the best bet is to open all the doors and windows, to equalize the pressure, and then get into the cellar.

This principle also explains the action of explosives. When a stick of dynamite is detonated, a very rapid chemical reaction generates hot gases at a high pressure. This pressure is so much greater than normal atmospheric pressure that the gases rush from the point of detonation

Fig. 13-8 A tornado is a region of extremely low pressure. If a building is tightly shut, the normal air pressure inside can blow it apart when a tornado approaches. (Courtesy NOAA).

and carry anything in the way with them. Thus a farmer can blast out a tree trunk, or a construction crew can alter the face of a hillside, both by exploiting the natural motion of fluids. Underwater explosions can also be used to form metals by pushing a metal sheet into a die. The process is called explosive forming. The advantage is that the water replaces the male die.

13-5 Fluid Friction

We saw in Chap. 6 that fluids have a natural tendency to resist flowing under their own weight. This resistance is called the fluid's viscosity, and it arises from cohesive forces between the molecules. Different fluids have different viscosities, and their viscosities change with the temperature.

Let's look at how viscosity affects fluid flow. Figure 13-9 shows a schematic of a pipeline filled with a fluid under pressure. In the upper drawing the valve is closed, so the pressure is the same through the entire length of the pipe. This, remember, was Pascal's principle (Chap. 6): a pressure exerted on a confined fluid is the same in all parts of that fluid. But with the valve opened, the situation is quite different. Now the pressure drops as the fluid moves away from its source. This pressure drop along the length of the pipe is due mainly to the fluid's viscosity. It comes about because the fluid is rubbing against the inner wall of the pipe, and layers of the moving fluid are rubbing against other layers. The

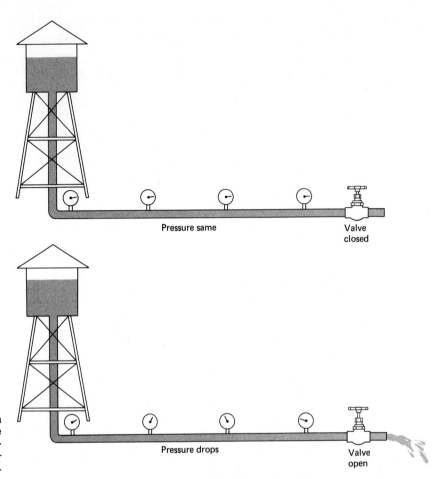

Fig. 13-9 The pressure of a moving fluid drops as distance from the pressure source increases. If the fluid is not moving, there is no pressure drop.

fluid in the very center is moving with the least restriction. The pressure pushing the fluid along is decreased by the cumulative effect of all this fluid friction.

There is something else that affects fluid friction. You have probably noticed what happens when a fast-moving stream (a creek or a river) encounters obstacles like large rocks or fallen trees: the stream gurgles and splashes and becomes very turbulent around the obstacle. This, in turn, acts to slow the overall fluid motion. The same thing happens in a pipe if there are obstacles or rough spots inside. Turbulence also develops when the stream is forced to quickly change its direction, as beneath a waterfall. Again, sharp bends in a pipe produce the same effect.

Together, viscosity and turbulence contribute to *fluid friction*.

Definition: *Fluid friction* is a force opposing the motion of a fluid, arising from the fluid's viscosity and/or turbulence.

fluid friction

Fluid friction leads to pressure drops in pipelines and has a detrimental effect on the flow rate. If we need a large flow rate in a pipe, there are several things we might do:

1. Increase the source pressure. This, unfortunately, tends to be expensive: heavier-gauge pipe may have to be used, for instance. In the case of home plumbing, it may not even be possible to change the pressure.
2. Use larger-diameter pipe. This decreases the amount of fluid in contact with the pipe walls and gives a larger central core of moving fluid.
3. Shorten the pipe, if possible. Eliminate all unnecessary bends.
4. If the fluid is a liquid, increase its temperature. This, as we noted earlier, decreases the viscosity.
5. Add viscosity-reducing chemicals to the fluid. (Some fire departments are now doing this.)

13-6 Bernoulli's Principle

We have just seen that fluid friction can cause a fluid's pressure to decrease at points downstream from the source. But there are also some other factors that affect the fluid's pressure. If a fluid is flowing downhill, for instance, the decrease in elevation may compensate for the effect of fluid friction, and the pressure may actually go *up*. This, in fact, is what happens in city water systems that are gravity-fed.

Furthermore, we have seen that a fluid's velocity may be increased by restricting its cross section. An increased velocity means an increased kinetic energy, and energy doesn't come from nowhere. If a fluid's kinetic energy is increased, this must come at the expense of something else. This something else is the fluid's pressure.

In Fig. 13-10, we see a fountain connected to a water tank. At the top of the tank, the gauge pressure is zero. Just below the fountain nozzle, the pressure is due to the depth of the water at that point. This depth is called the gravity head.

Definition: The *gravity head* is the difference in vertical elevation between two points in a fluid.

gravity head

Just outside the nozzles, the water is moving very rapidly. We can describe this motion by talking about the fluid's velocity (or inertia) head. This quantity is about the same as the height of the fountain.

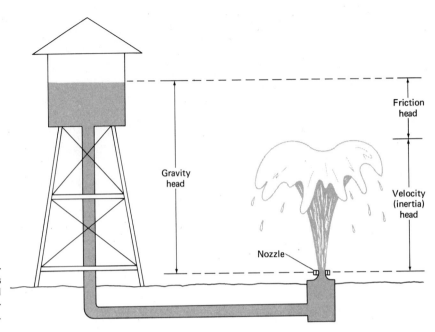

Fig. 13-10 A fountain. The gravity head inside the nozzle is transformed to a velocity head outside the nozzle. The difference is the friction head.

velocity (or inertia) head

Definition: The *velocity* (or *inertia*) *head* of a fluid is the decrease in elevation that would be needed to develop the fluid's velocity if there were no fluid friction.

Figure 13-11 shows how a fluid's velocity head increases with its fluid velocity. Notice that doubling the fluid velocity increases the velocity head by a factor of about 4.

Going back to Fig. 13-10, we see that the velocity head just outside the nozzle is not quite as big as the gravity head inside the nozzle. The difference is the friction head between the tank and the nozzle.

friction head

Definition: The *friction head* between two points of a moving stream is the amount of gravity or pressure head needed just to overcome the fluid friction between those points.

Although the water in the tank is at atmospheric pressure (Fig. 13-10 again), it doesn't have to be. If the tank is pressurized, the fountain will shoot higher. This additional height would be the result of the pressure head.

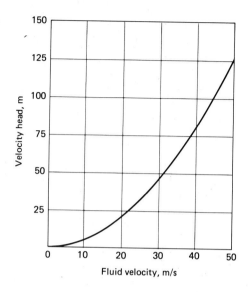

Fig. 13-11 Fluid velocity head versus fluid velocity. The same graph applies to all fluids, both liquids and gases. Note: 1 m/s = 3.6 kmph = 3.28 ft/s = 2.24 mph.

Definition: The *pressure head* in a confined fluid is the height to which the fluid would rise under the applied pressure, if allowed to.

pressure head

For instance, if the applied pressure is 1.0 atmosphere (gauge) and the fluid is water, the pressure head is 10 m.

The relationship between a fluid's pressure head, gravity head, velocity head, and friction head was first studied by an eighteenth-century Swiss scientist named Daniel Bernoulli. The principle he discovered may be stated like this:

Bernoulli's principle: At any point downstream from a fluid stream's source, the total head is the same as the total head at the source.

Bernoulli's principle

In other words, a fluid's velocity head can increase only if its pressure head or gravity head decreases and the friction head is too small to make up the difference. If you pour water from a jar, the drop in elevation (gravity head) causes the velocity head to increase. But if you pour cool molasses or honey from the jar, the decrease in gravity head goes into the friction head, and there will be little or no increase in the velocity head. Thus the stream of water speeds up as it flows downhill, but the stream of cool molasses or honey flows at a steady, sluggish rate.

But the most important applications of Bernoulli's principle involve cases where there is little or no change in elevation. In Fig. 13-12 we see a horizontal pipe whose diameter changes along its length.

Because of the continuity principle, the fluid speeds up as it passes through the narrow section. The fluid's increased velocity head here is compensated for by a drop in pressure head. As the fluid moves into the large section on the right, it slows down to its original velocity. The pressure then goes back up, but not quite as high as its original value (as a result of the friction head and the effect shown in Fig. 13-10). The point is this: *The faster a fluid moves, the lower its pressure becomes.*

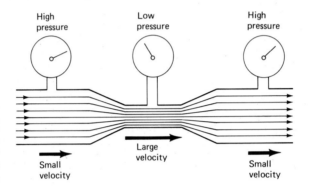

Fig. 13-12 Bernoulli's principle applied to a fluid moving in a horizontal pipe. When the fluid velocity increases, the pressure drops. When the fluid velocity decreases, the pressure goes up.

There are many ways to demonstrate this principle. Some simple ones are shown in Fig. 13-13. Hold a piece of paper in front of your mouth and blow hard over the upper surface. The paper rises because the fast-moving air on top is at a lower pressure than the stationary air underneath. Attach a string to a Ping-Pong ball and dangle it into a stream of water. The ball moves into the stream and stays there even if tugged sideways. Pop a Ping-Pong ball into a funnel as you blow through the spout. The ball stays up in the funnel as long as the air is moving around it. Turn a shower on full. The shower curtain moves into the stream of water.

Ships have been known to collide because of Bernoulli's principle. Figure 13-14 shows two deep-hulled ships that are traveling close to each other in the same direction. Water in the channel between them speeds up because of the continuity principle. This lowers the water pressure on the hull sides in the channel, and the higher water pressure on the out-facing hull sides forces the ships together. This effect must constantly be kept in mind by harbor and canal pilots. It is also good to keep in mind when canoeing fast-moving streams with many rocks.

The air-driven paint sprayer also makes use of Bernoulli's principle (Fig. 13-15). A source of compressed air is connected to the spray gun; its flow is controlled through a needle valve operated by a pistol grip. The flow of this air creates a low-pressure region in the mixing chamber. Atmospheric pressure forces the paint up the tube into this

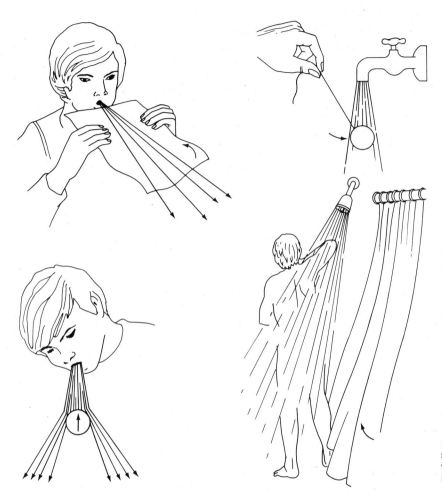

Fig. 13-13 Demonstrations of Bernoulli's principle. A piece of paper, a Ping-Pong ball, or a shower curtain will move into the low-pressure region created by a rapidly moving fluid.

low-pressure region, where it mixes with the air and is forced out the nozzle. This same basic arrangement is also used in the jet pump, the spray "atomizer," the aircraft airspeed indicator, and other applications.

Exercises

12. What is the velocity head of a stream of water with a fluid velocity of (a) 20 m/s, (b) 4$\overline{0}$ m/s?
     Answers: (a) 20 m, (b) 82 m
13. During a storm, a wind speed of 72 kmph [4$\overline{0}$ mph] is measured. (a) What is the wind's velocity head? (b) What is the velocity head if the wind speed is cut in half?
     Answers: (a) 20 m, (b) 5 m

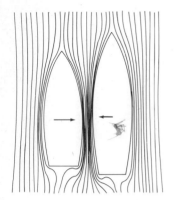

Fig. 13-14 Ships traveling too close together may be drawn into a collision because of the lowered fluid pressure between them.

14. In Fig. 13-12, the stream entering on the left has a pressure head of 115 m and a velocity head of 8.0 m. The actual fluid velocity of this stream (from Fig. 13-11) is 12 m/s. In the center section, the stream is constricted to one-fourth its original cross-sectional area. (a) What is the fluid velocity in the center "neck"? (b) What is the velocity head in the neck? (c) What is the pressure head in the neck?

Answers: (a) 48 m/s, (b) 120 m, (c) about 3 m

15. A crack develops on a seam near the bottom of a large gasoline storage tank. The level in the tank is 5.0 m above the leak. The friction head is very low. (a) What is the gravity head at the inside of the leak? (b) What is the gravity head at the outside of the leak? (c) What is the velocity head of the stream coming from the hole? (d) What is the fluid velocity of the stream?

Answers: (a) 5.0 m, (b) 0.0 m, (c) 5.0 m, (d) 10 m/s

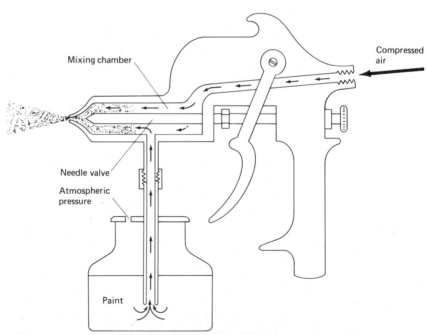

Fig. 13-15 A paint sprayer. The flow of compressed air creates a low-pressure region in the mixing chamber. Atmospheric pressure forces the paint into this low-pressure region, where it mixes with the air.

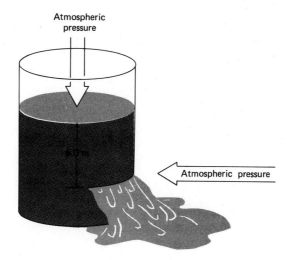

Except for space exploration, everything we do takes place in a body of air or water or both. This means that everything that moves must push a fluid out of the way in front of it. The action of the fluid in opposing the motion is called drag.

13-7 Motion through a Fluid

> Definition: *Fluid drag* is the force that a fluid exerts on an object moving through the fluid. This force always acts to retard any relative motion.

fluid drag

Notice that there is no such thing as fluid drag if there is no motion. But for moving objects, the drag increases rapidly as the object's speed through the fluid increases. This places an upper limit on the speed of a vehicle. As a car accelerates, the air drag it must overcome increases. Eventually, the car reaches a speed where most of the power output of its engine goes into pushing the air out of the way in front of it. At this point, the car can accelerate no further. It is at its top speed. The same thing happens with boats and aircraft. In fact, any moving object's top speed is determined by the point where fluid drag and other frictional forces balance the force propelling it.

One important factor governing fluid drag is the object's shape. Vehicles with smooth, streamlined shapes experience less drag than those that are box-shaped. Figure 13-16 on page 364 shows why. Because of its boxy shape, the trailer truck beats the air into eddies as it passes through. There are a lot of eddies behind the back of the trailer as the air abruptly flows in to fill the void left by the passing truck. These eddies act to hold back the trailer. With the sports car, on the other hand, the

Application: The Carburetor

The conventional internal-combustion engine uses the expansion of hot gases to drive its pistons. These hot gases come from burning a mixture of air and gasoline vapor in the engine's cylinders. The function of the carburetor is to provide this air-fuel mixture.

The diagram shows the carburetor's basic features. Its base is open to the engine's intake manifold, which is a piece of plumbing connected to each cylinder's intake valve port. The pumping action of the engine's pistons during their intake strokes keeps this region at a partial vacuum. Atmospheric pressure outside the air horn then forces air through the carburetor and toward this low-pressure region. The flow rate of this airstream is controlled by the throttle valve.

The air passage, or venturi, is purposely shaped like an hourglass, causing the airstream to speed up as it flows through. This increase in speed is accompanied by a decrease in pressure at the neck of the venturi (Bernoulli's principle). Now atmospheric pressure forces gasoline from the float chamber through a nozzle and into the airstream. The low pressure in this stream also causes the gasoline to evaporate very quickly, so it is fully vaporized before it enters the engine's cylinders.

The best fuel economy results from a mixture of 16 to 17 parts air (by weight) to 1 part gasoline. For cold starts, idling, and maximum acceleration, a somewhat richer mixture is needed: 1 part gasoline to 12 or 13 parts air is typical. Matters are complicated further because a rapid opening of the throttle valve tends to make the mixture very lean, which leads to a "flat spot" in the vehicle's acceleration. (This happens because the gasoline has a greater fluid inertia than the air, so its flow temporarily lags behind the increased airflow when the throttle is suddenly opened.) To enable the carburetor to supply the proper mixture over such a wide range of conditions, additional air and fuel passages are required. Although these passages are arranged differently in various models of carburetor, they all operate on these same basic principles: (1) fluids flow from regions of high pressure to regions of low pressure; (2) the fluid velocity increases when the flow is channeled into a smaller cross section; and (3) the fluid pressure drops when the fluid velocity increases.

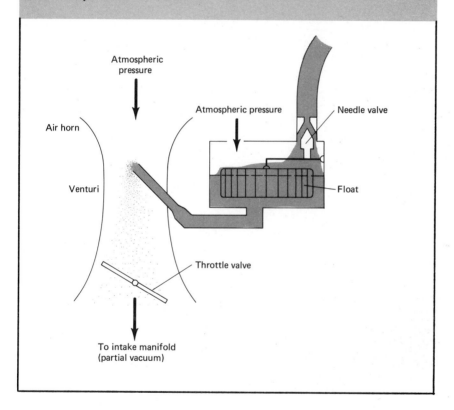

because streamlining lowers the demand on the engine, which decreases fuel consumption. But on racing cars, streamlining may lead to problems. In Fig. 13-16, the bottom of the sports car is relatively flat, while its top is curved. Above the car, then, we have a region where the air is being channeled into a smaller cross section. Because of the continuity principle, this channeled air speeds up, and its pressure

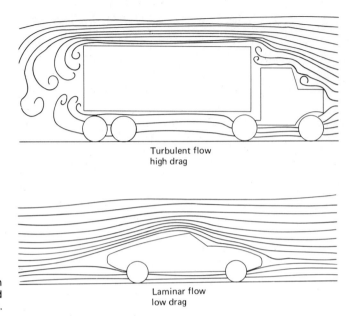

Turbulent flow
high drag

Laminar flow
low drag

Fig. 13-16 Fluid drag is high when the flow is turbulent and low when the flow is laminar.

drops. The air below the car then acts to lift the car into this low-pressure region, reducing tire traction. This is an extremely unsafe situation, since the only way the car is controlled is through the four little contact areas between the tires and the road. To combat this problem, designers of racing cars often put "spoilers" on their vehicles (Fig. 13-17). These break up the laminar flow above the car and reduce the lifting effect. Of course, they also increase the drag and reduce the top speed.

Fig. 13-17 A car with a "spoiler." This "spoils" the laminar flow and reduces the tendency of the car to lift off the road at high speeds.

This same principle holds for boats and aircraft. The fluid drag is low when the flow is laminar and high when the flow is turbulent. Because aircraft travel at such high speeds, it is particularly important that they be streamlined. But regardless of the shape of the vehicle, the flow eventually becomes turbulent if the speed gets high enough.

13-8 Fluid Inertia

We have already seen that a massive object like a boxcar or a ship is hard to get moving. Once the object is moving, great force is needed to stop it or change its direction. Because fluids also have mass, they exhibit this same property of *inertia*. This is why it is practically impossible to dam a stream while it is moving, or to carry a sheet of plywood in a windstorm.

The most important applications of fluid inertia have to do with changing the direction of flow. When a fluid's flow is diverted, the fluid exerts a force on the thing diverting it. The direction of this force is away from the center of curvature of the path which the fluid takes. The effect is exploited in many devices; the Pelton wheel shown in Fig. 13-18 is a

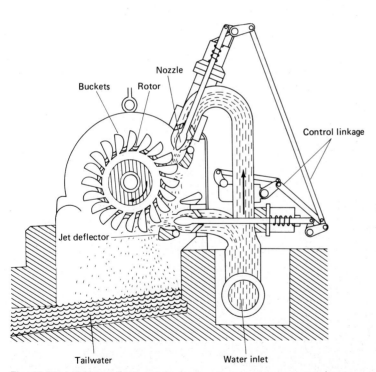

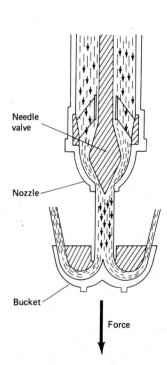

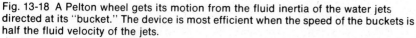
Fig. 13-18 A Pelton wheel gets its motion from the fluid inertia of the water jets directed at its "bucket." The device is most efficient when the speed of the buckets is half the fluid velocity of the jets.

typical example. The wheel is rimmed with a large number of curved buckets. Water under high pressure (usually from behind a dam) is channeled into high-speed streams directed at these buckets. As these streams are deflected, they force the wheel to rotate in the direction shown. Pelton wheels are commonly used to drive the dynamos at hydroelectric power plants.

This same principle allows a sailboat to sail into the wind (an accomplishment that is very mysterious to many people, even after they have seen it happen). Figure 13-19 shows how this happens. In the boat on the left, the wind strikes the single sail at a glancing angle. As it is deflected, this wind stream exerts a sideways force on the sail. The boat would move in this direction if it could; however, the boat's keel allows it to move only forward or backward. Since the force on the sail is more forward than it is backward, the boat sails upwind and heels over as it does. (Of course, it can't sail *directly* into the wind.) In the boat on the right, a front sail (jib) has been added. Besides increasing the total sail area, the jib provides a channel that increases the wind speed over the forward side of the mainsail. This reduces the pressure on the upwind side of that sail and increases the forward force on the boat.

In Chap. 12 we discussed the principles of flotation. In particular, we talked about *displacement-type hulls,* that is, hulls that float at a level

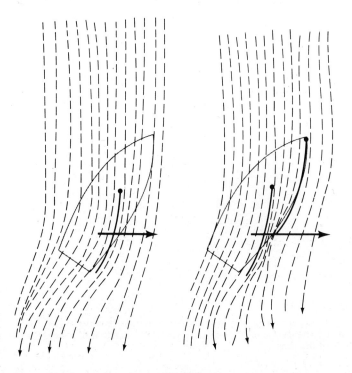

Fig. 13-19 By deflecting the wind with its sail, a boat can sail upwind. The addition of a front sail (jib) improves the boat's performance by reducing the pressure on the forward side of the mainsail. Forces shown are the effect of the wind on the sails. The boat also experiences a force to the left due to the action of water on the keel and rudder, as well as a rearward force of fluid drag acting on the hull.

where their weight equals the weight of the water they displace. Of course, every boat must do this when it is not moving. Power boats, however, are often designed to rise out of the water at high speeds and plane over its surface. We say that such boats have a *planing hull*. Aircraft may also be divided into these two categories: displacement type and planing type. The distinction is shown in Fig. 13-20. A hot-air balloon and an ore boat derive their flotation from the displacement principle. An airplane and a speedboat get at least some of their lift by planing.

The planing principle is based on fluid inertia. If a boat's hull is shaped to divert the water below it rather than around it, the boat rides up on its own bow wave and its hull climbs out of the water. This happens only because it is harder to move the water down than it is to move the hull up, in other words, because the water has more *inertia* than the hull. For the airplane, which is much heavier than the air around it, this effect by itself is not enough to sustain flight under normal conditions. It does, however, help quite a bit. And it has contributed to many student pilots' overshooting a landing as the airplane planes along on the layer of air trapped between the wing and the ground (a phenomenon sometimes called the *ground effect*).

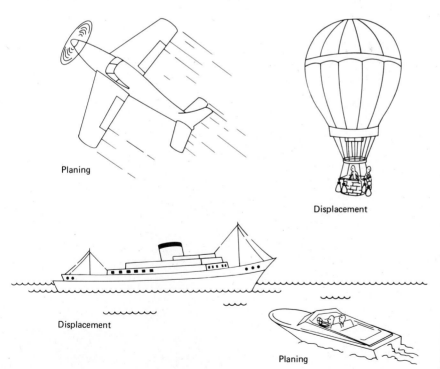

Planing

Displacement

Displacement

Planing

Fig. 13-20 Displacement-type and planing-type vessels in air and water.

The actual force developed by a fluid stream depends on the velocity head, the fluid's weight density, the frontal area of the object which the fluid strikes, and the angle at which it strikes. A very approximate formula for *fluid inertial force* may be written as follows:

$$\text{Fluid inertial force} = (\text{fluid weight density})(\text{velocity head})(\text{frontal area}) \qquad (13\text{-}4)$$

This formula does not consider turbulence or the fluid's compressibility, and it assumes that the fluid strikes a surface at a right angle. As a result, it gives values somewhat on the high side. Even so, it is sometimes used to make estimates when direct measurements are impossible or inconvenient.

Example 13-4 A Sluice Gate.

The gate is shown in Fig. 13-21. It allows water to enter a sluice (or open channel) possibly leading to a water turbine. The gate has an area of 4.5 m², and the stream above the dam has a velocity head of 0.35 m. What is the fluid force on the gate when it is shut?

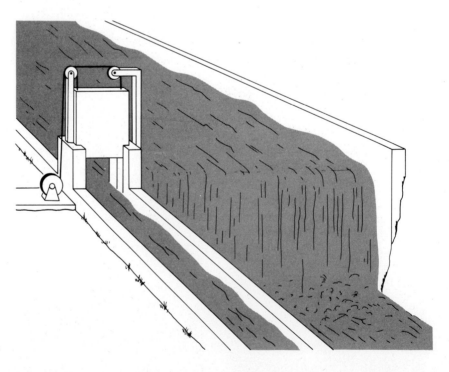

Fig. 13-21 A sluice gate.

Application: Water Hammer and Air Chamber

We have all heard the knocking sound that comes from suddenly shutting a faucet. It is particularly noticeable when an automatic washer solenoid valve closes. This effect is called "water hammer," and it places a great strain on pipe joints and valves. When a solenoid valve in an 18-in pipe at a power station accidentally tripped shut at full flow, some of the pipes jumped nearly 1 m, and welds were broken throughout the system.

  Water hammer comes about because of the fluid inertia of the moving water. If we suddenly shut a valve in a moving stream (where the pressure is low to begin with), the pressure has to jump as the fluid is brought to an abrupt halt. This pressure jump acts as a hammer against the inside walls of the pipe.

  An easy (and cheap) way to correct water hammer is to install air chambers. These are simply dead-end sections of pipe running vertically in the wall behind each faucet or valve. Because they run vertically, they are filled with trapped air that cannot escape. And since air is highly compressible, any sudden pressure changes are absorbed. The hammering now takes place against the air column rather than against the pipe walls.

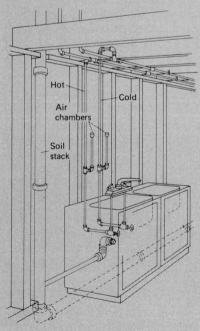

Application: The Airplane Wing

Although an airplane is designed to move through the air, its performance can be analyzed as if it were standing still and the air were doing all the moving. This, in fact, is why engineers can use a wind tunnel for design experiments. They direct an airstream from a huge fan over a design model, measure how the model responds, and then assume that the airplane would behave the same way if it were actually flying.

An airplane wing gets its lift from two sources: Bernoulli's principle and the fluid inertia of the air. In the top diagram, the wing is level. The curve of the wing's upper surface is designed to force the upper airstream into a smaller channel. This makes the air on top speed up, and so its pressure drops. The higher pressure on the underside of the wing then lifts the wing into this low-pressure region.

In the second diagram, the wing is angled slightly upward. This angle from the direction of the oncoming airstream is called the "angle of attack." Now part of the airstream is deflected off the underside of the wing, and this air's fluid inertia contributes to the lift.

The wing's total lift increases with its speed and its angle of attack—up to a point. If the angle of attack is made too large (typically around 15°), the flow becomes turbulent, as shown in the third diagram. Now the lift decreases and the drag increases tremendously. We say that the wing has *stalled*. To land an airplane, the pilot must intentionally stall it just a few feet above the runway.

The conventional screw propeller may be thought of as a rotating wing. Its cross section looks like the wing sections in the diagram, and it gets its forward thrust in the same way that the wing gets it lift. It may seem that the thrust generated by a propeller should increase with the frequency of rotation. Again, this is true only up to a certain point. If a propeller has a 1-m radius and rotates at 2500 rpm, its tip is traveling at about 70 percent of the speed of sound. At this speed, the compressibility of air begins to be important. Increasing the rate of rotation further does nothing but compress the air at the leading edge of the propeller and increase the drag. This requires additional power, which does not go into the forward motion of the aircraft.

For this reason, airplane engines are designed to operate at relatively low frequencies—typically 2500 rpm is tops. To further increase the plane's thrust without going to multiple engines, the plane may be equipped with a *variable-pitch propeller*. This allows the angle of attack of the propeller blade to be controlled by the pilot.

The wing area has a great deal to do with a plane's performance. Because a biplane (a two-winger) has so much wing area, it can climb and maneuver very quickly. It is also very slow, because of the fluid friction (drag) on all this wing surface. These features combine to make the biplane good for crop dusting and aerobatics, but poor for transportation.

We said that the pressure on the underside of a wing is high while the pressure on top is low, and this pressure difference supports the moving plane. Because fluids flow from regions of high pressure to regions of low pressure, there is a rapid flow of air around the wing tips, as shown, The result is called *wing-tip vortex*. The vortex (plural: vortices) generated by a large plane trails some distance behind it, creating a hazard for light aircraft. This is one reason that light planes are discouraged from landing at major commercial airports.

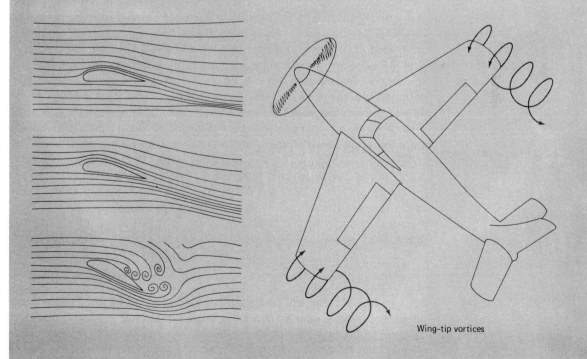

Wing-tip vortices

From Table 6-2 we see that the weight density of water is 0.998 kg_f/L, or 998 kg_f/m³. Then using Eq. (13-4), we have

Fluid inertial force = (fluid weight density)
(velocity head)(frontal area)

$$= \left(998 \ \frac{kg_f}{m^3}\right) (0.35 \ m)(4.5 \ m^2)$$

$$= 1600 \ kg_f$$

Although this result is, as we mentioned, a bit on the high side, we still are going to have a large force to reckon with. The gate here is really not that large, as sluice gates go. The force comes from the fluid inertia of the stream, which is flowing at about 2.6 m/s [9.4 kmph, or 5.9 mph]. ◄

13-9 Fluid Power

There are three basic ways of using fluids to produce mechanical power:

1. Use the fluid's pressure head to drive a piston, vane, or rotor.
2. Use the fluid's gravity head to produce an imbalance that rotates a wheel.
3. Use the fluid's velocity head (inertia head) to drive a propeller or rotor.

The first method is used in the air hammer (described in Chap. 12), in the reciprocating steam engine, and in various hydraulic motors. One such motor is shown in Fig. 13-22; here gear teeth act as the vanes.

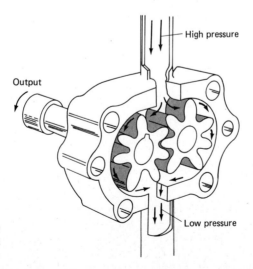

Fig. 13-22 A hydraulic gear motor. This device exploits the fluid's pressure head.

Application: The Hydraulic Ram

The hydraulic ram is a simple, automatic device that allows water to pump itself uphill. This is possible because a much larger amount of water flows to a lower level. Since the ram requires a gravity head to operate, it cannot pump water from a source lower than itself (it cannot be used with wells, for instance).

The ram's operation can be seen from the diagram. Water flowing down the inlet builds up speed until its inertia head slams the clack valve shut. Since this moving water cannot be stopped instantaneously, part of it flows through the check valve and up the delivery pipe. When the flow has slowed sufficiently, the check valve shuts and the clack valve reopens. The process then repeats, much like a jackhammer. Water flows into the inlet pipe in a series of bursts, but because of the air chamber the flow in the delivery pipe is fairly smooth.

It may still be surprising that water in the delivery pipe can be forced to a higher level than the original gravity head. But remember that the inlet flow is not continuous. It's a great deal like dropping a heavy stone into a pond and creating a splash that flies higher than the distance from which the stone was dropped. The ram uses a large amount of falling water to splash a small part of itself to a greater height.

With a gravity head of 2 m, a typical hydraulic ram requires 100 L/min of inlet flow to deliver 3 L/min to a height of 40 m. Most hydraulic rams operate at 25 to 100 cycles per minute (c/min). The efficiency may be as high as 65 percent.

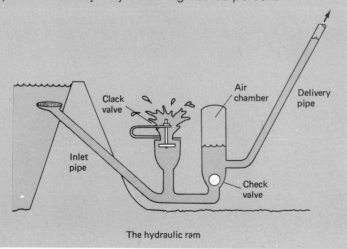

The hydraulic ram

The fluid pressure is typically around 135 kg$_f$/cm² [2000 psig]. This high pressure acting on the face of the gear teeth drives them in the direction shown. A motor of this design with a 19-kW [25-hp] output may require a fluid flow rate of 3.6 L/s [57 gpm] at a pressure of 135 kg$_f$/cm² [2000 psig]. The device is the size of a large electric motor.

The second method (using the fluid's gravity head) is seen in the overshot waterwheel in Fig. 13-23. Water pours over the wheel from above, throwing it off balance and causing it to turn. To get the same 19-kW output, the wheel must be around 9 m [30 ft] in diameter, with a flow rate of at least 630 L/s [10 000 gpm]. We don't see too many of these devices in use any more, mainly because they need to be so big. There is an advantage, however, in that the fluid need not be confined.

The third method (using the velocity head) was already seen in the Pelton wheel (Fig. 13-18). Although the gravity head also makes a small contribution here, mainly the velocity or inertia head drives the wheel. When the velocity head is relatively small but the flow rate is large, propeller turbines can be used effectively to produce mechanical power. The principle is that of a ship's propeller operating in reverse. Most windmills are propeller turbines.

Summary   The motion of a fluid can be described by either its fluid velocity or its flow rate. The flow rate is useful when we are filling or emptying reservoirs, transferring fluids through pipelines, circulating the fluid as a coolant, putting out fires, or in any other situation where the volume of the moving fluid is of concern. The fluid velocity is a useful quantity to know when navigating rivers, flying an aircraft, operating a wind machine, or in

Fig. 13-23 Overshot waterwheel. This device exploits the fluid's gravity head.

other cases where the fluid affects the motion of something else. The flow rate may be found from the fluid velocity, or vice versa, if the cross-sectional area of the flow is known.

When a fluid flows through a pipe or conduit whose cross section changes along its length, the fluid velocity increases at points where the flow is constricted. But the flow rate stays the same as long as the pipe doesn't branch. This is because the volume flowing in one end must equal the volume flowing out of the other end for an incompressible fluid. The effect is called the continuity principle.

When a fluid's velocity increases, either the pressure or the gravity head (or both) must decrease. This effect is described by Bernoulli's principle. In many applications a fluid is intentionally accelerated to lower its pressure.

Terms You Should Know

fluid velocity	friction head
anemometer	pressure head
fluid flow rate	fluid drag
continuity principle	turbulent flow
fluid friction	laminar flow
Bernoulli's principle	displacement-type hull
gravity head	planing hull
velocity (or inertia) head	fluid inertial force

Problems

1. In a certain oxyacetylene welding application, oxygen flows from the torch tip at a flow rate of 2.6 L/min. A full oxygen cylinder supplies 6200 L [220 ft³] of the gas at atmospheric pressure. Approximately how long will the tank last if used continuously?

2. A certain 3/4-in pipe supplies water to four 1/2-in pipes. If the flow rate in each of the smaller pipes is 10 gpm, what is the flow rate in the larger pipe that supplies them?

3. A certain 2-in-diameter pipe branches into four 1-in pipes. If the fluid velocity in each of the 1-in pipes is 12 ft/s, what is the fluid velocity in the 2-in pipe that supplies them?

4. A certain 2-in-diameter pipe narrows to a 1-in diameter. (a) If the fluid velocity in the 1-in section is 8.0 ft/s, what is the fluid velocity in the 2-in section? (b) If the flow rate in the 1-in section is 20 gpm, what is the flow rate in the 2-in section?

5. Gasoline is pumped at a flow rate of 0.50 L/s through a hose with an inside diameter of 2.0 cm. What is the fluid velocity of the gasoline?

6. Steam flows through an 18-cm-diameter pipe at a flow rate of 1.6 m³/s. Calculate the fluid velocity of the steam in (a) meters per second, (b) kilometers per hour, (c) miles per hour.

7. It takes 11 s to fill a 1-gal container with

water from a certain spigot. The supply line has a 1/2-in inside diameter. (*a*) What is the flow rate? (*b*) What is the fluid velocity in the supply line?

8. A cylindrical tank is 7.32 m high and has the same diameter. The tank is filled with fuel oil. What average flow rate would be needed to empty the tank in 1.0 week?

9. Water flows over the spillway of a dam at a fluid velocity of 2.5 m/s. The spillway is 12 m wide, and the stream is 42 cm deep. Calculate the flow rate in (*a*) cubic meters per second, (*b*) liters per second, (*c*) gallons per minute.

10. A fountain is to shoot 25 m into the air. (*a*) What is the minimum pressure head needed to do this? (*b*) What is the fluid velocity as the stream leaves the fountain's nozzle? (*c*) What is the approximate fluid pressure at the nozzle?

11. A certain storage tank can be filled through one of two pipes. One pipe alone can fill the tank in 3.0 h. The other pipe alone can fill it in 2.0 h. How long would it take to fill the tank using both pipes at the same time?

14
Waves

You have seen the waves that travel on bodies of water. Other types of waves cannot be seen directly, but they are waves nonetheless. In this chapter, we talk about sound waves and electromagnetic waves, as well as water waves.

Although these three kinds of waves are created in different ways, they have many things in common. For instance, they all carry energy at a predictable speed; they have a *frequency* and a *wavelength*; they may be absorbed, reflected, refracted, or diffracted; they can travel through one another without losing their identities; and their frequencies can be shifted by the motion of the source or receiver.

The study of waves can get very complicated indeed. In this short chapter, we limit ourselves to some of the basic features of waves and some of their more common applications.

14-1 Water Waves

When a gust of wind blows over the surface of a lake or ocean, the pressure reduction (Bernoulli's principle) causes portions of the water surface to rise. When the wind passes or stops, the region of higher water flows back into the low water around it, causing a disturbance much as if a very large rock or log had been dropped at that point. The result is a *wave crest* that travels away from the point of disturbance.

Water waves usually occur in clusters rather than singly. This can be seen very easily by dropping a pebble into a pond. An entire series of wave crests moves out from the source of the splash. These crests are fairly equally spaced, as with the wave in Fig. 14-1. The distance between two crests is called the *wavelength*.

wavelength   Definition: The *wavelength* of a wave is the distance between two successive crests or two successive troughs of the wave.

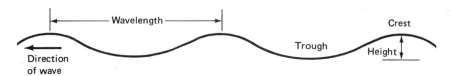

Fig. 14-1 Waves usually occur in clusters. The distance between two successive crests (or two successive troughs) is called the wavelength.

A wave is somewhat complicated to describe, since several things are happening at once. In Fig 14-2 we see a water wave traveling to the left. By this, we mean that the pattern of crests and troughs moves to the left, and energy is carried in this direction.

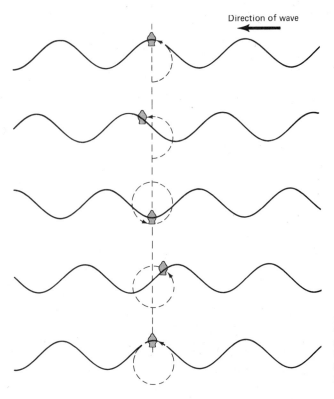

Fig. 14-2 A floating object moves in a circular path as a wave passes. The water itself also moves in circles. Although the wave carries energy in its direction of travel, there is no net fluid flow in this direction.

Now although the wave is moving to the left, the water is doing quite another thing. We can see this by putting a floating object on the water's surface. Figure 14-2 shows a floating cork, but a boat or person floating in the ocean would work just as well. As a wave crest approaches, the cork is lifted up and slightly forward. As the crest passes under it, the cork travels backward into the *wave trough*. The cork is then lifted by the next oncoming crest, and traces out a basically circular path. The water itself is doing the same thing. What we call the *wave*, then, is a pattern traveling on the water's surface, but not the water itself.

This, at least, is the case when waves are moving in deep water and the wave height is low. If a wind continues to blow against the wave, some of the wind's velocity head is changed to a gravity head in the wave, and the wave grows in height. If the wave gains so much energy that its height grows to more than one-seventh of its wavelength, the wave will "break." When this happens, the water at the crest is freed from its circular motion and is thrown violently in the direction of the wave. The wave height is then reduced to less than one-seventh of its wavelength, and the wave continues to progress as before. Of course, if a wave enters shallow water, it also "breaks" and throws a wall of water onto the beach. "Breakers" carry tremendous energy, and large ones have been known to destroy wharves and ships.

14-2 Wave Frequency

We have already used the word "frequency" to describe the rate of rotation of gears, pulleys, motors, and so on. Since the cork in Fig. 14-2 is moving in a circle, it also has a frequency of rotation. This frequency is the same as the frequency of the wave.

wave
frequency

Definition: *Wave frequency* is the same as the frequency of an object that is set into motion by the wave.

Now an object doesn't actually have to move in a circle to have frequency. As long as it traces the same path over and over at a well-defined rate, we can talk about its frequency. But instead of revolutions, we may have *cycles*. The frequency is then the number of cycles per minute (c/min) or cycles per second (c/s).

In the SI, the unit of frequency is the hertz (Hz).

hertz

Definition: The *hertz* (Hz) is the SI unit of frequency. It is equivalent to 1 revolution per second (r/s), or 1 c/s.

In the usual way, we can use the SI prefixes to get multiples of this unit: 1 kHz = 1000 c/s, and 1 MHz = 1 million c/s, for instance. Although water waves ordinarily have frequencies of only a fraction of 1 Hz, other types of waves often have frequencies in the kilohertz or megahertz ranges.

14-3 Wave Speed

Ocean waves created by a storm may cause damage hundreds of kilometers away. The waves carry some of the storm's energy across the surface of the water. The speed at which they do this is called the wave speed.

wave
speed

Definition: The *wave speed* is the rate at which energy is carried forward by a wave.

If you watch ocean waves or waves on a large lake rolling in toward shore, the speed of the crest and trough pattern you see is the same as the speed of the energy. But in deep water, the wave pattern travels at about twice the speed of the energy. How is this possible without having the wave get ahead of the energy it carries? Remember that the wave consists of an entire group of crests and troughs. As the leading crest moves ahead of the rest, it dies out and a new crest grows at the tail end of the group. The next leading crest then dies out, and another crest grows behind the rest. What we call the wave speed, then, is the forward speed of the entire group of the wave crests. This is also sometimes called the *group velocity*. The apparent forward speed of the individual crests in the group is called the *phase velocity*.

We are not particularly concerned about this phenomenon other than to point out that it sometimes happens. In *shallow* water, the group velocity and phase velocity are the same (there is no dying out of waves as they approach shore). The two velocities are also the same with sound waves and waves in strings. For light waves, we will not worry about the difference.

Now the wave speed of a wave depends on the properties of the medium, or the material through which it travels. It may also depend on other factors like the temperature or the pressure. But one thing it does *not* depend on is the speed of the source. The light beam coming from a car's headlamp travels away at the same wave speed regardless of whether the car is moving. Table 14-1 lists the wave speeds of some different kinds of waves.

Example 14-1 Distance to a Bolt of Lightning.

You have probably noticed that the thunder follows a flash of lightning by several seconds. Actually, the light wave and sound wave are produced at exactly the same time (Fig. 14-3). We see the light almost immediately because its wave speed is so great, but it takes awhile for the sound wave to reach us.

Suppose that we count a 5.0-s interval between the lightning and the thunder. How far away did the lightning strike?

TABLE 14-1   WAVE SPEEDS OF CERTAIN WAVES IN DIFFERENT MATERIALS

Type of wave	Medium and conditions	Wave speed, m/s
Electromagnetic waves	Free space (vacuum)	$2.997\ 924\ 62 \times 10^8$
Sound	Air, 0°C	331
	Air, 20°C	343
	Hydrogen, 0°C	1286
	Water, 15°C	1450
	Lead, 20°C	1230
	Aluminum, 20°C	5100
	Iron, 20°C	5130
	Granite, 20°C	6000
	Rubber, 20°C	54
Water	Shallow water	
	1.0 m deep	3.1
	4.0 m deep	6.3
	Deep water	
	10-m wavelength	4.0
	40-m wavelength	7.9
	Tsunamis	up to 200

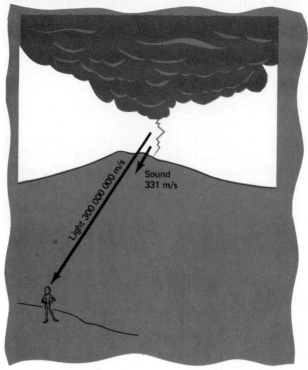

Fig. 14-3 Estimating the distance to a bolt of lightning.

In Table 14-1, we see that sound travels at 343 m/s at 20°C. The distance traveled in 5.0 s is then

$$\text{Distance} = (\text{speed})(\text{time})$$
$$= (343 \text{ m/s})(5.0 \text{ s})$$
$$= 1700 \text{ m}$$

We may also express the result as 1.7 km, or about 1 mi. Notice that each second corresponds to about 1/5 mi, or about ⅓ km. ◀

Exercises

1. Find the speed of jet plane in kilometers per hour if it is traveling at 1.2 times the speed of sound at 0°C.
    Answer: 1400 kmph
2. In a certain fireworks display, a flash of light is seen 2.0 s before the explosion is heard. The temperature is 20°C. How far away was the explosion?
    Answer: 690 m
3. How long does it take the sound in a large concert hall to travel from the stage to the last row of seats? The distance is 52 m.
    Answer: 0.15 s

4. A certain tsunami travels at an average wave speed of 150 m/s. How long does it take to travel halfway around the world, or about 20 000 km?
   Answer: 37 h
5. A floating buoy bobs up and down as a water wave passes. If it takes 2.0 s to make 1 c, what is the wave's frequency?
   Answer: 0.50 Hz
6. If the distance between a wave crest and the next trough is 5.5 m, what is the wavelength?
   Answer: 11 m

14-4 Relationship Between Frequency, Wavelength, and Wave Speed

We saw that the wave speed depends on properties of the medium. The wave's frequency, it turns out, depends on the *source* of the wave. If we strum a guitar string, the frequency of the sound wave depends on the tension, mass, and length of the string. But this frequency does *not* depend on the properties of the air the wave travels through. Similarly, the frequency of the radio wave from an antenna depends on the design of the transmitter. It does not depend on where the wave goes. If a radio station broadcasts at a frequency of 1.02 MHz, we can pick up the signal by tuning a receiver to this same frequency. The station may be in Pittsburgh, but the wave has the same frequency as it travels through St. Louis or Albuquerque.

We can calculate the wavelength if we know the frequency and the wave speed. The relationship is

$$\text{Wavelength in m} = \frac{\text{wave speed, in m/s}}{\text{frequency, in Hz}} \tag{14-1}$$

Example 14-2 Wavelength of an AM Radio Wave.

A certain AM radio station broadcasts at a frequency of 680 kHz. What is the wavelength?

From Table 14-1, we see that all electromagnetic waves travel at a wave speed of 3.00×10^8 m/s, or 300 000 000 m/s. Then using Eq. (14-1), we have

$$\text{Wavelength, in m} = \frac{\text{wave speed, in m/s}}{\text{frequency, in Hz}}$$

$$= \frac{3000\ 000\ 000 \text{ m/s}}{680\ 000 \text{ Hz}}$$

$$= 441 \text{ m}$$

Certainly this is a very long wavelength. Radio waves with higher frequencies (FM radio and television waves, for instance) have shorter wavelengths. Radar waves have such high frequencies that their wavelength is typically only a few centimeters. ◄

Of course, Eq. (14-1) may be rewritten in two other ways:

Wave speed, in m/s = (wavelength in m)(frequency, in Hz)　　(14-2)

$$\text{Frequency, in Hz} = \frac{\text{wave speed, in m/s}}{\text{wavelength, in m}} \qquad (14\text{-}3)$$

Exercises

7. A certain sound wave has a frequency of 1000 Hz. It travels from air at 20°C into water at 15°C. (a) What is its wavelength in the air? (b) What is its frequency in the water? (c) What is its wavelength in the water?
Answers: (a) 0.343 m, (b) 1000 Hz, (c) 1.45 m

8. What is the wavelength of (a) a 5.0-kHz sound wave in air at 0°C, (b) a 5.0-kHz electromagnetic wave?
Answers: (a) 6.6 cm, (b) $6\bar{0}$ km

9. A certain electromagnetic wave and a certain sound wave in 0°C air each have a wavelength of 1.00 m. (a) What is the frequency of the sound? (b) What is the frequency of the electromagnetic wave?
Answers: (a) 331 Hz, (b) $30\bar{0}$ MHz

10. Six wave crests strike a pier in $3\bar{0}$ s. The distance between crests is 24 m. (a) What is the wave's frequency? (b) What is the wavelength? (c) What is the wave speed?
Answers: (a) 0.20 Hz, (b) 24 m, (c) 4.8 m/s

14-5 Sound

Did you ever try to walk through a house without making a sound? People who try this late at night may be disappointed to find it is impossible. Just as you can't disturb the surface of a pond without creating a water wave, you can't disturb the air without creating a sound wave. Everything that moves creates sound.

Of course, there isn't any surface for the sound to travel on like our water waves do. Sound is a longitudinal wave rather than a transverse wave.

transverse
wave

Definition: A *transverse wave* is one in which the motion of the medium is transverse, or perpendicular, to the direction in which the wave travels.

Definition: A *longitudinal wave* is one in which the medium moves back and forth in the same direction the wave travels.

Figure 14-4 shows longitudinal and transverse waves in a long spring. One end of the spring is being set in motion by vibrating it at a steady rate. The longitudinal wave consists of regions of compression and tension that propagate to the right. Energy is also carried in this direction; but, of course, the spring itself doesn't move any great distance backward or forward.

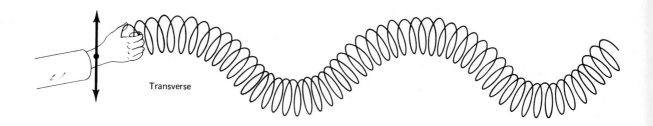

Transverse

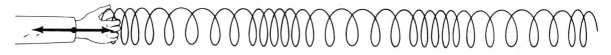

Longitudinal

Fig. 14-4 Transverse and longitudinal waves in a spring.

When sound travels through air, the same thing happens—a *sound wave* is created. Any disturbance causes some compression of the air near it, and this pressure disturbance propagates away at a rate that depends on the elasticity and density of the air. Figure 14-5 shows the pressure variation in a continuous sound wave coming from a speaker. The pressure pattern and graph are "frozen" at one instant in time. A fraction of a second later, the entire pattern is shifted to the right. The maximum gauge pressure in the wave is called its amplitude. Loud sounds have larger amplitudes than soft sounds.

Definition: The *amplitude* of a sound wave is the maximum gauge pressure in the wave. It is related to the loudness of the sound.

Application: Tsunamis—The Biggest Waves

Sometimes erroneously referred to as tidal waves, tsunamis are huge ocean waves that result from earthquakes or undersea landslides. They travel at fantastic speeds—often more than 650 kmph [400 mph]. In the open ocean their height diminishes to only 1 m or so, but as they enter shallow water they rise to 30 m [100 ft] or more. Because their wavelength is so long—usually several hundred kilometers—the volume of water in the wave crest is staggering.

The first sign of an approaching tsunami may be what appears to be an unusually low tide, uncovering many kilometers of previously unexposed ocean bottom and stranding fish and other sea creatures. This has the unfortunate effect of attracting throngs of the curious, who then find themselves unable to outrun the thundering breaker. With such a long wavelength, this breaker does not strike in a single wall, but pours into the shoreline continuously for several minutes.

Although several hundred great tsunamis have been recorded in history, very few eyewitnesses have survived to tell the tales. One group that did survive was the crew of the *U.S.S. Wateree*, a navy gunboat struck by a tsunami on August 13, 1868. Here is what happened.

Toward the end of the Civil War, the navy built several flat-bottomed, double-ended gunboats with the intention of invading the South via the Mississippi River. Since the war ended before they could be used for this purpose, the gunboats were sent on other missions. After a lengthy cruise, the *U.S.S. Wateree* found itself anchored in the harbor at Arica, Peru, while her boilers were being overhauled. A terrible earthquake struck and in a matter of seconds devastated what was then one of the richest cities in South America. The captains of other ships in the harbor recognized that a tsunami was likely, and they fired up their boilers to head out to sea (where the wave would not be as high). With his engines out of commission, the captain of the *U.S.S. Wateree* had no choice but to sit and watch helplessly. The ocean swelled, then quickly rolled back as far as the eye could see, as if someone had pulled a giant plug somewhere. Thousands of fish were left flipping around on the exposed bottom. The deep-hulled vessels fell onto their sides, some breaking up and some exploding. But the *U.S.S. Wateree*, because of her unusual flat-bottomed construction, landed upright and undamaged. For the next several hours the crew of the stranded vessel stood watch, scanning the horizon for signs of the approaching wave. The captain ordered everything movable tied down and the deck-level gunports opened to allow the water to run off in case they were swamped.

Night was approaching when the lookout shouted. A thin, phosphorescent line had appeared on the horizon. As the crew ran to watch, the line grew until it seemed to touch the clouds, roaring toward them with the thunder of a thousand breakers combined. It engulfed the ship, bore it up like a surfboard, then propelled it over the demolished city. By the time the surf ride ended, it had gotten too dark to see. Although there had been injuries, no one had been lost. The crew spent an uneasy night trying to sleep on the deck.

When morning broke, the amazed crew found nothing but sand around the ship. The wave had carried them some 5 km up the coast and 3 km inland. Again, the *U.S.S. Wateree* had landed upright and undamaged. Nearby were the remains of other ships with no survivors. One had its anchor chain wrapped around it as many times as it would go—showing that it had been rolled over and over by the wave.

There was no way to get the *U.S.S. Wateree* back to the water. The crew was rescued some three weeks later and the ship auctioned off to a hotel company. Later it became a hospital, then a warehouse, and finally an artillery target during the Chilean-Peruvian war.

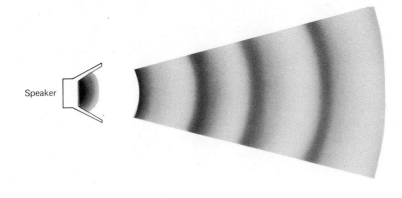

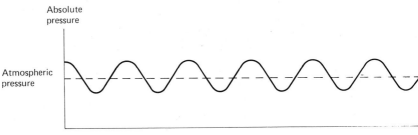

Fig. 14-5 Pressure variation in a
sound wave coming from a
loudspeaker.

The source of the sound may move in many ways. If there is a single violent motion, as in the explosion of a firecracker, the wave is a single pulse of high pressure, and we hear a "boom." A "rumble" is caused by a series of violent motions, as in a lightning discharge or a piece of sheet metal being rattled. A "hiss" is caused by continuous but random disturbances, such as water streaming from a hose, a worn-out phonograph record, or a fire. A musical tone is generated by a source that vibrates at a well-defined frequency: guitar and piano strings, reeds in wind instruments, and the metal bars in xylophones are examples of such sources. These four types of sound are shown in Fig. 14-6. The graphs show how the gauge pressure varies at the listener's ear.

A rumble, boom, hiss, gurgle, crackle, or pop has no single definite frequency and no definite wavelength either. A musical tone, on the other hand, has a fairly well-defined frequency and a fairly well-defined wavelength. The tone musicians call A_4 has a frequency of 440 Hz, and $F\#_5$ has a frequency of 739.99 Hz. The higher the frequency, the higher the *pitch*.

pitch   Definition: The *pitch* of a musical sound is the sensation of "high" or "low" associated with the tone. High-pitched sounds have higher frequencies than low-pitched sounds.

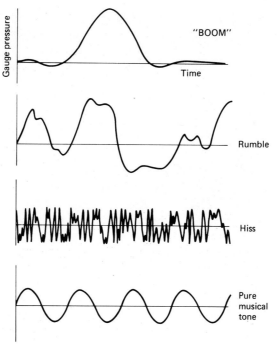

Fig. 14-6 Pressure variation in some different kinds of sound.

The human eardrum responds to sounds over a large range of frequencies. Although the actual range varies with the individual, we can say that in general the range of human hearing is from 20 Hz to 20 kHz. Frequencies above 20 kHz are called ultrasonic. Humans can't hear at ultrasonic frequencies, but some animals (dogs, for instance) can. "Silent" dog whistles have been designed on this principle.

> Definition: *Ultrasonic* waves are pressure waves with frequencies beyond the range of human hearing, or beyond about 20 kHz.

ultrasonic

When we discussed water waves, we said that the cork bobbed around at the same frequency as the wave traveling past it. The same principle holds for sound. A sound wave striking an object causes it to vibrate at the same frequency as the wave. This is how microphones and eardrums manage to sense sound. The principle is also used in ultrasonic cleaners (Fig. 14-7). Cleaning dirt and grime from small parts such as watch gears or diesel injectors can be difficult and tedious if done by hand. The ultrasonic cleaner uses ultrasonic waves to vibrate the dirt loose from the parts. In this case the wave passes through a cleaning solution surrounding the parts.

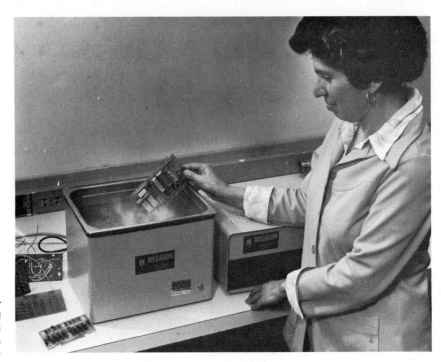

Fig. 14-7 An ultrasonic cleaner for removing dirt from small parts such as watch mechanisms, diesel injectors, carburetor parts, etc.

14-6 Electromagnetic Waves

Electromagnetic waves originate when an electric charge vibrates, rotates, starts moving, stops moving, or otherwise accelerates. If the charge's motion has a definite frequency, the wave has this same frequency. As we saw in Table 14-1, these waves travel at a very high speed: about 300 000 km/s [186 000 mi/s] in free space. They slow down a bit when traveling through air, glass, or other substances, but even here they are still traveling very fast.

But just what is an electromagnetic wave? Let's look at the electromagnet in Fig. 14-8. You may have seen such a magnet hanging from a crane and being used to lift scrap steel. When an electric current flows in the magnet's coils, the steel is attracted toward the magnet—even if they are separated a small distance. The magnetic field reaches across space to lift the steel. But when the magnet is first switched on, the steel scrap does not respond immediately. The very (*very*) short time lag is the time it takes for the magnetic field to travel across the space in between. This traveling magnetic field is one example of an electromagnetic wave.

Like water waves, electromagnetic waves are transverse. But since a magnetic field can reach through empty space, electromagnetic waves do not need a medium to carry them. As we know, they travel very well through even the vacuum of outer space.

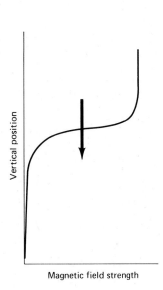

Fig. 14-8 An electromagnet. At the instant the current is switched on, an electromagnetic wave travels out from the magnet. The scrap does not "feel" the magnet until this wave reaches it.

In most cases (unlike the magnet example), electromagnetic waves are produced by electric charges that vibrate at a fairly definite frequency. Figure 14-9 shows the tremendous range of possible frequencies. Different frequency ranges, or different frequency *bands*, have been given different names.

At the lowest frequencies are the *long electric waves*. These are produced by the rotating armatures in motors and generators. *Radio waves*, which have higher frequencies, are generated by electric vibrations in electronic circuits. Super-high-frequency radio waves, which have very short wavelengths, are called *microwaves*. Still higher frequencies may be produced by the vibrations of atoms and molecules in materials that have been heated. This band is called *infrared radiation*, and it is the same as radiant heat.

Next is a rather narrow band of frequencies called *visible light*. This band may be divided into the colors of the visible spectrum, with red light having the lowest visible frequency (longest wavelength) and violet light having the highest visible frequency (shortest wavelength). Light is generated when electrons in atoms jump from one orbit to a smaller orbit. It may also be generated by the vibrations of atoms and molecules at extremely high temperatures.

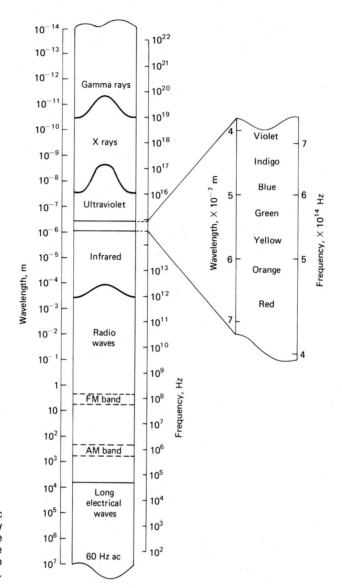

Fig. 14-9 The electromagnetic spectrum. Different frequency ranges, or frequency bands, are given different names. In some cases, there is a slight overlap between these bands.

At higher frequencies than violet light, we have *ultraviolet light*, or *ultraviolet radiation*. This is sometimes called "black light," since it cannot be seen directly. These waves are what cause sunburn. They can also cause serious damage to the eye by burning the retina. Welding produces large amounts of ultraviolet light because of the high temperatures involved, so the lenses in welding helmets are built to filter out

99.75 percent of these waves (Fig. 14-10). In doing so, they also filter out more than 99 percent of the visible light, and welders cannot see their work until they strike an arc.

Ultraviolet (UV) light also has its useful side. Certain minerals may be identified by the way they absorb UV and reemit visible light. The effect is known as *fluorescence*. Liquid dies made from such minerals may be used to detect surface flaws in machine parts and welds. The die is spread on the surface, the excess is wiped off, and then the part is held under a black light. Any die that has seeped into cracks or pits glows and reveals the irregularity.

Electromagnetic waves with frequencies higher than the ultraviolet band are called *X rays*. These very short waves originate when electrons in heavy atoms make high-energy orbit changes. They are very penetrating and can damage living tissue that has been exposed to too many of them. But they are also useful for such things as photographing broken bones or revealing internal defects in large castings. Very high-frequency electromagnetic waves that accompany nuclear processes are called *gamma rays*.

Fig. 14-10 Welding produces large amounts of ultraviolet light, which can damage the eyes. The filter in a welding helmet or mask removes 99.75 percent of the ultraviolet light that strikes it.

14-7 Absorption

As a wave travels through a solid, liquid, or gas, some of the wave energy is always changed to heat or other forms of energy. Eventually, the wave dies out. Of course, the wave may also be spreading out at the same time, which makes it weaken all the faster.

This is true for all types of waves. The sound from an indoor stereo bounces around the room and is eventually *absorbed* by the walls,

Application: The Oscilloscope

The shape of an electromagnetic wave or a sound wave cannot be seen directly. But for frequencies between about 1 Hz and 100 MHz, such waves can be displayed on the screen of an *oscilloscope*.

This instrument uses an electron beam to trace a graph on the face of a phosphor-coated tube. The tube is similar in construction to a television picture tube. Some oscilloscopes, like the one shown here, are equipped with a camera to make a permanent record of the graph.

The input to an oscilloscope must be an electromagnetic wave in a wire cable. To display an electromagnetic wave traveling through space, an antenna must be connected to the cable input. To show a sound wave, a microphone or transducer must be connected to the cable. Microphones and transducers convert sound (and ultrasonic waves) to electromagnetic waves in a wire.

The oscilloscope screen shows how the wave height changes in time. From this it is possible to estimate the wave amplitude and the wave frequency. The horizontal and vertical scales on the graph can be altered by the operator, so waves of different amplitudes and frequencies can be displayed.

Oscilloscopes have a wide variety of uses. They are used by electronic technicians to troubleshoot and adjust amplifiers and other electronic circuits. Automotive technicians use them to test and analyze ignition systems. Together with an ultrasonic source and transducer, an oscilloscope can be used to measure

carpets, furniture, and other objects, warming everything just a little. Turn off the lights, and a room quickly gets dark. The light waves have been absorbed by the surroundings, producing a small amount of heat.

The energy in a water wave is changed to heat when it washes up on a beach. Breakwaters are often built to absorb this wave energy before it travels into a harbor. Recently, there have been experiments in England to absorb wave energy from the ocean and use it to drive electric generators. The principle is shown in Fig. 14-11. A flexible raft, or hinged platform, is caused to flex and buckle by water waves passing underneath. This flexing operates hydraulic cylinders that pump fluid to

the thickness of a seamless pipe without cutting it open. And specially designed oscilloscopes are used in most sonar and radar displays.

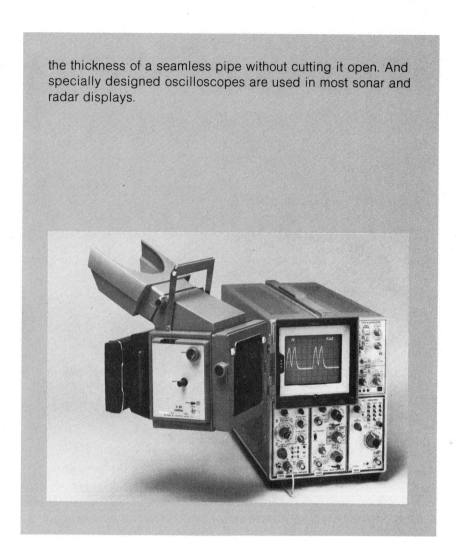

a hydraulic motor which turns a generator. Anchored offshore, a group of these rafts could generate a considerable amount of electricity while reducing the wave heights.

You are already familiar with the fact that waves bounce off things. Clap your hands in a valley or a large hall, and you will hear an echo; stand in front of a mirror, and you will see your reflection. Water waves do the same thing, as can easily be verified in a bathtub.

    Now a reflected wave never has 100 percent of the energy of the

14-8 Reflection

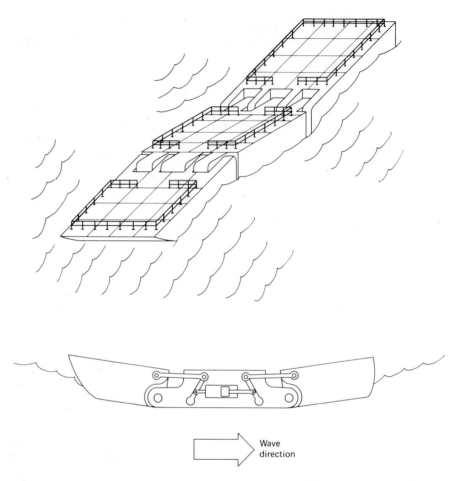

Wave direction

Fig. 14-11 Using a hinged floating platform to extract energy
from water waves.

original wave. The difference is usually absorbed. Sometimes, particularly with sound waves, a fair amount of the wave energy may travel right through the other substance. If you close a window to shut out street noise, some of the outside sound *reflects* off the window back toward the street. But quite a bit of it may still pass through the closed window and into the room.

So when a wave is reflected, this is seldom the *only* thing happening. And when a wave travels from one substance into another (light passing through a pane of glass, for instance), we can usually expect at least some *reflection* to occur as well. For instance, solar collectors (for heating) are covered with glass to limit the heat loss. A double pane of glass cuts the heat loss more than a single pane. But with additional thicknesses of glass, there is so much reflection at all the surfaces that this limits the wave energy entering more than it prevents additional heat loss.

Distance-measuring equipment often makes use of the reflection of waves. Sonar locates underwater objects by measuring the time between an outgoing sound pulse and the returning echo. The same principle is used in ultrasonic thickness gauges and the automatic focusing feature on some new cameras. Bats can navigate in the dark by emitting a high-pitched sound and listening for reflections. Their built-in sonar is so accurate that they can even catch insects in flight in the dark. Radar systems use the same idea, but with high-frequency radio waves instead of sound. Figure 14-12 shows two radar antennas; the one on the left transmits the wave, while the one on the right receives any reflections (from aircraft, for instance). The antennas here are actually fairly small. They are supported by the tripod structures and face back toward the "dishes," which act as reflectors.

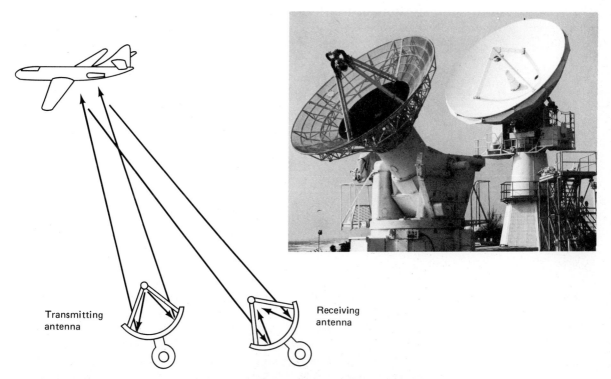

Transmitting antenna

Receiving antenna

Fig. 14-12 A radar system makes the use of the reflection of high-frequency radio waves. The object is located by measuring the time difference between the outgoing wave and the reflected wave. (Photo courtesy of RCA.)

14-9 Refraction

We said that the wave speed depends on properties of the medium. Light travels slower in water than in air, and slower yet in glass. Sound travels faster in water than in air. It also travels faster in warm air than in cold air. Water waves generally travel faster in deep water than in shallow water (although there are some other factors here, too).

When a wave travels from one substance into another where its wave speed is different, it tends to change direction. The process is called refraction.

refraction

Definition: *Refraction* is the bending of a wave as its wave speed changes.

Figure 14-13 shows a common sight on hot summer days: a *mirage*. The air just above the road surface is very hot, while the air farther up is a bit cooler. Light from the sky travels into the region of warm air, where it speeds up and bends away from the road surface. From a distance, the road appears to run into a large lake—when it is really the sky we are seeing. [Most optical devices (cameras, eyeglasses, etc.) make use of this effect. By using a properly shaped glass lens, we can bend light to make an object look larger or smaller than it otherwise would. We can also project an image onto a screen in this way.]

You can make a simple lens with clear plastic food wrap (or a cleaner's bag) and water, as shown in Fig. 14-14. This lens refracts direct sunlight and concentrates it at a small spot. But be careful—the spot gets *very* hot!

Fig. 14-13 A mirage.

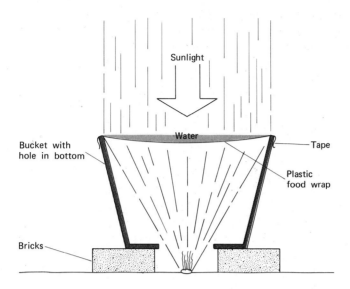

Sunlight

Water

Bucket with
hole in bottom

Tape

Plastic
food wrap

Bricks

Fig. 14-14 A water lens. Thin,
clear plastic food wrap is
stretched over the top of a
bucket and taped around the
sides. Water is poured on the
plastic to form the lens. The
bottom of the bucket has a hole
that lets the refracted light
through.

Waves also bend when they encounter obstacles in their path. The process is called diffraction.

14-10 Diffraction

> Definition: *Diffraction* is the bending of a wave around
> an obstacle in its path.

diffraction

Diffraction is most noticeable when the object is about the same
size as the wavelength of the wave. Sound waves, for instance, have
wavelengths between about 1.5 cm and 15 m in air. Door openings,
furniture, and other commonly encountered objects have sizes in this
same range. As a result, sound waves normally diffract a great deal as
they travel through our buildings, streets, and parking lots. Figure 14-15
shows how the sound from a ringing telephone bends as it travels through
an open door. We don't have to be standing in a direct line with the
telephone to hear it. Visible light, on the other hand, has wavelengths
between about 0.000 04 and 0.000 07 cm. Since this is very short
compared to the size of a door opening, we don't notice the *light* from the
telephone bending around the edge of the doorframe. In fact, we can't
see the telephone at all unless there is an unobstructed straight-line path
between the phone and our eyes.

Does light ever diffract at all? Yes. This is one reason why the
edges of shadows are "fuzzy" if we look at them closely. It is also why
conventional microscopes cannot magnify objects more than about
2500 times; very tiny objects diffract the light passing close to them, and

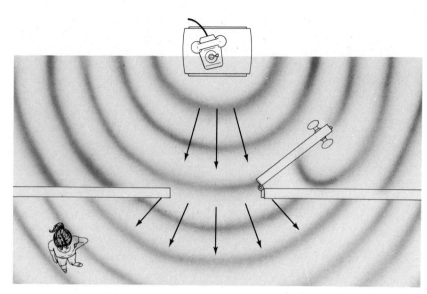

Fig. 14-15 Diffraction of sound. The telephone bell can be heard even though the door frame is between the listener and the phone.

there is no longer a distinct image. Figure 14-16 shows a nighttime photograph taken by a camera with a very small lens opening. The star effect results from the diffraction of light around the edges of the lens's six-sided diaphragm, or lens opening. If you were standing at this scene, the lights would not appear star-shaped to the eye. The construction of the camera causes this diffraction effect. Diffraction sets a limit on the resolution of many optical instruments.

You may also notice a fairly complicated diffraction pattern if you shine a bright light through a pinhole or a narrow slit. Another way of seeing this is to go out on a dark night and locate a streetlight far enough

Fig. 14-16 Diffraction of light. The streetlights appear star-shaped because the light bends around the edges of the camera's six-sided lens opening. Diffraction effects limit the resolution of many optical instruments.

Application: Sound Insulation

Sound insulation means preventing the propagation of unwanted sounds. There are two types of situation where we might want to do this. In auditoriums and other large rooms, too much sound reflected from the walls or ceiling combines with direct, unreflected sounds and may make them unintelligible. It is therefore desirable to reduce these *reflections*. But sometimes, particularly in small rooms and offices, the problem is sound transmitted in through the walls, floor, or ceiling. Here it becomes necessary to reduce the *transmission* of sound.

If the walls of a room are made of hard plaster, wood paneling, or any other rigid material, most of the sound striking the wall is reflected and very little is transmitted through. To reduce such reflections in a room, the inside surfaces should be covered with soft materials. Draperies, acoustical ceiling tiles, and carpeted floors all help reduce reflected sound. Entertainers are usually well aware that there are fewer reflections in an auditorium filled with people than in an empty auditorium; this is because bodies and clothing don't reflect sound very well.

But these same soft, light materials have little effect on sound traveling in *through* a wall. It is commonly believed that a wall filled with fiber glass is quieter than an empty wall, when, in fact, the fiber glass does very little to stop transmitted sound. It takes a dense, rigid material to do this. When sound enters a brick wall, for instance, it has to set a very large mass into vibration. This removes a large amount of energy from the sound wave, so the transmitted sound is greatly reduced. To isolate the inside of a building or room from outside noises, then, the walls must be made of dense structural materials.

away that it appears to be almost a geometrical point. Now hold your handkerchief (unfolded) at arm's length in front of the light. The single light now appears to be a bank of nine lights as the light waves diffract around the threads of the cloth.

Ordinarily, however, we may assume that light and other short-wavelength electromagnetic waves travel in fairly straight lines. Thus land surveys can be based on the idea that light entering the transit's telescope has traveled in a straight path. And sewer pipes are sometimes laid by using a narrow beam of laser light as a guide.

14-11 Superposition

We have talked about waves in a somewhat idealized way. If you look at the surface of a large lake or ocean, you see a very complicated wave pattern without a well-defined wavelength or wave height. The jumble of sounds at a football game and the light striking a movie screen are also very complicated wave patterns. In fact, very seldom do we find a wave having just one frequency and wavelength. A sound having a single frequency is a boring monotone. A light with a single frequency can be generated only with a laser, and it is such a pure color that it looks a bit strange.

Still, we may look at complicated wave patterns as the superposition of many waves having different frequencies. Visible sunlight appears white in color, but by using a glass prism we can split this light into its separate frequencies: the colors of the rainbow. So even though many different waves may be mixed in a complicated wave pattern, each separate wave still keeps its identity. This effect is sometimes called the superposition principle.

superposition principle

Definition: The *superposition principle* states that two or more waves can travel through the same space independently of one another.

The superposition principle also works for waves traveling in different directions. Two flashlight beams can cross without knocking each other off course. And sounds traveling in different directions can pass through one another and then continue on their separate ways. While two waves are crossing paths, there may be places where the combined wave height is very large. Tankers have been known to break in half because of a single large wave that suddenly rises from many smaller waves crossing one another. And offshore drilling platforms have been struck by 30-m waves when no one on shore has recorded a wave anywhere near this high. Such very high waves may suddenly appear from the superposition of several smaller waves. They are usually very short-lived, disappearing as quickly as they have appeared.

14-12 The Doppler Effect

We said that a wave has the same frequency as its source. There is one possible exception: if the source of the wave is moving or if its receiver is moving, the observed frequency will be different. This phenomenon is called the Doppler effect.

Doppler effect

Definition: The *Doppler effect* is the shift in a wave's frequency that results from the motion of the source or receiver.

You probably have heard a train whistle while you were stopped at a railroad crossing. As the train approaches, the whistle is high-pitched. But as the locomotive passes, the pitch of the whistle suddenly drops. This drop in pitch is due to a drop in frequency.

Why does this happen? Because the wave speed depends on properties of the medium rather than on the source. The sound from the moving train whistle travels no faster than sound from a stationary whistle. But with the moving whistle, the waves pile up in front; the wavelengths are shortened, and so the frequency is increased.

Figure 14-17 shows this same effect with a moving airplane. The airplane creates a great deal of noise from its engines, and the plane's own motion through the air creates sound as well. The sound wavelengths are shortened ahead of the moving plane and lengthened behind it. If the plane travels at the speed of sound, the forward-moving waves cannot get ahead of the plane. The wavelengths in front shrink to virtually nothing, and the total wave amplitude gets very large because of the superposition principle (the waves "pile up" on top of one another). The result is a region of very high pressure just in front of the plane and moving along with it. This high-pressure region is called the *sound barrier*. It takes a great deal of energy to accelerate the plane through this barrier.

If the plane travels faster than the sound it produces, we say that its speed is *supersonic*. Now all the pressure disturbances are left behind it. The waves expand in a cone-shaped region called the "Mach wedge." The leading edge of this wedge is a region of very high pressure, again resulting from the superposition principle. As this high-pressure region strikes points on the ground, a loud "boom" is heard and windows rattle or even break. Notice that this *sonic boom* is not a one-shot affair which happens just as a plane breaks the sound barrier. The sonic boom is actually a continuous high-pressure region that trails behind the plane as long as the plane is traveling faster than sound.

A Doppler shift can occur with any type of wave. Some radar systems use this principle to measure the speed of a moving object (a speeding car, for instance). If the car is traveling toward the wave, the reflected radar wave has a shorter wavelength and a higher frequency than the approaching radar wave. The higher the object's speed, the greater this frequency shift.

Summary

Our senses of sight and hearing are based on the interception of light and sound waves. But visible light accounts for only a small range of electromagnetic wave frequencies, and sounds may also be generated at frequencies beyond the response of the human ear. Waves that cannot

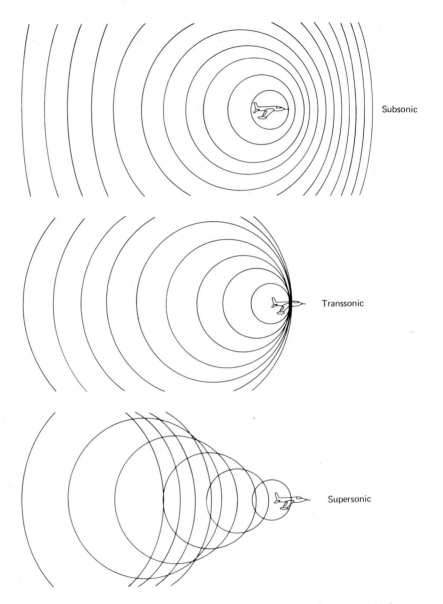

Subsonic

Transsonic

Supersonic

Fig. 14-17 The Doppler effect: sound waves created by a moving source. In the upper diagram, the plane's speed is less than the speed of the sound it creates. In the middle diagram, the plane is flying at the speed of sound. In the lower diagram, the plane is flying faster than sound. The sound itself travels at the same speed in all three cases.

be seen or heard are still sometimes very useful: radio and television, X-ray photography, ultrasonic cleaners and thickness gauges, radar, and many other applications make use of such waves.

Terms You Should Know

wavelength	microwave
wave crest	infrared radiation
wave trough	visible light
wave frequency	ultraviolet radiation
hertz	X rays
wave speed	absorption
transverse wave	reflection
longitudinal wave	refraction
sound wave	mirage
amplitude	diffraction
pitch	superposition principle
ultrasonic	Doppler effect
electromagnetic wave	sound barrier
long electric wave	sonic boom
radio wave	supersonic

Problems

1. A sonar depth finder uses a transducer to send a pulse of sound into the water. The same transducer picks up the reflection off the bottom. The time difference between the outgoing pulse and the reflected pulse can be measured with an oscilloscope. Suppose that this time difference is 32.2 ms. What is the water's depth?

2. An ultrasonic transducer sends out a wave with a frequency of 150 kHz. (a) What is the wavelength if the wave travels through air?

(b) What is the wavelength if the wave travels through water?

3. A piece of seamless aluminum pipe has a wall thickness of 3.1 mm. How long would it take a sound pulse to travel through the wall, reflect off the inside, and return to its starting point?

4. In what part of the electromagnetic spectrum do we find waves of the following wavelengths: (a) 3 m, (b) 5 picometers (pm), (c) 600 nanometers (nm), (d) 5 μm?

15
Heat and Temperature

Introduction We have seen that heat is a form of energy. In fact, it is a form of traveling energy, since it normally flows from hot places to cold places. There are ways to convert heat to other forms of energy, and there are ways to convert other forms of energy to heat. Some of these are listed in Tables 15-1 and 15-2.

If you think about it, you'll find many more ways of generating heat from other forms of energy than vice versa:

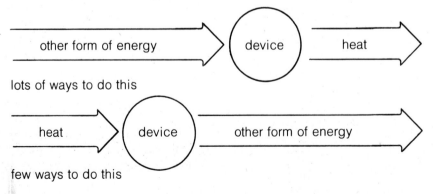

other form of energy → device → heat

lots of ways to do this

heat → device → other form of energy

few ways to do this

TABLE 15-1   *SOME WAYS OTHER FORMS OF ENERGY MAY BE CONVERTED TO HEAT*

Type of energy Converted to heat	Device
Electric	Resistance heater, light bulb, arc welder (or any other device where an electric current flows)
Mechanical	Saw, drill, grinder (or any other moving device where there is friction)
Light	Sunlight, spotlights, etc.
Sound	Sound converts to heat when it is absorbed, regardless of the absorbing material
Chemical	Solid, liquid, and gaseous fuels, explosives, and many (but not all) other chemical reactions
Elastic potential	Car and truck tires rolling without slipping at high speed

TABLE 15-2   *SOME WAYS OF GENERATING OTHER FORMS OF ENERGY FROM HEAT*

Converting heat to:	Device
Electricity	Thermocouple
Mechanical energy	Steam-turbine, or any other heat engine
Light	Fire
Sound	Lightning bolt
Chemical energy	Coke oven
Elastic potential energy	Boiler (production of high-pressure steam)

There is a general tendency for energy to run downhill, with heat being the "lowest" form of energy. And no matter what kind of energy conversion we deal with, we are always guaranteed that at least some heat is produced or wasted.

In this chapter, we talk mainly about how heat affects the properties of matter. In the next few chapters, we look into some of the important applications of heat flow.

15-1 Temperature and Heat Flow

In everyday language, we use the words "heat" and "temperature" to mean pretty much the same thing. In the sciences and the engineering fields, they are not the same thing at all. *Heat* is a form of energy. It can flow from one place to another, or it can be changed into some other form of energy. *Temperature* is a quantity that tells us which way and how fast heat flows. Unless we introduce some work (in the technical sense), heat flows from regions of high temperature to regions of low temperature. If we want to, we can think of temperature as the thing that pushes heat from one place to another.

Does this sound strange? It really isn't. We run into a similar situation in hydraulic systems and electric circuits (Fig. 15-1). In hydraulics, a fluid flows from a region of high *pressure* to a region of low pressure. We can think of the pressure as the thing that pushes the fluid from one place to another. In electricity, we have current and voltage. The voltage pushes, while the current is the flow of electricity that results.

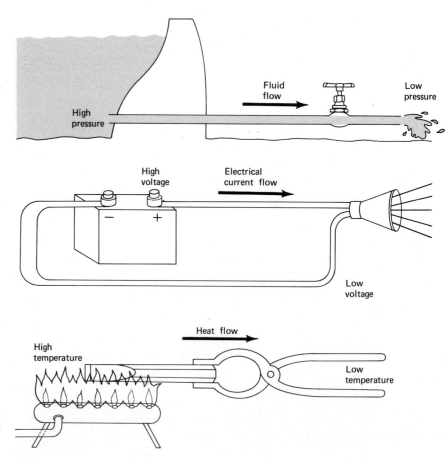

Fig. 15-1 Heat flow compared to fluid flow and electric current flow.

An electric current flows from a region of high voltage to a region of low voltage. It's the same with temperature and heat. Heat is the energy that flows from regions of high temperature to regions of low temperature.

Of course, it's perfectly possible to reverse the flow in all these examples. A hydraulic pump takes a fluid from a low-pressure region and pushes it to a high-pressure region. An electric generator takes electricity from the low-voltage terminal and "pumps" it to the high-voltage terminal. A refrigeration unit takes heat from a low-temperature container and "pumps" it outside to a high-temperature region. But in all these instances, outside work must be supplied. If no work is done, the natural course is from high pressure to low, high voltage to low, or high temperature to low.

Temperature, as we all know, can be measured with something called a thermometer. The most common type has some mercury or colored alcohol sealed in a tube. Yet there is nothing about mercury or alcohol itself that automatically tells the temperature. The expansion or contraction, as indicated on a scale, gives us the temperature reading. This is something to remember about all temperature-measuring devices: we never measure temperature directly; we measure only some secondary quantity that depends on temperature.

There are many such quantities: the electric voltage generated by a thermocouple, the electric resistance of a piece of wire, or the color of a hot, glowing object, to name a few. All these properties are dependent on temperature in a predictable way, and so they all can be used as the basis of a thermometer.

As we mentioned, the liquid thermometer makes use of the expansion properties of mercury or colored alcohol. Mercury freezes at −39°C, and it boils at 357°C. Ethyl alcohol freezes at −114°C and boils at 78°C. These freezing and boiling points limit the useful range of the instrument. But there are other problems with the liquid thermometer. For one thing, it is hard to read; for another, it is easy to break.

The *gas thermometer* is shown in Fig. 15-2. In principle, it works very much like the liquid thermometer. Gases, like liquids, expand when heated and contract when cooled. But if a gas (air, for instance) is confined in an airtight system, it has no place to expand to when it's heated. So instead its *pressure* increases as its temperature is raised. The system has a pressure gauge attached, and the pressure reading tells us the temperature. In fact, the scale on the pressure gauge may be calibrated to read the temperature directly. This type of thermometer is useful at very low temperatures, since air begins to liquefy only at −210°C. Unfortunately, the instrument is big and clumsy and not very portable, so it is seldom used outside the laboratory.

15-2 Measurement of Temperature

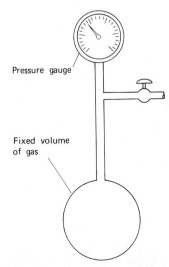

Pressure gauge

Fixed volume of gas

Fig. 15-2 The gas thermometer.

The *bimetallic-strip* thermometer is shown in Fig. 15-3. This instrument makes use of the fact that solids expand when heated and that different solids expand at different rates. Suppose that a strip of brass is laid on a strip of steel and then the two are wound into a coil and riveted together at the ends. The brass is on the inside. If the coil is heated, the brass expands more than the steel and the coil unwinds. If the coil is cooled, the brass contracts more than the steel and the coil winds up tight. One end of the coil is fixed, and the other end has a pointer attached so the pointer can indicate the temperature on a scale. This same principle is used in thermostats, except that instead of moving a pointer, the bimetallic strip throws a switch. (The strip may not be wound into a coil, but it is still a bimetallic strip.) In the thermostat, the actuating temperature can be raised by turning a screw or a dial.

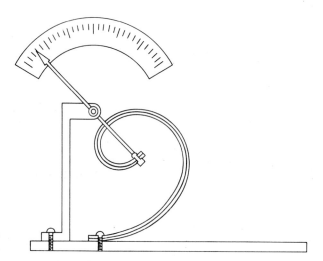

Fig. 15-3 The bimetallic-strip thermometer.

The *resistance thermometer* is based on the fact that the electric resistance of certain materials is strongly dependent on temperature. It's shown in Fig. 15-4. The material whose resistance varies is the sensor, and it is connected electrically to a resistance-measuring instrument

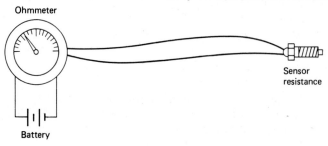

Fig. 15-4 The resistance thermometer.

called an ohmmeter. Again, the ohmmeter scale may be calibrated directly in units of temperature. This device is handy for giving remote readings and is therefore commonly used in automobiles.

Another thermometer based on electrical principles is called the *thermocouple thermometer*. It makes use of the fact that a small electric voltage can be generated at the junction between dissimilar metals and that this voltage is strongly dependent on temperature. The principle is shown in Fig. 15-5. Here the two metals are iron and constantan. There are two junctions: one is used as the temperature probe, and the other is maintained as a reference temperature. The voltmeter scale may be calibrated to read temperature directly. This device can be a very accurate thermometer.

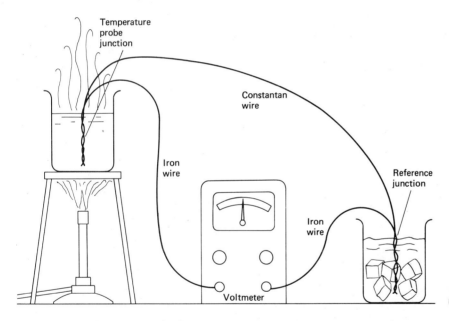

Fig. 15-5 Principle of the thermocouple thermometer.

For very high temperatures, which would melt ordinary thermometers, an *optical pyrometer* may be used. This device is based on the fact that very hot materials glow. The color (red, orange, yellow, white) is an indication of the temperature. The color of the hot material is compared to the color of a hot, glowing tungsten filament in the instrument. The device is shown in Fig. 15-6.

There are also other temperature-measuring devices, like direct-reading liquid-crystal displays and special crayons and paints that melt or change color at prescribed temperatures.

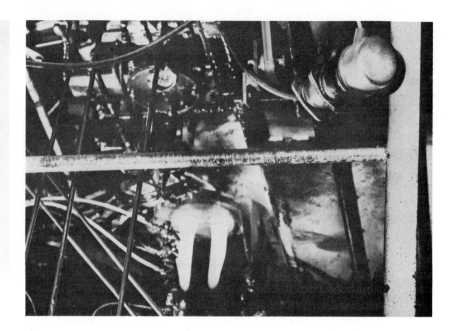

Fig. 15-6 An optical pyrometer measures temperature by comparing the color of a hot glowing object with the color of a glowing tungsten filament in the instrument. The pyrometer in the upper right-hand corner of the picture is measuring the temperature of a glass blob being sheared. A temperature table follows.

Light color	Approximate temperature, °C
Red-black (barely visible)	550
Red	700
Red-orange	850
Orange	1000
Orange-yellow	1200
Yellow	1400
White	1600

15-3 Temperature Scales

The molecules in any substance are in constant motion. Even in solids, where the molecules hold one another tightly, they are constantly vibrating about their equilibrium positions. A certain amount of kinetic energy is associated with this molecular motion, and temperature can be thought of as a measure of the average kinetic energy of an object's molecules. High temperatures correspond to a great deal of molecular motion, while low temperatures mean very little molecular motion.

Now measuring what individual molecules are doing is a very difficult thing. This is why thermometers always measure temperature indirectly—by measuring some secondary quantity which, in turn, depends on the temperature. Still, a mental picture of all these moving molecules can be very useful for describing temperature effects.

Unfortunately, the two temperature scales in common use were developed long before anything was known about molecules. In the year 1714, a Dutch instrument maker named Daniel Fahrenheit built a mercury thermometer that read the freezing point of water as 32° and the boiling point of water as 212°. Later, in 1741, a Swedish astronomer named Anders Celsius devised a new scale, the *Celsius scale,* which made 0° the boiling point and 100° the freezing point. Within a few years, this was turned right side up so 100° became the boiling point. Today, these two scales still go by the names "Fahrenheit" and "Celsius."

Since 180 Fahrenheit degrees (180 F°) span the same range as 100 Celsius degrees (100 C°), the Celsius degree is bigger by a factor of 1.8. It would be nice to use this as the conversion factor between the two scales, but unfortunately we can't. The problem is that the two scales have their zero points in different places. To calculate a Celsius temperature from a Fahrenheit temperature, we can use the formula

$$C = \frac{5}{9}(F - 32) \qquad (15\text{-}1)$$

And to convert degrees Celsius to degrees Fahrenheit, we can use

$$F = \frac{9C + 160}{5} \qquad (15\text{-}2)$$

In many cases, however, it is quicker and more practical to consult a conversion table such as Table 15-3. Notice that for temperatures between 250 and 600°F, the Celsius temperature is about half the Fahrenheit temperature.

There seems to be no limit to how high a temperature can be, while there *is* a limit to how low it can be. How slow can a molecule move? If it is stopped, it can't move any slower. Now it's true that scientists have discovered that molecular motion can never actually be stopped completely. But when molecules are moving as slow as they can go, their temperature is as low as it can get. This temperature is called absolute zero.

Definition: *Absolute zero* is the coldest possible temperature.

absolute zero

TABLE 15-3   FAHRENHEIT-CELSIUS EQUIVALENTS

°F	°C	°F	°C	°F	°C
−250	−157	900	482	2100	1149
−200	−129	950	510	2200	1204
−150	−101	1000	538	2300	1260
−100	−73	1050	566	2400	1316
−50	−46	1100	593	2500	1371
0	−18	1150	621	2600	1427
50	10	1200	649	2700	1482
100	38	1250	677	2800	1538
150	66	1300	704	2900	1593
200	93	1350	732	3000	1649
250	121	1400	760	3100	1704
300	149	1450	788	3200	1760
350	177	1500	816	3300	1816
400	204	1550	843	3400	1871
450	232	1600	871	3500	1927
500	260	1650	899	3600	1982
550	288	1700	927	3700	2038
600	316	1750	954	3800	2093
650	343	1800	982	3900	2149
700	371	1850	1010	4000	2204
750	399	1900	1038	4100	2260
800	427	1950	1066	4200	2316
850	454	2000	1093	4300	2371

If an object could be cooled to absolute zero, no more heat could possibly be extracted from it.

Now absolute zero is a very good place to begin a temperature scale. There are two such *absolute temperature scales*. The *Rankine scale* has a degree the same size as the Fahrenheit degree; the freezing point of water then becomes 491.67°R. The *kelvin* scale has a degree the same size as the Celsius degree; it places the freezing point of water at 273.15 K. It is customary to omit the degree symbol when writing kelvin temperatures. The relationships between the Fahrenheit, Celsius, Rankine, and kelvin scales are listed in Table 15-4.

Exercises

1. Tin melts at 45$\overline{0}$°F. Express this in (a) degrees Celsius, (b) degrees Rankine, (c) kelvin.
   Answers: (a) 232°C, (b) 91$\overline{0}$°R, (c) 505 K

TABLE 15-4   *COMPARISON OF THE FOUR TEMPERATURE SCALES*

	Temperature Scales			
Temperature of:	Fahrenheit	Rankine	Celsius	Kelvin
Boiling water*	212.00°	671.67°	100.00°	373.15
Freezing water*	32.00°	491.67°	0.00°	273.15
Absolute zero	−459.67°	0.00°	−273.15°	0.00

* At standard atmospheric pressure.

2. Steam entering a large turbine engine is at a temperature of 1050°F. Express this in (a) degrees Celsius, (b) degrees Rankine, (c) kelvin.
    Answers: (a) 566°C, (b) 1510°R, (c) 839 K
3. Molten iron tapped from a blast furnace has a temperature of 2400°F. (a) What is this temperature on the Celsius scale? (b) What is the iron's color?
    Answers: (a) 1300°C, (b) orange-yellow
4. A mixture of dry ice (frozen CO_2) and acetone comes to a temperature of −78.5°C. Express this temperature in (a) kelvin, (b) degrees Fahrenheit, (c) degrees Rankine.
    Answers: (a) 194.7 K, (b) −109.3°F, (c) 350.4°R
5. The temperature on a certain winter night in the Canadian plains drops to −40.0°C. Express this temperature in (a) kelvin, (b) degrees Fahrenheit, (c) degrees Rankine.
    Answers: (a) 233.2 K, (b) −40.0°F, (c) 419.7°R
6. The lowest outside temperature ever recorded in the United States was −79.8°F on January 23, 1971, at Prospect Creek Camp in northern Alaska. The highest outside temperature in the United States was 134°F in Death Valley, California, on July 10, 1913. Express these temperatures in degrees Celsius.
    Answers: −62.1°C and 56.7°C

15-4 Thermal Expansion of Solids

When a solid is heated the motion of its molecules increases, and this usually makes the solid expand. The amount of expansion depends on the substance. Aluminum expands more than brass (for the same temperature change), and brass expands more than steel.

   The expansion characteristics of a substance are described by its coefficient of linear expansion.

coefficient of linear expansion

Definition: A substance's *coefficient of linear expansion* is its fractional increase in length for each degree of temperature increase.

Table 15-5 lists coefficients of linear expansion for some common materials. For instance, steel expands by 11.4 *millionths* of its original length for each degree Celsius of temperature increase. Brick expands by about 6 millionths of its length for each degree Fahrenheit increase.

TABLE 15-5   *COEFFICIENTS OF LINEAR EXPANSION FOR SOME COMMON MATERIALS*

Substance	Millionths per F°	Millionths per C°
Aluminum	12.34	22.21
Brass	9.57	17.2
Brick	6	11
Bronze	9.86	17.7
Concrete	10–14	18–25
Copper	8.87	16.0
Glass, ordinary	5	9
Glass, heat-resistant	1.8	3.3
Gold	7.86	14.1
Iron, cast	5.56	10.0
Lead	15.71	28.28
Silver	10.79	19.42
Steel	6.36	11.4
Wood, across grain	20–35	35–60
Wood, along grain	2–4	3–6

To get the actual amount of expansion, then, we can use the formula

$$\begin{matrix} \text{Length} \\ \text{expansion} \end{matrix} = \begin{pmatrix} \text{coefficient of} \\ \text{linear expansion} \end{pmatrix} \begin{pmatrix} \text{original} \\ \text{length} \end{pmatrix} \begin{pmatrix} \text{temperature} \\ \text{increase} \end{pmatrix} \quad (15\text{-}3)$$

Example 15-1 Expansion of a Steel Rail.

A certain rail is 29.871 m in length when the temperature is −10°C [14°F]. On a bright summer day, the rail can reach temperatures as high as 72°C [162°F]. How much does the rail expand when its temperature increases this much?

From Table 15-5, we see that the coefficient of linear expansion of steel is 11.4 millionths per C°. Then using Eq. (15-3), we have

$$\begin{array}{c} \text{Length} \\ \text{expansion} \end{array} = \left(\begin{array}{c}\text{coefficient of} \\ \text{linear expansion}\end{array}\right) \left(\begin{array}{c}\text{original} \\ \text{length}\end{array}\right) \left(\begin{array}{c}\text{temperature} \\ \text{increase}\end{array}\right)$$

$$= (11.4 \times 10^{-6} \; C°) \, (29.871 \text{ m}) \, (82C°)$$

Notice that the temperature increase here is 82C° (−10 to 72°C). Multiplying this through gives

$$\text{Length expansion} = 0.028 \text{ m}$$

or 2.8 cm [1.1 in]. While this is a small amount compared to the rail's original length, it is certainly enough expansion to be easily noticed.

What if the rail is cooled from 72 to −10°C? Obviously, it will shrink or contract by this same 2.8 cm. Equation (15-3) and Table 15-5 describe cases of contraction as well as expansion. ◄

Allowances need to be made for thermal expansion in structures. If a solid is heated but there is no place for it to expand to, thermal stresses build up. In extreme cases, these thermal stresses can exceed the material's elastic limit or even its ultimate strength. Roads and masonry structures can crack, and steel structures can buckle or twist out of shape. Figure 15-7 shows some schemes used to prevent stress buildup on expansion. Bridges are built on rollers or rockers, power lines are strung with a certain amount of slack, steam lines exposed to the elements are built with vertical "jogs," and large masonry walls need vertical expansion joints filled with flexible caulking. You have probably noticed the steel expansion joints on roads crossing bridges. You may also have noticed a series of clicking sounds when running hot water from a spigot. This happens when the hangers holding the hot-water line to the joists (or studs) are too tight and the pipe cannot smoothly expand in length. The clicks may be heard again when the hot water is shut off and the pipe contracts as it cools.

If solids expand when heated, what happens to the size of a hole in a solid plate? Many people think that the hole should shrink as the solid expands. But this doesn't happen. The plate in Fig. 15-8 can be considered to be a very fat ring (or a series of rings nestled in one another). And, of course, the hole in a ring always expands as the ring is heated, and its circumference increases. So, just like the material around it, a hole always expands when heated. Certainly it can never contract.

What if the plate is rigidly fastened so its outer edges can't move? Then heating the plate will probably cause it to buckle. The hole will either pop forward or backward. Either way, the hole gets bigger.

Fig. 15-7 Allowance must be made for expansion and contraction in structures exposed to outside temperature variations.

Exercises

7. A 100-m steel tape measure is accurate when the temperature is 20.0°C. How far is it off when the temperature is (a) −10.0°C, (b) 42.0°C?
    Answers: (a) The tape is 3.42 cm short. (b) The tape is 2.51 cm long.
8. How big an expansion joint is needed between two 30-ft sections of brick wall if the expected temperature rise is no more than 60F°?
    Answer: At least 0.13 in
9. An aluminum wire 1.4 km long is strung between two towers for the transmission of high-voltage electricity. How much does its

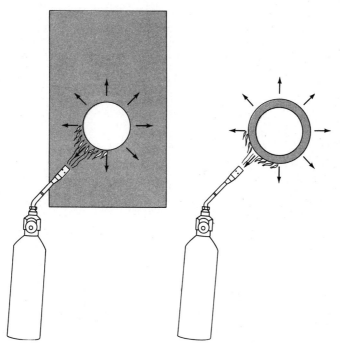

Fig. 15-8 A solid ring expands when heated. So does the hole in a solid plate.

length change when the temperature goes from −10.0 to 40.0°C?
Answer: 1.6 m

10. A suspension bridge has a main span 3800 ft long. The girders supporting the road are made of steel. On a cold winter day the temperature may reach −5.0°F, while in the summer a temperature of 95°F may be reached. How much does the bridge's length change as a result of thermal expansion?
Answer: About 4.1 ft

Like solids, liquids usually expand when heated. This volume increase causes a decrease in density. As a result, hot oil floats on cold oil, and hot mercury floats on cold mercury.

The only important exception to this behavior is water between 0 and 4°C [32 and 39°F]. As water is heated from 0 to 4°C, it actually contracts, and its density increases. Figure 15-9 shows the actual density variation of water at low temperatures. As a result, 4°C water will sink in 3°C water, which sinks in 2°C water, and so on. At 4°C, water is as dense as it gets. Water at any other temperature will float on 4°C water.

15-5 Thermal Expansion of Liquids

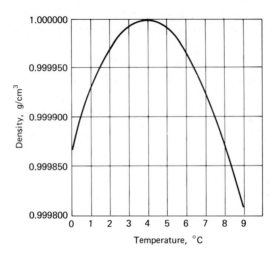

Fig. 15-9 Density of water at
low temperatures.

15-6 Charles' Law

If a gas is confined and heated, it can't very well expand since there is no place for it to go. In this case the gas pressure increases as the temperature goes up. This pressure increase is similar to the thermal stress buildup in solids that are not allowed to expand when heated. This pressure increase can be predicted from Charles' law.

Charles' law

Charles' law: In a confined gas at temperatures high compared to the substance's boiling point, the absolute pressure is proportional to the absolute temperature.

In other words, if we double the absolute temperature, we expect the absolute pressure to double. Increase the absolute temperature by 50 percent, and the absolute pressure increases by 50 percent. An so on. This may be written as a formula:

$$\begin{array}{c}\text{New absolute}\\\text{pressure}\end{array} = \left(\begin{array}{c}\text{original abso-}\\\text{lute pressure}\end{array}\right) \times$$

$$\left(\frac{\text{new absolute temperature}}{\text{original absolute temperature}}\right) \qquad (15\text{-}4)$$

Notice that both the temperatures and pressures must be absolute.

Example 15-2 Tire Inflation and Temperature.

A certain tire is inflated to a pressure of 2.0 kg$_f$/cm², gauge [28 psig] when the temperature is 0.0°C [32°F]. What is the tire pressure when the temperature goes up to 30°C [86°F]?

First we have to calculate the absolute pressure and temperatures. By using 1.03 kg$_f$/cm² [14.7 psia] as atmospheric pressure, the tire's original inflation pressure is

$$\text{Original absolute pressure} = 2.0 \text{ kg}_f/\text{cm}^2 + 1.0 \text{ kg}_f/\text{cm}^2$$

$$= 3.0 \text{ kg}_f/\text{cm}^2$$

The original temperature, based on the definition of the kelvin scale in Table 15-4, is

$$\text{Original absolute temperature} = (0.0 + 273.16) \text{ K}$$

$$= 273 \text{ K}$$

And the new temperature, expressed in kelvin, is

$$\text{New absolute temperature} = (30 + 273.16) \text{ K}$$

$$= 303 \text{ K}$$

Then using Eq. (15-4), we have

$$\begin{matrix} \text{Absolute} \\ \text{pressure} \end{matrix} = \left(\begin{matrix} \text{original abso-} \\ \text{lute pressure} \end{matrix} \right) \times \left(\frac{\text{new absolute temperature}}{\text{original absolute temperature}} \right)$$

$$= \left(3.0 \; \frac{\text{kg}_f}{\text{cm}^2} \right) \left(\frac{303}{273} \right)$$

$$= 3.3 \text{ kg}_f/\text{cm}^2$$

Subtracting standard atmospheric pressure then gives the new tire pressure of 2.3 kg$_f$/cm² gauge [33 psig]. The temperature increase has caused the tire to be overinflated by about 5 psig.

This is a problem in Northern climates. People forget to recheck their tire pressure when spring temperatures begin to rise, and this leads to overinflation and uneven tire wear. Perhaps even worse is that the reverse happens when temperatures drop in the fall. Now the tires run underinflated, which affects vehicle handling and fuel economy as well as tire wear. ◀

Exercises

11. When the temperature is 15.0°C, the pressure in a certain oxygen cylinder is 133 kg$_f$/cm², gauge. What is the gauge pressure if the cylinder's temperature is increased to 35.0°C?
   Answer: 142 kg$_f$/cm², gauge

12. The superheated steam generated in a certain boiler has a pressure of 750 psig when its temperature is 350°F. What is the steam pressure if it is heated to 1000°F without being allowed to expand?
Answer: 1360 psig

13. Just before ignition, the fuel-air mixture in a certain car engine is at a pressure of 10.8 atm (absolute) and a temperature of 185°F. The mixture burns rapidly enough that the piston doesn't move very far during combustion. If the temperature goes up to 4140°F at the beginning of the power stroke, what is the cylinder pressure at this point?
Answer: 77.2 atm, absolute

15-7 Thermal Expansion of Gases

Gases that are not confined expand when heated. Thus hot air rises from a fire or from an asphalt road baking in the sun. The decreased density of hot air can even provide enough buoyant force to lift a balloon.

When a gas is used in a refrigeration system, an engine, or a pneumatic system, usually it is partially confined. Heating such a gas may result in a volume increase and a pressure increase at the same time. It stands to reason that if a gas's volume is allowed to increase as it's heated, its pressure will not increase as much as Eq. (15-4) predicts.

15-8 Specific Heat Capacity

There have probably been hot spring days when you went for a swim and found the water quite cold. You may have gone back to your car and found that the car's body surface had gotten very hot while sitting in the sun. Even though the car body and the water were both exposed to the same sunlight, the steel got much hotter than the water.

Now it's quite possible that the car and the water absorbed different amounts of heat energy. But the point is this: If 1 kg of steel absorbs 100 kJ of heat and 1 kg of water also absorbs 100 kJ of heat, then the steel will get much hotter than the water. Water tends to store heat energy with very little temperature increase, while the temperature of steel and of other metals increases very rapidly as heat is absorbed. The ability of a material to store heat is described by its specific heat capacity.

specific heat capacity

Definition: The *specific heat capacity* of a substance is the amount of heat energy required to produce a 1° temperature increase in a unit amount of the substance.

Since heat is a form of energy, it may be measured in units of joules, foot-pounds, kilowatthours, electronvolts, or any other energy unit mentioned in Table 9-3. But most of the time heat is measured in joules or British thermal units. It takes 4184 J of heat energy to raise 1 kg of water 1 C°. This is the same as saying that the specific heat capacity of water is 4184 J/(kg·C°). And it takes 1 Btu of heat to raise 1 lb_m of water 1 F°. This is the same as saying that the specific heat capacity of water is 1.00 Btu/(lb_m·F°). For aluminum, the corresponding values are 895 J/(kg·C°) and 0.214 Btu/(lb_m·F°). The lower values reflect the fact that aluminum is much easier to heat than water.

Until recently, heat was commonly measured in units of calories and kilocalories in metricated countries. Dieticians still use the kilocalorie (incorrectly calling it the calorie), and this unit is also occasionally found in technical publications. But in the vast majority of cases, heat is now expressed in terms of the British thermal unit or the joule and its SI multiples. Because of the trend toward phasing out the calorie, we will not use it further in this book.

Table 15-6 lists the specific heat capacities of some common materials. In looking over this table, keep in mind that a large specific heat capacity means that the substance stores a great deal of heat with only a small increase in temperature. Substances with large specific heat capacities are very good for cooling. Water is commonly used for cooling in engines and many industrial processes. High-voltage electric circuit breakers are often filled with oil because the hydrogen released by its chemical decomposition quickly cools any plasma arcs that may form.

Provided that there is no phase change (melting or boiling), the heat required to raise an object's temperature is given by

$$\text{Heat} = \left(\begin{array}{c} \text{specific heat} \\ \text{capacity of substance} \end{array} \right) \times \left(\begin{array}{c} \text{mass of} \\ \text{substance} \end{array} \right) \times$$

$$\left(\begin{array}{c} \text{temperature} \\ \text{change} \end{array} \right) \qquad (15\text{-}5)$$

This same formula also gives the heat energy released by a substance as it cools off. Let's look at an example.

Example 15-3 The Hot-Water Tank.

A certain hot-water tank is filled with 40 gal of cold water at 62°F. The water is to be heated to 165°F. How much heat energy must the water absorb?

From Table 15-6, we see that the specific heat capacity of water is 1.000 Btu/(lb_m·F°). The first step, then, is to find the mass in pound-mass of 40 gal of water.

Table 6-2 tells us that each 1 ft³ of water has a mass of 62.3 lb_m.

TABLE 15-6   SPECIFIC HEAT CAPACITIES OF SOME COMMON MATERIALS

Material	Specific heat capacity	
	Btu/(lb$_m$ F$^\circ$)	J/(kg C$^\circ$)
Solids		
Aluminum	0.214	895
Brass	0.094	390
Brickwork	0.20	840
Copper	0.094	390
Glass	0.194	812
Gold	0.031	130
Ice	0.504	2 110
Iron, cast	0.130	544
Lead	0.031	130
Sand	0.195	816
Silver	0.056	234
Steel	0.116	485
Stone (typical)	0.20	840
Wood, oak	0.57	2 400
Wood, pine	0.467	1 950
Liquids		
Aluminum	0.25	1 000
Copper	0.101	423
Alcohol, ethyl	0.58	2 400
Alcohol, methyl	0.60	2 500
Gold	0.032 7	137
Lead	0.038	160
Mercury	0.033	140
Nitrogen	0.474	1 980
Oxygen	0.394	1 650
Oil, machine	0.400	1 670
Silver	0.068 5	287
Water	1.000	4 184
Gases (confined)		
Air	0.168	703
Ammonia	0.399	1 670
Hydrogen	2.412	10 090
Nitrogen	0.173	724
Oxygen	0.155	649
Steam	0.346	1 450

Note: There may be some variation in these figures, depending on the actual temperature, the presence of impurities, etc.

Table 3-3 gives us the conversion 1 ft^3 = 7.480 5 gal. So the mass of 40 gal of water is

$$\text{Mass of water} = (40 \text{ gal})\left(\frac{1 \text{ ft}^3}{7.480\ 5 \text{ gal}}\right)\left(62.3 \frac{\text{lb}_m}{\text{ft}^3}\right)$$

$$= \frac{40\ (62.3)}{7.480\ 5} \text{ lb}_m$$

$$= 330 \text{ lb}_m$$

The water is heated from 62 to 165°F. This gives a temperature change of 103 F°. Then using Eq. (15-5), we have

$$\text{Heat absorbed} = \left(\begin{array}{c}\text{specific heat}\\\text{capacity of water}\end{array}\right) \times \left(\begin{array}{c}\text{mass of}\\\text{water}\end{array}\right) \times \left(\begin{array}{c}\text{temperature}\\\text{increase}\end{array}\right)$$

$$= \left(1.000 \frac{\text{Btu}}{\text{lb}_m \cdot \text{F°}}\right)(330 \text{ lb}_m)(103 \text{ F°})$$

$$= 34\ 000 \text{ Btu}$$

If the water is to be heated in 1 h, then the burner or heating element must put out 34,000 Btu/h. Actually, the heat production must be a bit greater than this since the efficiency of the heater will not be 100 percent. ◀

Example 15-4 A Hot Ingot.

A steel ingot (Fig. 15-10) has a typical mass of 13.5 t. How much heat does it give up in cooling from 1150°C [2100°F] to 20°C [68°F]?

From Table 15-6, we see that the specific heat capacity of steel is 485 J/(kg·C°). This means that each 1 kg gives up 485 J of heat in cooling off 1 C°. Using Eq. (15-5), we have

$$\text{Heat lost} = \left(\begin{array}{c}\text{specific heat}\\\text{capacity of steel}\end{array}\right) \times \left(\begin{array}{c}\text{mass of}\\\text{steel}\end{array}\right) \times \left(\begin{array}{c}\text{temperature}\\\text{decrease}\end{array}\right)$$

$$= \left(485 \frac{\text{J}}{\text{kg} \cdot \text{C°}}\right)(13\ 500 \text{ kg})(1130 \text{ C°})$$

$$= 7.40 \times 10^9 \text{ J}$$

By using the conversion factors in Table 9-3, this may also be expressed as 7 million Btu. Notice that this is a tremendous amount of heat. You don't have to stand very close to the hot ingot to be aware that this heat is being lost.

Let's say that the ingot, which is standing in the open air, cools off

Fig. 15-10 A stripper crane removing the mold from a hot
steel ingot.

in 16 h. (This is a fairly typical value.) A rough estimate of the average power lost is

$$\text{Power} = \frac{\text{energy transformed}}{\text{time}}$$

$$= \frac{7.40 \times 10^9 \text{ J}}{16 \text{ h}} \left(\frac{1 \text{ h}}{3600 \text{ s}} \right)$$

where the last quantity is the conversion factor from hours to seconds. This gives

$$\text{Power} = \frac{7.40 \times 10^9}{16\,(3600)} \frac{\text{J}}{\text{s}}$$

$$= 1.3 \times 10^5 \text{ W}$$

$$= 130 \text{ kW}$$

This is equivalent to the power developed by a good-sized car engine running at full load.

　　To make this story complete, we should say something else about

Application: Thermal Expansion in Internal-Combustion Engines

A conventional car engine contains hundreds of parts that must function over a range of temperatures. In very cold weather, all these parts contract and, since they are made of different materials, the clearances change. When the engine warms up to operating temperature, all these parts expand by different amounts. The engine must be designed with tolerances close enough to prevent excess wear and loss of lubricants when cold, yet loose enough that no parts ever seize when hot.

Most aluminum pistons are cam-ground so they have a slightly oval shape when cold. When the piston heats up to operating temperature, most of the expansion takes place parallel to the piston pin, and the piston becomes circular in cross section. The piston may also be tapered so that the top is smaller in diameter. This is because the top of the piston runs much hotter than the bottom, and there is more expansion on top. As a result, there is usually a poor fit when the piston is cold; but at normal operating temperatures, the piston assumes the same shape as the cylinder in which it slides.

An engine's valve stems and pushrods also expand as the engine warms up. Thus a certain amount of tappet clearance must be allowed when adjustments are made on a cold engine with mechanical lifters. Hydraulic lifters are designed to compensate automatically for this thermal expansion.

The bolt holes in an engine's block expand slightly more than the head bolts, because the block gets slightly warmer than the bolts—therefore the bolt holes are slightly larger than the bolt diameters. If the bolts are not torqued to specification on the cold engine, they may loosen when the engine warms up. A blown head gasket can result.

Many engines have aluminum heads and even aluminum blocks. As we have seen, aluminum has the advantage of being lightweight. Unfortunately, its ultimate strength (even when alloyed) is much less than that of steel. If an aluminum engine is overheated and the head bolts and other parts keep it from expanding the way it wants to, the resulting thermal stresses can easily exceed the aluminum's elastic limit. Mechanics say that the head has "warped." The remedy is very costly: removing the head and planing it or replacing it altogether.

steel ingots: the 7.4 GJ of heat loss we calculated here would have to be resupplied to get the cold ingot ready for rolling into slabs or other forms. This is no small amount of heat. When possible, the time and fuel expense of reheating is avoided by storing the hot ingot in a furnace called a "soaking pit." When the ingot's temperature is uniform throughout, it is withdrawn from the pit and rolled. ◄

Exercises

14.   How much heat energy, in megajoules, is needed to raise 10.0 kg of water from 20 to 80°C? How much heat energy is needed to do the same thing with 10.0 kg of brass?
      Answers: 2.5 MJ for the water, 0.23 MJ for the brass
15.   A cast-iron engine block has a mass of 422 lb_m. (a) How much heat, in British thermal units, is needed to raise its temperature from 42 to 205°F? (b) Express this result in joules.
      Answers: (a) 8900 Btu, (b) 9.4 million J, or 9.4 MJ
16.   Pure lead melts at 330°C. How much heat, in joules, is needed to raise 15 kg of lead from 20°C to its melting point?
      Answer: 6̄00 000 J
17.   A certain swimming pool holds 48 000 gal of water. How much heat, in British thermal units, is needed to raise the water's temperature from 62 to 71°F?
      Answer: 3.6 million Btu

15-9 Freezing

If you ask someone to tell you the freezing point of water, she or he will probably answer correctly that it is 32°F [0.0°C]. Now follow up by asking for the melting point of water, and most people get confused. Is it 33°F? Or 32.5°F? No—it is still 32.0°F. Water, and most other pure chemical compounds, freezes at exactly the same temperature as it melts at.

Okay, but suppose that you have some H_2O at exactly 0.00°C. Is it liquid or solid? The answer is that it can be either. It all depends on whether you are taking out heat or putting in heat. If heat is coming out, more and more of the water turns solid until eventually it has all frozen. At this point, if we are still taking out heat, the ice begins to cool to temperatures below 0°C.

If we put heat into a mixture of ice and water, more and more of the water turns liquid but the temperature remains at 0.00°C as long as there is still some ice left. When the last of it has finally melted, any additional heat goes into warming the water above 0°C.

Figure 15-11 shows how the temperature of water is affected by heat flowing in or out. For pure substances, the phase transition from solid to liquid (or vice versa) usually takes place at a constant temperature. So does the phase transition from liquid to gas.

For a mixture, the transition is not as abrupt. Figure 15-12 compares the freezing of pure substances like water with the freezing of mixtures like plumber's solder, gasoline, and so on. (Yes, molten metals do freeze.) The pure substance's temperature remains constant during the phase-change period, while with mixtures the temperature drops as they go through a "slushy" stage between the solid and liquid phases. With bronze (90 percent copper and 10 percent tin), freezing begins at 1000°C and ends when the alloy has cooled to about 850°C. Care must be taken not to disturb molten alloys as they cool through the liquid-solid transition region. Otherwise, mechanical and metallurgical flaws may develop. This should always be kept in mind when soldering or brazing.

We all know that water expands on freezing, and this is why ice floats on water. It is also why we need to use antifreeze in a car's cooling system in the winter: the expansion can crack the engine block if the coolant freezes. But water is unusual in this respect. For most other substances, the solid sinks in the liquid. When molten metal is poured into a mold, it freezes from the bottom up and from the outside in. The metal's contraction on freezing can leave empty cracks or holes in metal castings, which, or course, can lead to serious problems. This is why machine parts are usually forged rather than cast.

Pressure has a slight effect on a substance's freezing point. If the substance expands on freezing, an increased pressure will lower the freezing temperature. Thus ice will melt when subjected to pressure, as under the wheels of a parked truck or under the blade of an ice skate. Substances that contract on freezing behave in just the opposite way. Pressurizing such substances causes them to freeze at temperatures higher than normal. This effect is exploited in the injection molding of plastics.

There is another way to change a substance's freezing point, and that is by mixing it with something else. This mixing itself changes the temperature. If finely crushed ice (or snow) is mixed with salt, the temperature of the resulting slush may drop as low as −18°C [0°F]. If calcium chloride is mixed with ice or snow, temperatures as low as −42°C [−44°F] may result. This principle was commonly used in ice cream freezers back in the days before household refrigeration. Today it is still used when icy city streets are "salted" in the winter. The salt-ice slush may not actually melt (after all, it is very cold), but then it isn't frozen solid either. With snow tires, cars can move around on it.

Application: Sweat Fits on Shafts

Gears, pulleys, sprockets, fly cutters, and other rotating parts are often attached to their drive shafts with setscrews. Although this allows for convenient assembly and replacement, it does have a disadvantage. By tightening the setscrews to take up the clearance between the part and the shaft, the part is forced slightly off center. The eccentrically mounted part then wobbles as it rotates. This is obviously something to be avoided in applications like power transmissions and work rolls for forming metals.

One alternative is to use a sweat fit. The sleeve of the rotating part is made of bronze or some other metal with a large coefficient of expansion. The hole is machined slightly too small to accept the shaft. To mount the part, the sleeve is heated until the hole expands sufficiently to slide over the shaft. On cooling, the sleeve centers itself on the shaft, and thermal stress holds it in place.

Removing the part is more difficult than putting it on. This is because heating the mounted part also heats the shaft, and both expand together. But if the mounting sleeve is bronze and the shaft is steel, the bronze expands at a greater rate; eventually a temperature is reached where the part can be slipped off.

When a pickle jar is sealed at the processor's, it is usually very hot. Subsequent contraction produces an effective sweat fit between the metal top and the jar. The cold jar is then very hard to open (particularly if it has been refrigerated). A simple way of loosening the cap is to run hot water over it. Because the metal expands more quickly than the glass, the fit relaxes and the hot cap can easily be twisted off.

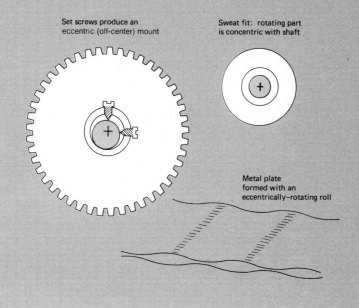

Set screws produce an
eccentric (off-center) mount

Sweat fit: rotating part
is concentric with shaft

Metal plate
formed with an
eccentrically-rotating roll

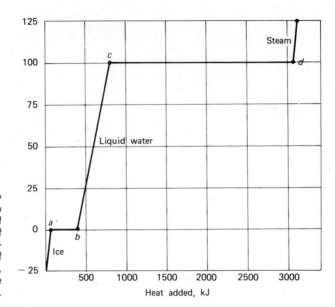

Fig. 15-11 Heat required to raise 1 kg of ice from −25 to 125°C. Point a is the onset of melting, b is the completion of melting, c is the onset of boiling, and d is the completion of vaporization. Beyond point d, the water is completely in the form of steam.

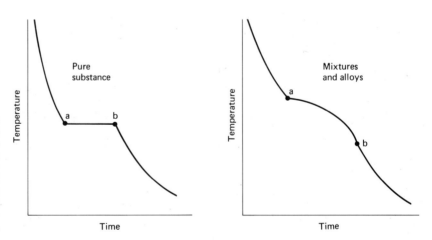

Fig. 15-12 Pure substances usually remain at a constant temperature as they freeze, while alloys and other mixtures drop in temperature as they go through the liquid-to-solid phase change. Points a on both graphs are the onset of freezing while points b are the completion of freezing.

15-10 Vaporization

We have all been told that the boiling point of water is 212°F [100°C]. Actually this is true only when the pressure is 1 atm. If the pressure is increased, the boiling point also increases. At 2.0 atm pressure, for instance, water must be heated to 250°F [121°C] before it boils.

Now if you boil water in an open pan on a stove, there is no way you can get the water any hotter than 100°C. As soon as the water reaches this temperature, any additional heat transforms some of the water to steam that escapes. Turning up the burner makes the water boil faster, but this does not make it get any hotter. But if the water is pressurized in a pressure cooker or a boiler, it can be heated well beyond 100°C. The limit is 374.1°C [705°F], which is called the water's *critical point*. It takes a pressure of 218.3 atm to keep water in the liquid state at this temperature.

Definition: A substance's *critical point* is the highest temperature and pressure that will permit it to remain in the liquid state.

critical point

Most other substances exhibit this same general behavior; only the numbers are different. Liquid oxygen boils at −183°C when exposed to the atmosphere, but it remains liquid at −118.4°C if confined at a pressure of 50.1 atm. This is the oxygen's critical point, and at higher temperatures there is no way to keep it liquid. Table 15-7 lists the critical points of some other substances with commercial or industrial importance. Freon is used as a refrigerant, since its boiling point can be drastically altered by changing the pressure. Helium, on the other hand, is normally a gas since it boils at such low temperatures, even under pressure. Methane is one of the main ingredients in natural gas; it boils at very low temperatures, even when pressurized. Octane is one of the

TABLE 15-7 *CRITICAL POINTS OF SOME IMPORTANT SUBSTANCES*

Substance	Critical temperature, °C	Critical pressure, atm	Normal boiling point, °C
Carbon dioxide	31.0	72.85	*
Freon-12	111.7	39.4	−29.8
Helium	−267.9	2.26	−268.9
Methane	−82.1	45.8	−161.5
Octane	296	24.8	125.6
Oxygen	−118.4	50.1	−183
Propane	96.8	42	−42.1
Water	374.1	218.3	100.0

* Carbon dioxide is never liquid at atmospheric pressure.

ingredients in gasoline. Propane normally boils at −42.1°C, but under pressure it remains liquid at temperatures as high as 96.8°C. Propane is the main ingredient in l.p. (liquefied petroleum) gas.

When any liquid vaporizes (boils), it expands a great deal. If 1 L of water (about 1 q) is boiled away under atmospheric pressure, it generates nearly 1900 L of steam! Thus if any confined liquid is heated beyond its critical temperature, a violent explosion is likely to result. This is why CO_2 tanks need to be kept out of the sun. The carbon dioxide is in the liquid state in such tanks, but beyond 31.0°C [87.8°F] it boils rapidly and ruptures the container, or else pops the pressure relief valve if the tank is equipped with one.

Figure 15-13 shows the variation in the boiling points of some common coolants at different pressures. Most car and truck radiators are equipped with pressure caps to allow the coolant to get very hot without boiling. With pure water in the radiator, a 5-psig cap releases at 226°F [108°C], or a 10-psig cap at 115°C [239°F] and a 15-psig cap at 121°C [nearly 250°F]. With ethylene glycol in the water, these temperatures can be higher yet. What happens if we suddenly remove the radiator cap on a hot engine? If the coolant is at a temperature above its normal boiling point, it instantly begins to boil violently, and we may have a geyser. The effect is shown in Fig. 15-14. Many people who didn't understand the principle of the pressurized cooling system have been badly scalded when this happened.

Some years ago, various water-alcohol mixtures were used as antifreeze coolants for winter driving. The problem was that they quickly boiled over under summertime conditions, and so the coolant had to be changed every fall and spring. Today we use permanent-type antifreeze solutions, which are usually mixures of ethylene glycol and water. Not

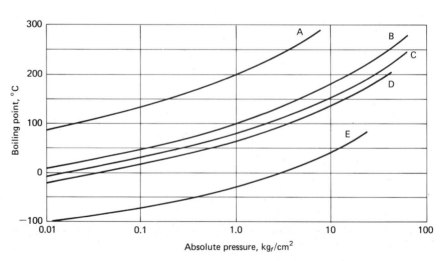

Fig. 15-13 Effect of pressure on the boiling points of some coolants: (a) ethylene glycol, (b) water, (c) ethyl alcohol, (d) methyl alcohol, (e) Freon-12. Note: 1 atm = 1.03 kg_f/cm^2.

Fig. 15-14 It can be very dangerous to remove the radiator cap on a hot engine with a pressurized cooling system. (Courtesy Shell Oil)

only does this suppress the freezing point, but it also raises the boiling point. A cooling system filled with permanent-type antifreeze has less tendency to boil over than one filled with pure water.

Since raising the pressure on a liquid raises its boiling point, it follows that lowering the pressure on a liquid will allow it to boil at a lower temperature. If the pressure is lowered to 17.54 mmHg [0.023 atm] water actually boils at 20°C [68°F], or normal room temperature. And at a pressure of 47 mmHg [0.062 atm] water boils at body temperature [37°C, or 98.6°F]. This is why the cabins of military aircraft must be pressurized for high-altitude flights. At altitudes above 19 km [62 000 ft], the atmosphere pressure is low enough that an exposed pilot's blood would boil. Of course, the cabins of commercial aircraft are also pressurized even though they do not reach such high altitudes. This solves the practical problem of requiring each passenger to wear an oxygen mask.

It is common practice in distillation processes to lower the pressure to get the liquid to boil at a lower temperature. This principle is used in refineries to separate oils and waxes from crude-oil residue after the gasoline has been distilled off. The principle has also been used to distill fresh water from seawater and in the manufacture of some distilled liquors.

15-11 Latent Heat

We have seen that heat absorbed by a substance can do one of two things: it can cause a temperature increase, or it can cause a phase change. The heat required for a phase change alone is called *latent heat*. For instance, melting 1 kg of ice at 0°C into 1 kg of water at 0°C requires 335 kJ of heat. We call this 335 kJ latent heat because it is stored in the phase change rather than in a measurable temperature increase. If another 335 kJ is absorbed by the 1 kg of water, its temperature will rise all the way to 80°C [176°F]. We see, then, that this latent heat is a fairly significant quantity.

We need to distinguish between the latent heat in the solid-liquid phase change and that in the liquid-gas phase change.

latent heat of fusion

Definition: A substance's *latent heat of fusion* is the amount of heat needed to melt a unit amount of the substance when it is already at the melting temperature.

latent heat of vaporization

Definition: A substance's *latent heat of vaporization* is the amount of heat needed to vaporize the substance when it is already at its boiling point.

The SI unit of latent heat is joules per kilogram (J/kg). In the old USCS units, latent heat is expressed in British thermal units per pound-mass (Btu/lb$_m$). Table 15-8 lists the conversion relations between these units. Also listed, for reference, is calories per gram (cal/g).

Table 15-9 lists the latent heats of some substances, along with their normal melting and boiling temperatures. Copper, for instance, melts at 1080°C [1976°F], and it takes 180 kJ of heat to melt 1 kg of copper that is already at this temperature. To melt 100 kg of copper, then, requires (100)(180 kJ), or 18 000 kJ. To get the total heat involved in the phase change, we simply multiply the latent heat by the mass of the substance.

TABLE 15-8   CONVERSION FACTORS FOR LATENT HEAT

	$\dfrac{\text{J}}{\text{kg}}$	$\dfrac{\text{cal}}{\text{g}}$	$\dfrac{\text{Btu}}{\text{lb}_m}$
1 joule per kilogram =	1	2.3901 × 10^{-4}	4.3021 × 10^{-4}
1 calorie per gram =	4184.0	1	1.800 0
1 Btu per pound-mass =	2324.4	0.555 56	1

Note: 1 Btu = 1054.4 J
These units and conversions are also used for heat of combustion (Chap. 16).

TABLE 15-9   *LATENT HEATS AND NORMAL BOILING AND MELTING POINTS OF SOME COMMON SUBSTANCES*

Substance	Normal melting point, °C	Latent heat of fusion, kJ/kg	Normal boiling point, °C	Latent heat of vaporization, kJ/kg
Alcohol, ethyl	−114	100	78	854
Alcohol, methyl	−97	67	65	1120
Aluminum	658	390	2057	8340
Ammonia	−77.7	351	−33.35	1370
Copper	1080	180	2310	7360
Freon-12	−155	—	−29.8	167.2
Gold	1063	67	2500	1870
Iron	1535	96−140	3000	6800
Lead	330	25	1170	732
Mercury	−39	12	357	300
Oxygen	−219	14	−183	210
Silver	961	110	1950	2300
Tungsten	3400	—	5830	4940
Water	0.00	335	100.0	2260

Note: $1 \dfrac{kJ}{kg} = 1 \dfrac{J}{g}$

Example 15-5 Melting Aluminum.

Some aluminum castings are to be made, and this requires the melting of 18 kg [40 lb_m] of the metal. The aluminum ingots are initially at a room temperature of 20°C [68°F]. How much heat must the aluminum absorb before it all melts?

From Table 15-9 we see that aluminum has a latent heat of fusion of 390 kJ/kg. So to melt 18 kg, we need

$$\text{Heat for melting} = (390 \text{ kJ/kg})(18 \text{ kg})$$

$$= 7020 \text{ kJ, or } 7.02 \text{ MJ}$$

But before the aluminum can be melted, it has to be heated to 658°C (its normal melting point). Since the metal starts out at 20°C, its temperature increase is 638C°. Using the specific heat capacity of solid aluminum as 895 J/(kg·C°), we can calculate the heat required to raise the aluminum to its melting point from Eq. (15-5):

$$\text{Heat} = \left(\begin{array}{c}\text{specific heat}\\\text{capacity of aluminum}\end{array}\right) \times \left(\begin{array}{c}\text{mass of}\\\text{aluminum}\end{array}\right) \times \left(\begin{array}{c}\text{temperature}\\\text{change}\end{array}\right)$$

$$= \left(895 \ \frac{J}{kg \cdot C°}\right)(18 \text{ kg})(638C°)$$

$$= 10.3 \text{ million J, or } 10.3 \text{ MJ}$$

The heat required for the entire process is then 7.02 MJ + 10.3 MJ, or about 17 MJ. By using the conversion factors in Table 9-3, this may also be expressed as 16 000 Btu. ◀

It doesn't take long to heat a teapot of water to its boiling point, but after that it takes a very long time to completely boil off the water. This is because water has such a high latent heat of vaporization. It takes a long time to supply 2260 kJ to each 1 kg of water, even when the burner is turned up high.

Because water absorbs so much heat when it vaporizes, it is commonly used in quenching, or quickly cooling, hot metals. If we plunge a red-hot chisel into a bucket of water, the water warms up just a little while the chisel cools off drastically. Most of the heat is carried away in the escaping steam. The same principle is used in spraying water on the redhot steel coming out of a hot-strip mill (Fig. 15-15). As the water turns to steam, each 1 kg of it absorbs 2260 kJ of heat from the hot steel. The steel can be cooled several hundred degrees in a matter of seconds.

Example 15-6 Quenching a Hot Metal.

A certain slab of steel has a mass of 3800 lb_m and is at a temperature of 1600°F. (This steel is red-hot.) How much water is needed to cool it to 600°F?

First let's calculate the amount of heat given up by the steel as it cools. Using Eq. (15-5) and 0.116 Btu/(lb_m·F°) as the steel's specific heat capacity, we have

$$\text{Heat lost} = \left(\begin{array}{c}\text{specific heat}\\\text{capacity of steel}\end{array}\right) \times \left(\begin{array}{c}\text{mass of}\\\text{steel}\end{array}\right) \times \left(\begin{array}{c}\text{temperature}\\\text{loss}\end{array}\right)$$

$$= \left(0.116 \ \frac{Btu}{lb_m \cdot F°}\right)(3800 \text{ lb}_m)(1000 \text{ F°})$$

Fig. 15-15 Quenching in a hot-strip mill. As the water boils off, it absorbs its heat of vaporization from the hot steel. This lowers the steel's temperature to a point where it can be coiled.

$$= 440\,000 \text{ Btu}$$

Now when water is sprayed on steel this hot, it very quickly boils off. Most of the heat lost by the steel then goes into vaporizing the water. From Table 15-9, we see that when 1 kg of water vaporizes, it absorbs 2260 kJ of heat. To express this in British thermal units per pound-mass, we can use the conversion factor from Table 15-8:

$$2260\ \frac{\text{kJ}}{\text{kg}} = \left(2260\ \frac{\text{kJ}}{\text{kg}}\right)\left(\frac{1000\ \text{J}}{1\ \text{kJ}}\right)\left(\frac{1\ \text{Btu/lb}_m}{2324.4\ \text{J/kg}}\right)$$

$$= 972\ \frac{\text{Btu}}{\text{lb}_m}$$

This is to say that 972 Btu of heat is absorbed when 1 lb$_m$ of vaporizes. The mass of water needed to absorb 440 000 Btu is

$$\text{Mass of water} = \frac{440\,000\ \text{Btu}}{972\ \text{Btu/lb}_m}$$

$$= 450\ \text{lb}_m$$

g cooled.

Notice that this is much less than the mass of

Application: Casting Steel

Today most steel is made in basic oxygen furnaces (BOFs), which can produce as much as 300 t of the alloy in less than 0.5 h. The molten steel is poured from the furnace into a large ladle suspended from a crane. The crane carries this liquid steel to a railroad platform, where it is cast into ingot molds, or to a continuous casting machine, where it is cast directly into slabs or billets.

Casting steel into ingots leads to a number of problems. Since the steel in a mold freezes from the bottom up and the outside in, and since it contracts as it freezes, a void (or hole) may be left in the center of the solid ingot. This void may remain in the form of cracks or other irregularities after the ingot is rolled to its finished form. Another problem is that an ingot must have a uniform temperature throughout before it can be rolled. But if it is just left standing in the open air, its outside cools faster than its inside, and the temperature is anything but uniform. So steel companies place their hot ingots into special ovens, called soaking pits, for several hours before they are rolled. This leads to handling and scheduling problems, as well as additional fuel expenses.

A more modern way to make high-quality steel products is through continuous casting. This eliminates the soaking pits and produces slabs or billets directly from the molten steel. In a typical installation, the ladle of molten steel is hoisted to the top of a 12-story building where it is emptied into a reservoir called a "tundish." At the bottom of this tundish are two to six nozzles that direct the molten metal into the same number of water-cooled copper molds. Rolls withdraw the solidified steel continuously from the bottoms of these molds. At the bottom of the building, the hot slabs or billets are bent so they move horizontally rather than vertically. A "flying shear" then cuts them into manageable lengths while they are still moving.

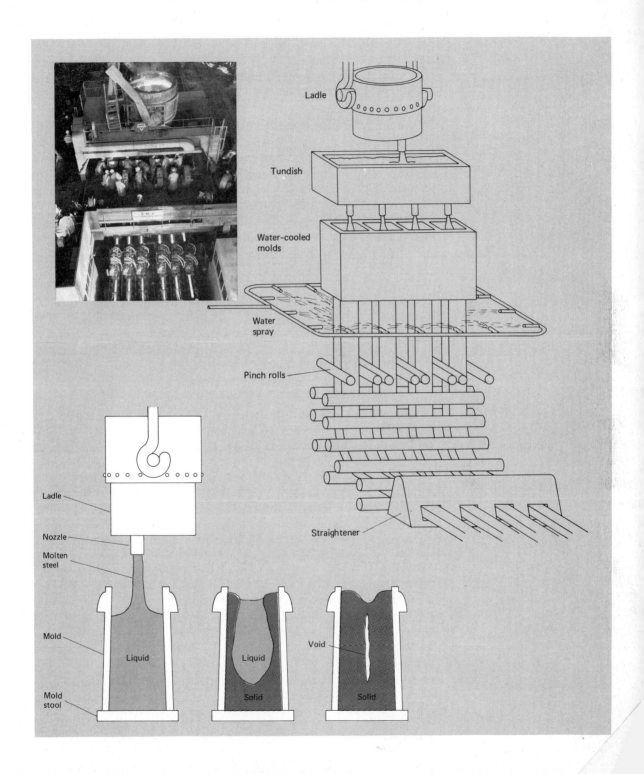

Ladle

Tundish

Water-cooled
molds

Water
spray

Pinch rolls

Straightener

Ladle

Nozzle

Molten
steel

Mold

Mold
stool

Liquid

Liquid

Solid

Void

Solid

Now the water absorbs heat not only when it vaporizes, but also when it heats up to its boiling point. So 450 lb$_m$ of water is *more* than enough to cool this steel by 1000 F°. This would allow for some loss because of splashing, for instance. ◀

One more thing needs to be said about latent heat. When steam condenses into water, it gives up its latent heat to its surroundings. This is why steam burns are so serious. One gram of boiling water gives up 260 J of heat if it is spilled on the skin and cools to body temperature. But 1 g of steam gives up an additional 2260 J, for a total of 2520 J. Thus the steam gives about 10 times as much heat to cause bodily damage as the same amount of boiling water at the same temperature. Extreme care should always be taken when working with steam.

Exercises

18. How much heat, in kilojoules, is needed to melt a 25-kg block of ice?
    Answer: 8400 kJ
19. How much heat, in British thermal units, is needed to melt a 100-lb$_m$ block of ice?
    Answer: 14 400 Btu
20. The condenser on a certain steam engine must convert 950 g of steam to water each second. (a) How many kilojoules of heat are given up by the steam each second? (b) If the temperature of the cooling water can increase by $5\bar{0}$C°, how many grams of cooling water must be circulated through the condenser each second? (c) What is the cooling water's flow rate in liters per second?
    Answers: (a) 2147 kJ, (b) 10 300 g, (c) 10.3 L/s
21. In a certain refrigeration system, 5.8 g of Freon-12 vaporizes each second. (a) How many joules of heat are absorbed by this Freon? (b) Calculate the cooling power in watts.
    Answers: (a) 970 J, (b) 970 W
22. Approximately how much water (in pound-mass) is needed to cool 121 lb$_m$ of aluminum from 968 to 212°F [520 to 100°C]? Assume that most of the cooling takes place when the water vaporizes.
    Answer: 20.1 lb$_m$
23. Approximately what volume of water (in liters) is needed to cool 10.7 kg of copper from 765 to 100°C? Assume that most of the cooling takes place when the water vaporizes.
    Answer: 1.24 L

We have seen that vaporization is a technical term for boiling. For a given pressure, there is only one temperature at which a given liquid will boil. But a liquid can *evaporate* at any temperature and pressure.

15-12 Evaporation

Definition: *Evaporation* is the process in which a liquid loses molecules from its surface.

evaporation

When a liquid boils, bubbles of gas form throughout it and float to the surface. But evaporation is strictly a surface phenomenon. Thus a liquid evaporates faster if it has a large exposed surface. This is why a wet car dries off very quickly out of doors, but an open can of beer (that someone has forgotten about) can sit around many days without much evaporation loss.

Evaporation also depends on whether the environment is open or closed. Figure 15-16 shows three jars filled with the same amount of alcohol. One jar has a lid, the second sits inside a larger inverted jar, and the third is open to the air. Obviously the alcohol open to the air continues to evaporate until it is gone. The alcohol inside the inverted jar evaporates some, then stops. But there is practically no evaporation in the covered jar.

We said earlier that the molecules in a liquid (or any other substance) are in constant motion. But it turns out that the molecules don't all move at the same speed. In Fig. 15-17 we see how the molecular speeds are distributed in a liquid at two different temperatures. At the higher temperature, the average speed is higher, of course. But at the lower temperature, at least a few of the molecules are traveling as fast as the average speed at the higher temperature.

When a liquid evaporates, these fast-moving molecules escape from the surface. If the environment is closed, these molecules bounce

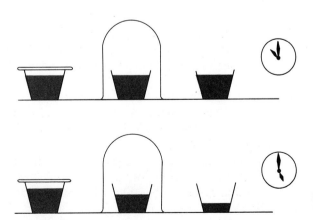

Fig. 15-16 Evaporation depends on whether the environment is open or closed.

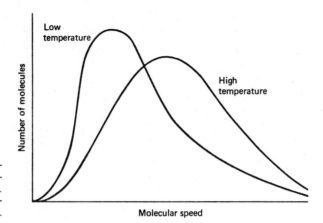

Fig. 15-17 At any given temperature, the molecules in a substance have a range of speeds. At higher temperatures, the average speed is higher.

around for awhile and eventually find their way back into the liquid. But if the environment is open, they escape forever.

We can see, then, why the surface area is important in evaporation: the surface molecules have the best chance of getting away without hitting another molecule and bouncing back in. We can also see why temperature affects the rate of evaporation: at higher temperatures, the average surface molecule is moving faster, and so there is a better chance that it will get away. One other factor affects the rate of evaporation. If the gas above the liquid is constantly moving, the escaped molecules are dragged away and will not get back into the liquid. This is why wet things dry off more quickly when the wind is blowing.

Since the fast-moving molecules are the ones that escape from a liquid's surface, evaporation lowers the average molecular speed. This means that the liquid's temperature is also lowered. If a hot liquid is allowed to evaporate, it cools off much more quickly than if it is covered and no evaporation occurs. This principle is used in the water bag (Fig. 15-18). Desert travelers often hang a canvas bag filled with water on the front of their jeeps. Capillary action keeps the outside of the bag wet, and there is a high rate of evaporation from this large surface—particularly when the vehicle is moving. This evaporation carries away the "hot" molecules (the fast-moving ones), so the remaining water is left much cooler than the desert air.

If you need more convincing that this *evaporation cooling* really happens, you can try the simple demonstration in Fig. 15-19. Wrap the bulb of a thermometer with a wet rag and blow on it. You will see the temperature drop a bit. If the rag is wet with alcohol or ether, you can get the temperature to drop farther because these liquids evaporate more quickly.

Fig. 15-18 The water bag. Water in a canvas bag is kept cool by inducing some of it to evaporate.

Summary

Temperature is related to the average kinetic energy of the molecules in an object. When the molecules are moving fast, the temperature is high. When they are moving slowly, the temperature is low. When the molecules are moving as slowly as they can, the temperature is absolute zero. No lower temperature is possible. On the other hand, there appears to be no limit to the highest temperature; plasma temperatures of several million degrees have been generated in research laboratories.

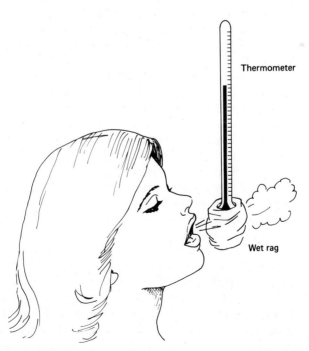

Thermometer

Wet rag

Fig. 15-19 A simple demonstration of evaporation cooling. A bigger temperature drop is generated if the rag is wet with alcohol or ether.

Application: Cavitation in Liquid Streams

When a liquid moves through a nozzle or other constriction in its path, it speeds up and its pressure drops (Bernoulli's principle). In extreme cases, this pressure reduction may be enough to cause the liquid to boil at ordinary temperatures. The resulting rapid formation and collapse of vapor bubbles is called *cavitation*.

This effect is often experienced with fire hoses. If the nozzle opening is reduced to shoot the stream a great distance, the production of rapidly expanding vapor bubbles may cause the stream to disperse and not travel very far at all. On the other hand, if the intention is to spread the water around, then cavitation may be desirable. Cavitation is also a frequent cause of structural damage in ship propellers, water turbines, and pump parts.

It is difficult and impractical to measure temperatures by measuring molecular kinetic energy. Instead, temperature is always measured by monitoring some other quantity that depends on temperature. Some such quantities are the length of a solid, the volume of a liquid, the pressure of a gas, the color of a hot glowing object, and electrical properties like resistance and thermocouple voltage.

Heat normally flows in response to temperature differences. The flow of heat can also change the temperature: if there is no phase change, an object's temperature drops when heat flows out and increases when heat flows in. An object's specific heat capacity tells us how much heat it stores in a 1° temperature increase.

Heat may also cause phase changes. An object's latent heat tells us how much heat is needed to change its phase.

In this chapter, we have seen some of the effects of heat flow. In the next chapter, we look in greater detail at how heat flows from one place to another.

Terms You Should Know

heat	gas thermometer
temperature	bimetallic strip

resistance thermometer
thermocouple thermometer
optical pyrometer
Celsius scale
absolute zero
kelvin
Rankine scale
coefficient of linear expansion

Charles' law
specific heat capacity
critical point
latent heat of fusion
latent heat of vaporization
evaporation
evaporation cooling

Problems

1. A cylindrical stone-lined well is filled with ice. The well measures 1.34 m in diameter and 4.56 m in depth. (a) How much heat, in joules, is needed to melt all the ice? (b) If heat flows into the well at a rate of 2.0 MJ/h, how long does it take to melt the ice?

2. A desalinization plant produces fresh water from seawater by boiling it and condensing the steam. This process leaves the salt and other minerals behind. To supply a small community, the plant must produce 500 000 L of pure water each day. Assume that the seawater is initially at 20°C and that its thermal characteristics are the same as those of fresh water. (This is not quite true, but it will serve as an approximation.) (a) How much heat is needed to produce 1 L of pure water? (b) How much heat does the plant require each day? (c) Calculate the thermal power requirement in kilowatts.

3. Oxygen in a cylinder is under a pressure of 2200 psig when the temperature is 65°F. The test pressure of the cylinder is 3300 psig. At what temperature will the oxygen pressure exceed the cylinder's test pressure?

4. A certain solar heating system must store enough heat for a 3-d operation with no sun. Suppose that an average 30 000 Btu/h of heat is needed to heat the building. The collectors can produce hot air or hot water at 180°F. The building is to be maintained at 68°F. (a) What mass of stone (in pound-mass) would be needed to store the heat required for a 3-d operation? (b) What mass of water would be needed to store this much heat? (c) Approximately what volume of stone would be needed? (d) Approximately what volume of water would be needed? (Note: This problem has to do with the storage of the heat; its collection is another matter.)

5. The cooling system in a certain engine holds 16 qt of coolant. The engine is started on a day when the outside temperature is −10°C. How much heat must the coolant absorb to warm up to an operating temperature of 100°C? (Assume that the coolant is mostly water.)

16
Heat Transfer

Introduction   We have seen how an object's temperature, phase, and other properties can change when it absorbs or gives off heat. Now it's time to look at how heat is generated and transferred from one place to another.

There are three different processes by which heat is transferred: conduction, convection, and radiation. Conduction takes places mainly in solids, convection is the transfer of heat through fluids, and radiation transfers heat through gases as well as some liquids and solids and even a vacuum. Often, two or all three of these processes take place at the same time. But regardless of how heat is generated, and how it is transferred, it is still measured in the same energy units we've already discussed: kilowatt-hours, British thermal units, or joules, for instance.

16-1 The First Law of Thermodynamics   Since heat is a form of energy, all heat flow must conform to the law of conservation of energy. Figure 16-1 gives us a convenient way of looking at this. The dotted line represents the boundary of some system of interest: the water in a boiler, a building that is being heated, a compressed-gas cylinder, an engine, or anything else where heat flow is important.

Now there is always some energy stored in the system's atoms and molecules. If the temperature is high, the molecules move very fast and have a large amount of stored kinetic energy. If the system contains an unstable chemical compound (gasoline, for instance), the molecules

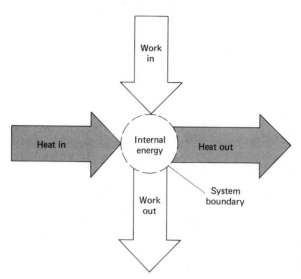

Fig. 16-1 A system can have heat flowing in and heat flowing out while work is done on it and it does work itself. At the same time, its internal energy may be changing. The first law of thermodynamics is a relationship between these quantities.

have a large potential energy. Energy is also associated with the system's phase: the molecular energy in steam at 100°C is greater than in water at 100°C. The sum of these molecular energies is called the system's internal energy.

> Definition: A system's *internal energy* is the total energy stored in its atoms and molecules.

internal energy

Although it is difficult to say just how much internal energy a given system actually has, we will find it easy (and useful) to say how much the internal energy *changes* in certain kinds of processes.

In some processes heat flows into the system, and in others heat flows out. Sometimes, it does both. At the same time, work may be done on the system, or work may be done by the system, or both. Because energy cannot be created or destroyed, any change in a system's internal energy must be balanced by a flow of heat, or the performance of work, or both. This relationship is described by the first law of thermodynamics.

> *First Law of Thermodynamics*: The net heat flowing into any system is completely accounted for in the net work done by the system and the change in the system's internal energy. Written as a formula:

first law of thermodynamics

(Heat in) − (heat out) = (work out) − (work in) + (final internal energy) − (original internal energy)   (16-1)

Now there is really nothing complicated or mysterious about the first law of thermodynamics. It really just amounts to the law of conservation of energy applied to heat flow. You can't have heat flowing into something without it being accounted for in the system's internal energy or work output. And you can't have heat flowing out of something except at the expense of internal energy or work.

Example 16-1 The Human Engine.

A medium-sized person typically gives off 430 000 J, or 430 kJ, of heat each waking hour. At least this is the case when the air temperature is around 20°C. At colder temperatures or in water, the rate of heat loss is greater.

Of this 430 kJ, about half is sensible heat that warms the air around the person's body. The other half is latent heat that vaporizes the body's perspiration. When this person is sleeping, the total heat loss drops to about 280 kJ/h. The person can also do around 260 kJ of work in 1 h (when awake, of course). Based on this, what can we say about this body's internal energy if the person goes a day without food?

The person is shown in Fig. 16-2. He or she has no work or heat input. In one day, this person's work output is around 2100 kJ (8 h at about 260 kJ/h). The heat output is 430 KJ/h for 16 h, plus 280 kJ/h for 8 h, for a total of about 9100 kJ. All this heat and work must come at the expense of internal energy. This person's internal energy therefore decreases by 11 200 kJ in one 24-h day. If this person is not to quickly lose weight, she or he must restore this 11 200 kJ through chemical

Fig. 16-2 The first law of thermodynamics applied to the human engine. People work and generate body heat at the expense of their internal energy. Every so often, they have to restore the lost internal energy by eating.

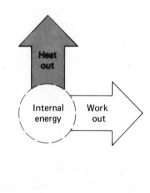

energy in food. This 11 200 kJ is equivalent to 2700 kcal or "food calories."

We might mention that these figures were based on a body mass of about 70 kg [155 lb$_m$]. A heavier person will generate more heat and will also do more work (even just in moving around). Thus a heavy person needs more than 11 200 kJ of food energy to balance the body's energy losses in a day. ◄

As Example 16-1 showed, each time we eat we are refueling a biological engine. Between meals, the engine operates by depleting its own supply of internal energy.

But in something like an electric motor, there is no stored fuel. The flow of electricity does work on the motor, and the motor's rotating shaft does work on something else. At the same time, some heat is generated. This is shown in Fig. 16-3. Ideally, the motor should warm up to an acceptable operating temperature and then remain at this temperature. This requires that the work and heat output balance the electrical work

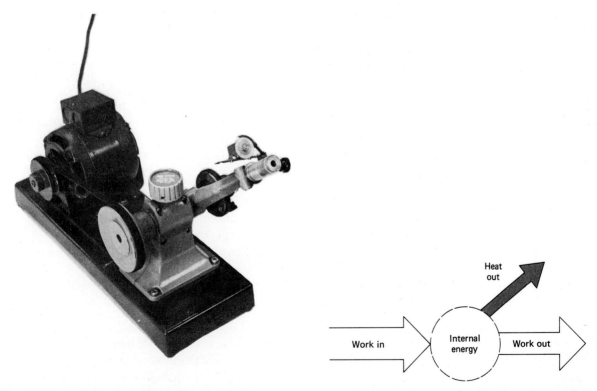

Fig. 16-3 Energy flow in an electric motor. If provision is not made to allow the heat to flow out, the motor's internal energy will increase until it burns up.

input. But if provision is not made to carry the heat away, the motor's internal energy must increase. As it increases, the motor's temperature increases. Eventually, a point is reached where the motor's insulation melts, and the motor burns up. The motor runs at a steady temperature only if there is enough ventilation to carry away the heat generated.

Exercises

1. On a cold winter day, a certain home loses 120 000 Btu of heat to the outside in 1 h. (*a*) How much heat must be supplied by the furnace if the home's temperature is not to decrease during this hour? (*b*) What happens if the furnace supplies 150 000 Btu of heat during the hour? (*c*) What happens if the furnace supplies 90 000 Btu of heat during the hour?
   Answers: (*a*) 120 000 Btu, (*b*) home gets warmer, (*c*) home cools off
2. An electric motor has a power output of 0.75 hp and an efficiency of 95 percent. (*a*) How many British thermal units of heat does it generate in 1 h of continuous operation at full load? (*b*) Express this heat in kilojoules.
   Answers: 100 Btu, (*b*) 106 kJ

16-2 Expansion Cooling; Compression Heating

When a gas is quickly compressed, its temperature goes up. When a compressed gas is allowed to expand quickly, its temperature drops—this is called *expansion cooling*. These effects are a direct consequence of the first law of thermodynamics.

Figure 16-4 shows a small CO_2 cartridge of the type used in pellet guns. The nozzle is sealed with a lead plug. When the plug is punctured, the high-pressure carbon dioxide drives itself out of the cartridge at high speed. Considerable work is done in expanding the gas (remember, work is done when a force acts through some distance). But we have put practically no work into the system, and we have certainly put in no heat. Thus the work output can come only at the expense of the gas's internal energy. This causes the gas's temperature to drop drastically. Frost forms on the outside of the cartridge, and some of the CO_2 itself may freeze.

When a gas is compressed, just the opposite happens. Most of the work done in compressing the gas goes into its internal energy, and its temperature rises. This principle is used in the diesel engine (Fig. 16-5). When a cylinder full of air is quickly compressed into about 5 percent of its original volume, the temperature gets high enough to ignite an oil spray. Thus the diesel engine needs no spark plugs. This process is called *compression heating*.

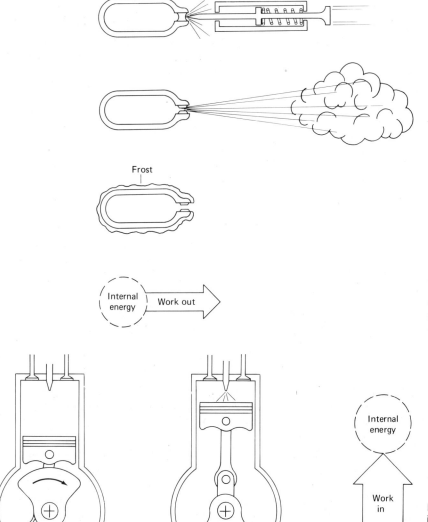

Fig. 16-4 Puncturing a CO_2 cartridge. Rapid expansion of a gas decreases its internal energy and lowers its temperature.

Fig. 16-5 Compression stroke in a diesel engine. Rapid compression of the air increases its internal energy. This makes its temperature increase enough to ignite an oil spray.

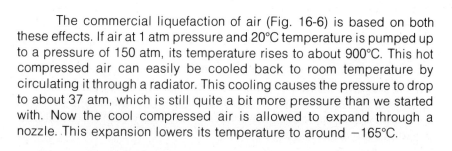

The commercial liquefaction of air (Fig. 16-6) is based on both these effects. If air at 1 atm pressure and 20°C temperature is pumped up to a pressure of 150 atm, its temperature rises to about 900°C. This hot compressed air can easily be cooled back to room temperature by circulating it through a radiator. This cooling causes the pressure to drop to about 37 atm, which is still quite a bit more pressure than we started with. Now the cool compressed air is allowed to expand through a nozzle. This expansion lowers its temperature to around −165°C.

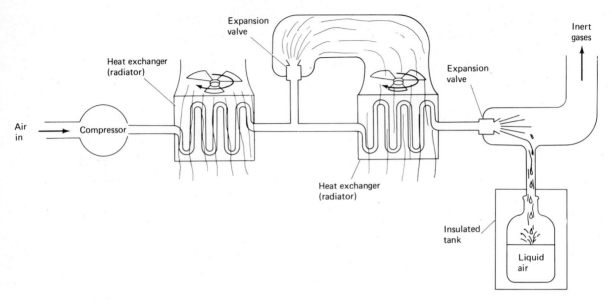

Fig. 16-6 Production of liquid air. Expansion cooling allows the air to act as its own refrigerant.

Of course, −165°C is not quite cold enough to liquefy air. But it gives us quite a head start if we use it to cool some room-temperature compressed air before it expands. In this second stage, compressed air that has been cooled by the first stage again expands through a nozzle. This causes such a decrease in temperature that the oxygen and nitrogen in the air liquefy. It is interesting to note that the air acts as its own refrigerant.

A certain part of the air is never liquefied by this procedure: a mixture of the low-boiling inert gases remains. These gases can be separated and bottled for commercial use. The liquid air then consists of nitrogen and oxygen. These can be separated by controlling the temperature so the nitrogen boils off, leaving only liquid oxygen. Almost all oxygen for industrial use is manufactured in this way.

Exercises

3. A steam hammer does 650 kJ of work in forming a large turbine shaft at a forge. If 250 kJ of heat is conducted away, what happens to the shaft's internal energy? What happens to its temperature?
    Answer: The internal energy increases by 400 kJ, which causes the shaft's temperature to increase.

4. Steam entering a large turbine engine is at a high temperature and pressure. The engine does work at a rate of 600 kW, and waste heat is carried away by the condenser and other cooling provisions at a rate of 1000 kW. (*a*) Does the steam get hotter or cooler as it travels through the engine? (*b*) What is the rate at which the steam loses internal energy?
   Answers: (*a*) cooler, (*b*) 1600 kW

When we burn a fuel, the stored chemical energy turns into heat and work. Why work? Because combustion produces hot gases that naturally expand. This expansion powers our car engines, for instance. In Fig. 16-7, the expansion does work in propelling a rocket.

    We can burn fuel oil in a furnace with the intention of producing heat for a home, or we can burn the same oil in a diesel engine with the intention of producing work. The *total* energy (heat plus work) given off by each kilogram of fuel is about the same in both cases. The difference between the engine and the furnace lies in how this total energy is split up. In the engine the oil produces more work but less heat than in the

16-3 Heat of Combustion

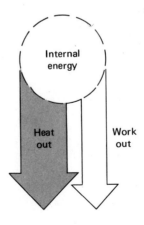

Fig. 16-7 A rocket. When any fuel is burned, heat and work are produced at the expense of the fuel's internal energy. (Photo courtesy NASA)

furnace. Thus the available internal energy per kilogram of fuel can be a very useful number. This number is called the fuel's heat of combustion.

heat of combustion

Definition: The *heat of combustion* of a fuel is the amount of internal energy that can be converted to heat and work by burning a unit amount of the fuel. For solid and liquid fuels, heat of combustion is expressed in British thermal units per pound-mass, or joules per kilogram and its SI multiples. For gaseous fuels, it is often expressed on a volume basis: British thermal units per cubic foot, or kilojoules per liter, for instance.

Table 16-1 lists the heats of combustion of some solid and liquid fuels. One kilogram of gasoline, for instance, has about 44 MJ of available internal energy that can be released when it burns. This is equivalent to about 19 000 Btu from each pound-mass of gasoline. Also listed is the air-fuel ratio needed for complete combustion. For instance, 14.2 kg of air is needed to burn 1 kg of crude oil, and 14.2 lb_m of air is needed to burn 1 lb_m of crude oil. In furnaces, the amount of air supplied should always exceed this figure. Otherwise, fuel will be wasted and the flue gas (chimney exhaust) will contain poisonous carbon monoxide.

Table 16-2 lists the heats of combustion of some gaseous fuels. Since fuel gases are usually metered according to volume (cubic feet, liters, etc.), it's useful to know their heating value on a volume basis. We see, for instance, that burning 1 ft³ of natural gas releases up to 1000 Btu,

TABLE 16-1   *TYPICAL HEATS OF COMBUSTION OF SOME SOLID AND LIQUID FUELS.**

Fuel	Heat of combustion		Air-fuel ratio for complete combustion, lb_m/lb_m or g/g
	Btu/lb_m	MJ/kg	
Butane	20 000	46	15.4
Gasoline	19 000	44	14.8
Crude oil	18 000	42	14.2
Fuel oil	17 500	41	13.8
Coke	14 500	34	11.3
Coal, bituminous	13 500	31	10.4

* Actual values will vary slightly depending on geographical origin. Values do not include latent heat in any water vapor formed as a product of combustion.
Note: 1 MJ/kg = 1000 kJ/kg = 1000 J/g

TABLE 16-2   TYPICAL HEATS OF COMBUSTION OF SOME GASEOUS FUELS.*

| Fuel | Heat of combustion | | | | Air-fuel ratio for complete combustion, ft³/ft³ or L/L |
| | On mass basis | | On volume basis | | |
	Btu/lb$_m$	MJ/kg	Btu/ft³	kJ/L	
Hydrogen	61 400	143	275	10.3	2.4
Methane	23 800	55.2	900	33	9.6
Propane	22 200	51.5	2400	88	13.7
Natural gas	23 600	54.8	1000	37	11.9
Coke gas	23 000	53.6	1300	50	14

*Data applies to gases at 1.0 atm pressure and 15.6°C [60°F]. Actual values vary slightly depending on geographical origin. Values do not include latent heat in any water vapor formed as a product of combustion.

Note: 1 MJ/kg = 1000 kJ/kg = 1000 J/g
      1 kJ/L = 1 J/cm³ = 1 MJ/m³

and complete combustion of this gas requires at least 11.9 ft³ of air. The assumption is that these volumes are measured at standard atmospheric pressure. Natural gas lines are normally maintained at pressures only slightly higher than atmospheric.

Hydrogen offers interesting possibilities as a fuel. When burned, it produces nothing but water vapor. In terms of volume, we see that the heat of combustion of hydrogen is quite low, but in terms of mass it is very high. If a practical way could be found to keep hydrogen safely liquefied until it was actually needed, the large amount of energy per unit mass would be very attractive. And hydrogen is abundantly available in our oceans and streams. What we need are efficient methods to extract it.

Example 16-2 Fuel Consumption of a Furnace.

A certain oil furnace has a rated output of 95 000 Btu/h and an input of 130 000 Btu/h. These figures are usually stamped in a metal tag riveted to the furnace. (Note that this furnace has an efficiency of 73 percent.) What is the rate of fuel consumption when the furnace is operating at capacity?

From Table 16-1, we see that fuel oil can supply 17 500 Btu for each 1 lb$_m$ that is burned. We will assume that the burner has been adjusted so enough air is supplied. To put in 130 000 Btu in 1 h, we then need

$$\text{Mass of oil} = \frac{130\ 000\ \text{Btu}}{17\ 500\ \text{Btu/lb}_m}$$

$$= 7.43\ \text{lb}_m$$

If we want to find the volume of this oil, we can consult Table 6-2 to find that the density of fuel oil is about 58 lb_m/ft^3. Then our 7.43 lb_m of oil has a volume of 7.43/58 ft^3, or 0.13 ft^3. Since there is 7.48 gal in 1 ft^3,

$$\text{Volume} = 0.13\ \text{ft}^3 \left(\frac{7.48\ \text{gal}}{1\ \text{ft}^3} \right)$$

$$= 0.96\ \text{gal}$$

(This is the U.S. liquid gallon.) Remember that this was all based on one hour's operation of the furnace. If the furnace operates at full capacity for a 24-h day, it then uses about 23 gal of fuel. You may be interested in calculating the cost of this one day's heating, based on the local price of fuel oil. ◄

Exercises

5. How much heat, in British thermal units, is released by the complete combustion of 1 ton of coal?
   Answer: 27 million Btu

6. One metric ton of coal is burned. (*a*) How much heat is released, in joules? (*b*) How many metric tons of air (minimum) are needed to burn this much coal?
   Answers: (*a*) 31 billion J, or 31 GJ; (*b*) 10.4 t

7. Some 850 m^3 of natural gas is burned. How much heat is released in gigajoules?
   Answer: 3.2 GJ

8. What volume of propane gas would have to be burned to release 830 MJ of heat?
   Answer: 9.5 m^3

9. A certain furnace has an input capacity of 40.7 kW. (*a*) How many cubic meters of natural gas will it consume in one day, operating at full capacity? (*Note:* 1 kW = 1 kJ/s.) (*b*) How many cubic meters of air are needed to support this combustion?
   Answers: (*a*) 94 m^3, (*b*) 1100 m^3

16-4 Thermal Conduction   When one end of a metal bar is placed in a furnace, the other end soon gets very hot (Fig. 16-8). This is because metals are good conductors of heat. The conduction process amounts to a sort of chain reaction in the

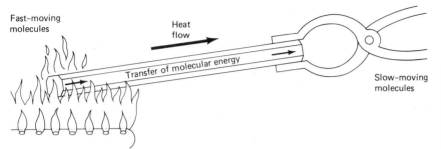

Fast-moving molecules

Heat flow

Transfer of molecular energy

Slow-moving molecules

Fig. 16-8 Heat is conducted as fast-moving molecules transfer energy to slow-moving molecules. Metals are good conductors because they have free electrons that also take part in this process.

molecules. The flame has a high temperature, which is the same as saying that its molecules (and atoms, ions, and electrons) are moving very fast. As these fast-moving particles bounce off the bar, they transfer some energy to the molecules of the metal. Of course, the metal molecules aren't free to move very far unless the metal melts. But they do begin to vibrate quite rapidly. This increase in the vibration rate is transferred to the adjacent molecules that are not in the flame. As these molecules begin to vibrate faster, they transfer energy to the next molecules down the line, and so on. Thus there is a net transfer of energy from the hot end of the bar to the cold end. This is called *thermal conduction*.

Since all substances have molecules, anything can conduct heat in this way. Why, then, do some things conduct heat better than others? There are several reasons. One has to do with how massive the molecules are—light molecules speed up quicker, so they tend to transfer energy faster. But more importantly, some atoms and molecules do not hold all their electrons very tightly. In a metal, vast numbers of free electrons swim around from atom to atom. These electrons also move faster when heated; and being unattached, they transfer energy very quickly (the fact that the electrons are very light also helps). It is these same free electrons that make metals good conductors of electricity. When we call a material a good *conductor*, then, this usually means for electricity as well as heat. Materials that are good *insulators* do not conduct either electricity or heat very well.

The ability of a substance to conduct heat is described by its thermal conductivity.

Definition: A substance's *thermal conductivity* is the amount of heat energy per unit time that will flow through a sample of unit area and unit thickness under a temperature difference of 1°.

thermal conductivity

We can appreciate this definition better by looking at Fig. 16-9. Here is a sample of material 1.00 m thick and 1.00 m² in face area. Since

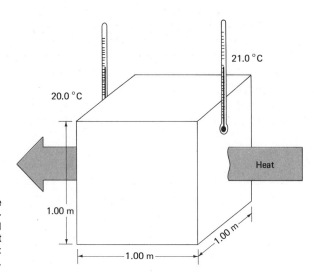

Fig. 16-9 Standard sample size for expressing thermal conductivity in SI units. The resulting SI unit of conductivity is the watt per metre, per Celsius degree: W/(m•C°).

one side is at a temperature of 21°C while the other is at 20°C, the temperature difference across the sample is 1.0 C°. As a result of this temperature difference, heat flows through the sample. If the sample is aluminum, it conducts heat at a rate of 220 W. If it is concrete, the rate of heat flow is about 1.3 W. These values are the thermal conductivities of aluminum and concrete.

In USCS units, the rate of heat flow is expressed in British thermal units per hour. Now the standard sample size is 1 ft² in area by a thickness of 1 in, and the temperature difference is in Fahrenheit degrees (F°). This standard sample size is shown in Fig. 16-10.

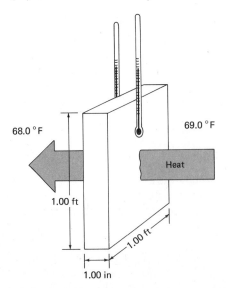

Fig. 16-10 Sample size for expressing thermal conductivity in USCS units. The unit of conductivity is Btu-inches per hour, per square foot, per degree Fahrenheit: Btu·in/(h·ft²·F).

Table 16-3 lists the thermal conductivities of some common substances in terms of these two standard sample sizes. Notice the tremendous range of values between the high value for silver and the low value for trapped air. Of course, a vacuum conducts no heat at all since there are no molecules to vibrate.

Now very seldom do we have an object exactly the same size as our standard samples. And very seldom is the temperature difference

TABLE 16-3   TYPICAL THERMAL CONDUCTIVITIES OF SOME COMMON SUBSTANCES

Substance	Thermal conductivity	
	$\dfrac{W}{m \cdot C°}$	$\dfrac{Btu \cdot in}{h \cdot ft^2 \cdot F°}$
Air, trapped*	0.025	0.17
Aluminum	220	1500
Brass	$10\overline{0}$	730
Brick, common	0.72	5.0
Celotex	0.049	0.34
Concrete	1.3	9.0
Copper	385	2670
Fiber glass	0.037	0.26
Glass	0.79	5.5
Gold	320	2200
Oak	0.147	1.02
Pine, white	0.11	0.78
Rock wool	0.040	0.28
Sawdust	0.059	0.41
Silver	420	2900
Steel, 0.1 percent carbon	59	410
Steel, 1.0 percent carbon	46	320
Stone	1.7	12
Urea-formaldehyde foam	0.030	0.21
Water, trapped*	0.63	4.4
Vacuum	0	0

* No fluid flow.

Note: $1 W = 1 \dfrac{J}{s} = 3.414\,4 \dfrac{Btu}{h}$

$1 Btu = 1.054\,4 \text{ kJ}$

exactly 1°. To find the actual heat flow in all cases, we can use the following formula:

$$\text{Rate of heat flow} = \frac{\left(\begin{array}{c}\text{thermal}\\\text{conductivity}\end{array}\right)\left(\text{area}\right)\left(\begin{array}{c}\text{temperature}\\\text{difference}\end{array}\right)}{(\text{thickness})} \qquad (16\text{-}1)$$

Notice that the heat flow increases if the area or temperature is increased but is lower when the material is thicker.

Example 16-3 Heat Transfer through a Brick Wall.

The solid brick wall shown is part of a building. The inside surface of the wall is at 12.0°C while the outside surface is at 5.0°C. Let's find the rate of heat conduction through the wall. Using SI units, this result is expressed in watts (W).

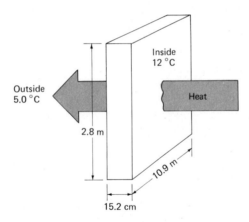

To use Eq. (16-1) with SI units, we need to express the area of the wall in square meters, the thickness in meters, and the temperature difference in Celsius degrees. The area is

$$\text{Area} = 2.8\ \text{m} \, (10.9\ \text{m})$$

$$= 30.5\ \text{m}^2$$

Expressing the thickness in meters gives

$$\text{Thickness} = 15.2\ \text{cm} \left(\frac{1\ \text{m}}{100\ \text{cm}}\right)$$

$$= 0.152\ \text{m}$$

And the temperature difference is 12.0°C − 5.0°C, or 7.0°C. Then

$$\text{Rate of heat flow} = \frac{(\text{thermal conductivity})(\text{area})(\text{temp. difference})}{\text{thickness}}$$

$$= \frac{[0.72 \text{ W/(m} \cdot \text{C°)}](30.5 \text{ m}^2)(7.0 \text{ C°})}{0.152 \text{ m}}$$

$$= 1\bar{0}00 \text{ W}$$

This answer may also be expressed as 1.0 kW. Or, using the fact that 1 W = 3.414 4 Btu/h, we can express it in USCS units as 3400 Btu/h. ◀

Example 16-4 Heat Transfer through a Steel Tank.

The steel holding tank has a total surface area of 51.8 ft². It is made of 0.1 percent carbon steel 0.132 in thick. The pressurized water in the tank is at 210°F while the tank's outside surface is at 206°F. What is the rate of heat loss from the tank?

      Using Eq. (16-1) with old USCS units, we express the tank's surface area in square feet, the thickness of the steel in inches, and the temperature difference in degrees Fahrenheit. Substituting the values into the equation, we get

$$\text{Rate of heat flow} = \frac{(\text{thermal conductivity})(\text{area})(\text{temp. difference})}{\text{thickness}}$$

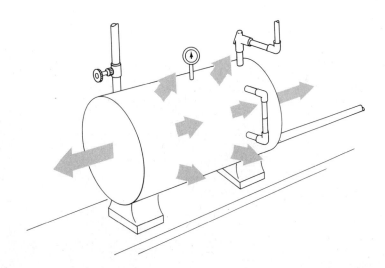

Application: Heat Pipes

Transferring heat from one place to another can be a difficult job, requiring a closed fluid circuit, pumps, valves, thermostats, and regulators. But for distances less than about 20 m where the heat is to be transferred uphill, heat pipes offer a cheap and simple solution.

A heat pipe is a hollow tube, sealed at both ends, with an inside diameter of a few centimeters. Inside is a small amount of water, Freon, acetone, or some other fluid. When the lower end of the tube is heated, the liquid boils and its vapor travels upward to the cooler end, where it condenses and gives up the heat it absorbed at the hot end. The condensed liquid then flows downhill back to its starting point, and the cycle repeats continuously.

Since the liquid must boil at the hot end, it may seem as if heat pipes are useful only at high temperatures. This is not true. By using low-boiling liquids under reduced pressures, heat pipes can be built to operate when the hot end is 0°C or colder. Of course, the cold end must be at an even lower temperature.

Large numbers of heat pipes are used on the Alaskan pipeline. The oil must be warm enough to allow it to flow easily, but this heat cannot be allowed to flow through the pipeline supports to the ground. If it did, the permafrost would melt into a bog, the pipeline would sink, and there would be an ecological disaster. So the heat flowing through the support structure must be conducted away before it enters the ground. In the photograph, the "antlers" on the pipeline supports are the finned cold ends of heat pipes. Inside the structure are the hot ends, whose temperature is still below 0°C.

One interesting feature of gravity-operated heat pipes is that they pass heat in only one direction, something like a fluid check valve. Because of this property, they are being installed in some solar heating units. When the solar collector plate is hot, the heat pipe transfers heat into a water reservoir for storage. When the collector plate is cold (at night, for instance), the heat transfer stops and no heat is lost from the reservoirs to the outside. The really nice thing is that this is done without valves, pumps, or thermostats.

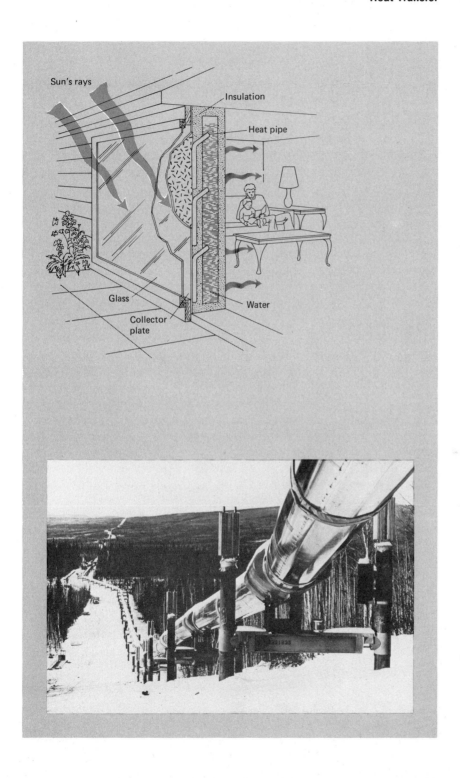

Sun's rays

Insulation

Heat pipe

Glass

Collector plate

Water

$$= \frac{(410)(51.8)(4)}{0.132} \frac{Btu}{h}$$

$$= 640\,000 \ Btu/h$$

Notice that this is a very high rate of heat transfer. (A typical home furnace has a capacity of around 100 000 Btu/h.) This high heat loss results from a combination of factors: steel is a good conductor, the tank has a large surface area to pass the heat, and the metal is fairly thin. ◀

Example 16-4 showed that thin materials may conduct a large amount of heat with only a small temperature difference. In fact, there can never be more than a few degrees temperature difference between the two sides of a thin plate. Thus touching the outside of a water pipe gives a good indication of how hot or cold the water is inside.

When a kettle of water is heated on a stove, the bottom of the kettle is never much hotter than the water—even though the outside of the kettle is in contact with a flame. The same holds true for the water tubes in industrial boilers. A simple demonstration of this is shown in Fig. 16-11.

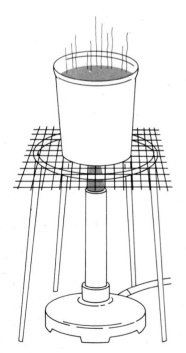

Fig. 16-11 Boiling water in a paper cup. Because the cup is so thin, there can never be more than a few degrees temperature difference between the inside and the outside of the cup's bottom.

If a paper cup full of water is placed over a burner, the cup simply cannot get hot enough to burn until all the water has boiled off.

Exercises

10. The walls of a garage are made of pine sheathing 0.75 in thick. There is a total of 960 ft² of wall area. The inside wall surface is at 45°F, and the outside surface is at 35°F. (a) Find the rate of heat loss resulting from conduction through the walls. (b) How would this result be affected if the wall thickness were doubled?
    Answers: (a) $1\bar{0}$ 000 Btu/h, (b) it would be cut approximately in half

11. A certain windowpane measures 52 cm by 68 cm and is 0.32 cm thick. With the window closed, the outer surface is at 32°C [90°F] while the inside surface is at 29°C [84°F]. Notice that the building is being air-conditioned. Calculate the rate of heat flow into the building through the window, in watts.
    Answer: 260 W

12. A brass rod is 50 cm long and 2.5 cm² in cross-sectional area. One end is heated to 120°C [248°F]. If the other end is at a temperature of 50°C [122°F], what is the rate of heat conduction through the rod from one end to the other?
    Answer: 3.5 W

13. How is the answer to Exercise 12 affected if (a) everything else is the same but the rod has twice the cross-sectional area; (b) everything else is the same but the rod is twice as long; (c) everything else is the same but the hot end is at 190°C; (d) everything else is the same but the rod is made of aluminum?
    Answers: (a) double the heat transfer, or about 7.0 W; (b) half the heat transfer, or about 1.8 W; (c) double the heat transfer, or about 7.0 W; (d) double the heat transfer, or 7.0 W

16-5 Convection

Convection is the transfer of heat by a moving fluid. With *forced convection,* the fluid is pumped from someplace hot to someplace cooler. As it moves, the fluid carries heat along with it. A home hot-water heating system transfers heat by the forced convection of water from the boiler into the radiators. The cooling system of a modern car engine also uses forced convection, since it depends on the water pump to circulate the coolant from the hot engine block to the cooler radiator. Just as with conduction, the direction of heat flow is from high-temperature regions to lower-temperature regions.

But convection may also take place without a pump. In this case, it is called *natural convection*. This is an extremely important effect. In fact, when we hear the word "convection" used by itself, we can be pretty sure it means natural convection.

How does this happen? We have already seen that most substances expand when heated. If just part of a fluid is heated, the hot part becomes less dense than the cold part, and so it tries to float upward. The cooler and denser part of the fluid sinks to take its place. This process may develop upward and downward currents in the fluid (as with hot, smoky air rising from a chimney), or it may just create a general turbulence. Either way, heat is transferred from the hot part of the fluid to the cold part. This is why when we heat a pan of water on the bottom, all the water comes to pretty much the same temperature.

Early automobiles (the Model T Ford, for instance) relied on natural convection currents to move the engine coolant from the engine block to the radiator and back again. There was no water pump. Of course, the coolant circulated rather slowly, but then the engine wasn't very large and didn't have a very big cooling requirement.

A more modern application of the same idea is the solar hot-water heater shown in Fig. 16-12. The water is heated in the solar collector panel, where it expands and rises into the water tank. Here it displaces the cooler and denser water, which sinks into the return pipe and enters the bottom of the collector. A continuous flow is maintained with no pump, and heat is transferred from the collector to the water tank.

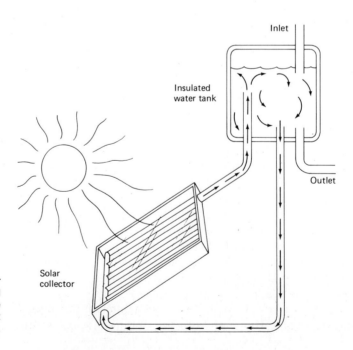

Inlet

Insulated
water tank

Outlet

Solar
collector

Fig. 16-12 A solar hot-water heater that depends on natural convection currents to circulate the water. The water tank must be above the collector.

But pipes aren't always necessary for natural convection. In Fig. 16-13, we see how natural convection helps circulate the warm air from a furnace. The blower pressure forces warm air through the ducts; but once it enters the room, this air is on its own. Here it rises, displacing colder air that drops to the floor and eventually enters the cold-air return. Unfortunately, the warm air keeps right on rising—all the way up to the roof, if it can. This is why attics should be well insulated. If the home is air-conditioned, just the opposite happens: the cooled air wants to drop. Ideally, most of the air conditioning in a multistory building should be on the upper floors.

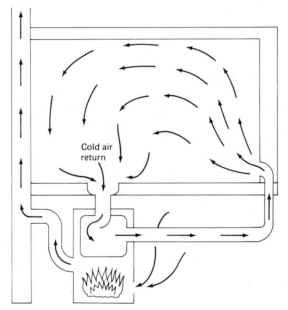

Cold air return

Fig. 16-13 Convection and home heating.

Many factors affect the rate of natural convection. If there are pipes, their sizes and arrangement obviously make a big difference. But even when there are no pipes or ducts, it is still difficult to predict the rate of convection. The best we can do is to carefully note some of the important factors.

First, the rate of fluid flow is important. The faster a fluid moves, the greater is its cooling effect. This has led the U.S. Environmental Science Services Administration to publish the *wind-chill index* shown in Table 16-4. The heat loss from the surface of the human body is greater at lower temperatures, but it is also greater when the wind is blowing. From this table we can see, for instance, that the cooling effect of a 30-mph wind at 10°F is the same as that of still air at −33°F. And if the air is moving faster, its effective temperature is even lower, as far as body

TABLE 16-4   WIND-CHILL INDEX IN EFFECTIVE DEGREES FAHRENHEIT FOR VARIOUS COMBINATIONS OF ACTUAL TEMPERATURE IN DEGREES FAHRENHEIT AND WIND SPEED IN MILES PER HOUR

Wind speed (mph)	Actual temperature (°F)																		
	45	**40**	**35**	**30**	**25**	**20**	**15**	**10**	**5**	**0**	**−5**	**−10**	**−15**	**−20**	**−25**	**−30**	**−35**	**−40**	**−45**
4	45	40	35	30	25	20	15	10	5	0	−5	−10	−15	−20	−25	−30	−35	−40	−45
5	43	37	32	27	22	16	11	6	0	−5	−10	−15	−21	−26	−31	−36	−42	−47	−52
10	34	28	22	16	10	3	−3	−9	−15	−22	−27	−34	−40	−46	−52	−58	−64	−71	−77
15	29	23	16	9	2	−5	−11	−18	−25	−31	−38	−45	−51	−58	−65	−72	−78	−85	−92
20	26	19	12	4	−3	−10	−17	−24	−31	−39	−46	−53	−60	−67	−74	−81	−88	−95	−103
25	23	16	8	1	−7	−15	−22	−29	−36	−44	−51	−59	−66	−74	−81	−88	−96	−103	−110
30	21	13	6	−2	−10	−18	−25	−33	−41	−49	−56	−64	−71	−79	−86	−93	−101	−109	−116
35	20	12	4	−4	−12	−20	−27	−35	−43	−52	−58	−67	−74	−82	−89	−97	−105	−113	−120
40	19	11	3	−5	−13	−21	−29	−37	−45	−53	−60	−69	−76	−84	−92	−100	−107	−115	−123
45	18	10	2	−6	−14	−22	−30	−38	−46	−54	−62	−70	−78	−85	−93	−102	−109	−117	−125

Source: National Oceanic and Atmospheric Administration.

heat loss is concerned. This same effect also works in reverse: a hot fluid warms an object more quickly if the fluid is moving. This idea is used when frozen water pipes are thawed with a portable hair dryer.

Second, fluids with high thermal conductivities are also good convectors generally. It is very easy to set up convection currents in air; but even when moving, the air still transfers heat fairly slowly. Water is better. Molten metals, because of their high thermal conductivities, are excellent convectors. This is why molten sodium metal is used as a heat-transfer medium in some nuclear reactors.

Third, the surface area in contact with the fluid makes a big difference. The "fins" on radiators, oil coolers, brake drums, and power transformers are designed to increase the surface area in contact with the fluid (usually air). This increases the rate at which the fluid carries away heat. Figure 16-14 shows an air-cooled aircraft engine that uses fins to encourage convective heat loss.

Fourth, the rate of convective heat transfer is greater when there is a big temperature difference between the object and the fluid. This works both ways: when the fluid is hotter than the object and when the object is hotter than the fluid. Thus the air circulating through a car radiator carries away more heat when the radiator is very hot. If an engine is designed to operate with a high cooling-system temperature, its radiator needn't be

Fig. 16-14 An air-cooled aircraft engine. By increasing the outside area of the cylinders, the fins encourage convective heat loss.

very big. This is why radiator sizes shrunk when car manufacturers went to pressurized cooling systems.

Finally, the hot object's orientation affects the rate of convection around it. On a cold day, outside air convection carries away a large amount of heat from a window in a vertical wall. But if the same window is set into the roof as a skylight, the rate of convective heat loss is even greater. The rate of convection is greatest when the window is horizontal, all other factors being equal.

It is fortunate that convection is not a more effective mechanism of heat transfer; otherwise it would be practically impossible to keep homes warm in cold climates. Suppose that we have a building with an excavated foundation and concrete-block basement walls. In the winter, the soil outside the wall may be at 5°C (Fig. 16-15). The basement air is at about 20°C. If the wall itself were at 20°C on the inside, there would be a tremendous heat loss as a result of conduction. But in practice, if you touch the wall, you find it is quite cold, certainly much less than 20°C (but higher than 5°C). The rate of conduction through the wall depends on the actual temperature difference between its inner and outer surfaces, and this is nowhere near the difference between the inside air temperature

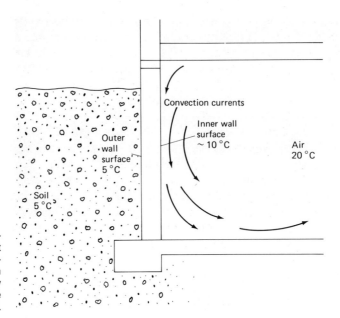

Convection currents

Inner wall
surface
~ 10 °C

Air
20 °C

Outer
wall
surface
5 °C

Soil
5 °C

Fig. 16-15 The inside of a concrete basement wall will not warm up to the inside air temperature, because air convection cannot supply heat rapidly enough. The effect limits the heat loss through the wall.

and the outside soil temperature. Why isn't the inside of the wall at the same temperature as the air? Because convection just can't transfer heat to the wall very quickly. The heat that is transferred by convection travels through the concrete and does not allow the inside of the wall to warm up. And this self-regulating effect keeps the total rate of heat transfer (conduction plus convection) relatively low.

Okay, but suppose now that the wall is well insulated. The inside will no longer be cold to the touch. But this has been achieved only by reducing the wall's conductivity. Now there is even less total heat loss.

Windows are always a great source of heat loss in cold climates. Still, if the inside of the windows were actually at the temperature of the inside air, conductive losses would be so great that it would take an industrial boiler to keep a home warm. In fact, the inside of a single-thickness window gets quite cold in subfreezing temperatures, and it is not unusual for ice to form on the *inside*. This happens because the rate of convective heat transfer from the warm inside air to the glass is quite low. And on the outside, the rate of convective heat transfer from the glass to the outside air is also rather low.

16-6 Radiation

Radiation is the third method of heat transfer. If you've ever stood near a fire on a cold day, you probably noticed that you could warm just one side of your body at a time. The radiation travels along a "line of sight." Thermal radiation can pass through a vacuum (in certain light bulbs, for instance), air, water, glass, or other transparent materials. It also travels fairly easily through some opaque materials like tar, wood, ceramics, and plastics—which is the same as saying that these materials are transparent to thermal radiation. But it does not travel through metals at all.

Thermal radiation is the same thing as infrared radiation, and it is made up of many infrared waves traveling together. This radiation originates in electromagnetic disturbances caused by the rapidly vibrating atoms in a hot object. These disturbances set off the waves that travel through space at the speed of light: about 300 000 km/s [186 000 mi/s].

Now any object at a temperature above absolute zero sends out at least some thermal radiation. The hotter the object (that is, the higher its temperature), the greater is its rate of heat loss by radiation. Normally, this radiation consists of a jumble of infrared waves having a whole range of wavelengths. When the radiating object is hotter, its radiation not only is more intense but also contains shorter wavelengths. This effect is shown in Fig. 16-16.

When an object is only slightly warm, as with the human body, the wavelengths of the thermal radiation are too long to be visible. Thus we can't see other people in the dark. But if we use an infrared detector,

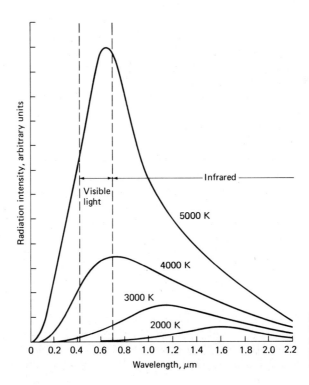

Fig. 16-16 Characteristics of the thermal radiation from a hot object. Hotter objects emit much more radiation, particularly at the shorter wavelengths.

Application: Colonel Pleasant's Air Pump

In June 1864, federal troops laid siege to the city of Petersburg in southeastern Virginia. Both armies constructed complicated lines of breastworks and trenches running for miles in a great arc south of the city, and there they stood at a stalemate.

Colonel Henry Pleasants commanded the 48th Pennsylvania regiment, which contained many coal miners. Pleasants himself had been a mining engineer before the war. During their idle hours, some of the miners got an idea: why not tunnel under the Confederate lines, set an explosive charge, and blow a hole in the breastworks that would give them the opening they needed to take the city? The higher brass approved of the idea (probably just to keep the men busy), and the project was begun under Pleasants' supervision.

But before the digging got very far, work was halted by the problem of ventilation. No mechanical pumps were available; and, of course, it was out of the question to sink vertical air shafts in view of the Confederates. The colonel's solution is shown in the diagram (not to scale). The tunnel's entrance was covered with an airtight leather door (which, of course, could be opened and closed). A fire was maintained just inside the entrance, and smoke and hot air escaped through a vertical flue. Leading from the outside of the tunnel to the end being excavated was an air pipe. The pipe was actually made of wood, and extra sections were added as the digging progressed. The fire's draft lowered the air pressure in the tunnel, which caused fresh outside air to flow in through the pipe.

At 4:45 A.M. on July 30, 1864, a huge explosion blew a hole in the Confederate breastworks. But this technological

which is sensitive to these longer wavelengths, the thermal radiation from a person will give him or her away.

For hotter objects, like the firebrick in a fireplace whose fire has just gone out, we can sense the thermal radiation just by holding a hand close. If the object is really hot, like a ladle of molten steel, the radiation is so intense that we can't even get very close. At the same time, some of the wavelengths are short enough to be visible. When a solid or liquid is so hot that it glows, we say it is *incandescent.* In an incandescent light

success surprised the Union Army almost as much as the Confederates. Unprepared to take full advantage of the opening, the bluecoats lost the "Battle of the Crater," and the siege didn't finally end until March 1865.

More recently, this same idea has been used to ventilate solar-heated buildings in the summer. Heat rising through hot-air solar collectors sets up a flue draft that draws in fresh outside air.

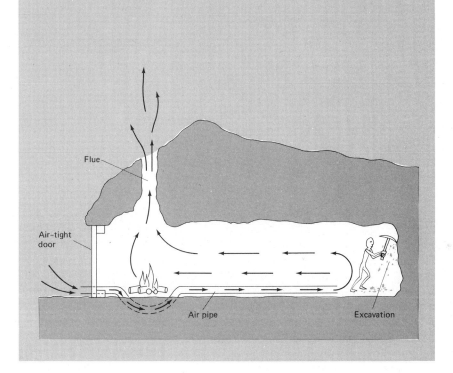

Flue

Air-tight
door

Air pipe

Excavation

bulb, an electric current heats a tungsten filament to a high enough temperature that it glows very brightly. Still, most of the radiation from a light bulb is in the form of heat rather than visible light.

In Chap. 15 we mentioned that hot objects glow red at first, then orange, yellow, white, and finally blue-white as their temperature is increased. Now we see why this happens: at higher temperatures, the thermal radiation contains shorter wavelengths.

16-7 The Greenhouse Effect

You have probably had the experience of going to your parked car on a chilly but sunny day. When you open the door, you find that the inside air has gotten very warm, even though the outside air is cold (Fig. 16-17). How does this happen?

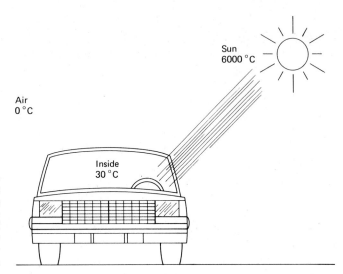

Fig. 16-17 The greenhouse effect heats a car's inside on a chilly but sunny day. Glass is transparent to most of the thermal radiation from the sun, but it will not pass the thermal radiation from the much cooler inside air.

The surface of the sun is at a temperature of about 6 000°C [over 10 000°F]. Thus the sun's thermal radiation not only is very intense, but also contains some fairly short wavelengths. These short wavelengths easily pass through the car's closed windows. Inside, the radiation is absorbed by the car's dashboard, upholstery, and so on. From these solids the heat is transferred to the inside air by convection.

Now the warm things inside the car also radiate heat. But here the temperature is so low that the thermal radiation has very long wavelengths. The window glass is not very transparent to these very long wavelengths, and most of this secondary radiation is reflected back into the car. Thus the car gains heat by radiation but loses heat by conduction and convection. The car's inside temperature rises to where the rate of heat gain balances the rate of heat loss.

Of course, the *greenhouse effect* gets its name because this is what keeps greenhouses warm during the day in cold climates. It also explains how the earth manages to stay warm when space is so cold: the atmosphere acts like window glass in trapping the longer-wavelength radiation from the earth's surface. The greenhouse effect is also exploited in solar heat collectors.

When a hot object is placed in cooler surroundings, heat flows until the object and surroundings are at the same temperature. This heat flow is a combination of conduction, convection, and radiation. Thus the cooling process can be fairly difficult to describe in detail.

16-8 Newton's Law of Cooling

As an approximation, however, we can use Newton's law of cooling:

> *Newton's Law of Cooling*: A warm object cools off at a rate proportional to the temperature difference between it and its surroundings.

Newton's law of cooling

Suppose that we place a bucket of water at 80°C into a freezing compartment at 0°C, and we refrigerate the compartment so it stays at 0°C. It may take 10 min for the water to cool to 40°C. If so, it takes about another 10 min to cool to 20°C. Then it takes an additional 10 min to cool to 10°C. After another 10-min period, the water has cooled to 5°C. In the next 10 min it cools by only 2.5 C°. The cooling rate decreases as the water's temperature gets closer to the temperature of the compartment. Obviously, this water will never quite reach 0°C, and so it never freezes. If we want ice to form, we have to refrigerate the compartment to a temperature *below* 0°C [32°F].

Figure 16-18 shows typical cooling curves for two identical hot objects. Neither object changes phase (if there are phase changes, the curves look like Fig. 15-12). The surroundings are at 20°C [68°F]. At the start, one object is at 160°C while the other is at 90°C. In 1 h, the hot object cools from 160 to 90°C, or a change of 70 C°. But the object starting at 90°C cools at only half the rate, so it loses only 35 C° and ends up at 55°C at the end of the hour. After 5 h, the two objects are at pretty much the same temperature.

Figure 16-19 shows a home whose inside is warmer than the outside. Because of the temperature difference, heat naturally flows from the inside to the outside. The function of the furnace is to keep heat flowing in at the same rate at which it flows out.

16-9 Thermal Resistance ("R Value")

The heat flows out through a combination of conduction, convection, and radiation. But if the home is well insulated and its doors and windows are weather-stripped, most of the heat flows by conduction.

We have already seen how to estimate the rate of conductive heat transfer with the help of Table 16-3. In Example 16-3, for instance, we estimated the rate of heat transfer through a certain brick wall as 1.0 kW. If a building has four such walls, they conduct a total of 4.0 kW. If the roof loses another 2.5 kW, the total rate of heat loss comes to 6.5 kW.

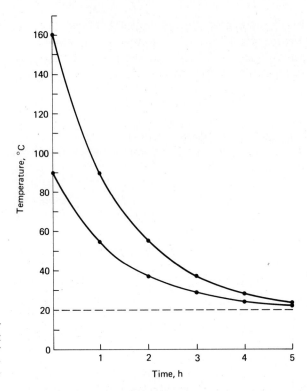

Fig. 16-18 Cooling curves for two identical objects starting at different temperatures. Neither object changes phase. The surroundings are at 20°C.

This, then, is the rate at which heat must be supplied to the building to maintain its inside temperature.

The problem is that most walls are "sandwiches" of several different materials. There may be a brick veneer on the outside, then

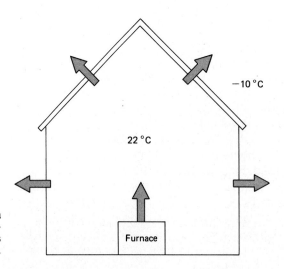

Fig. 16-19 The function of a furnace is to balance a building's heat gains and heat losses in cold weather.

insulating sheathing, then fiber glass batts, then wallboard. The thermal conductivity of the wall is different from the conductivity of any of these substances alone. Heating engineers discovered long ago that the easiest way to deal with this problem is to calculate something called the thermal resistance.

Definition: A substance's *thermal resistance* is the ratio of its thickness to its thermal conductivity:

thermal resistance

$$\text{Thermal resistance} = \frac{\text{thickness}}{\text{thermal conductivity}} \qquad (16\text{-}2)$$

Example 16-5 Thermal Resistance of Fiber Glass.

A fiber glass batt for wall insulation is 4.0 in thick. What is its thermal resistance?

   In old USCS units, we measure heat flow rate in British thermal units per hour, temperature difference in Fahrenheit degrees, wall area in square feet, and material thickness in inches. In terms of these units, the thermal conductivity of fiber glass is about 0.26 (Table 16-3). Then using Eq. (16-2), we have

$$\text{Thermal resistance} = \frac{\text{thickness}}{\text{thermal conductivity}}$$

$$= \frac{4.0 \text{ in}}{0.26 \text{ Btu} \cdot \text{in}/(h \cdot ft^2 \cdot F°)}$$

$$= 15 \ \frac{ft^2 \cdot F°}{Btu/h}$$

   These units probably look a bit confusing. What they mean is this: If there is a 1 F° temperature difference across this insulation, 15 ft² of area conducts heat at a rate of 1 Btu/h. Or, if there is a 15 F° temperature difference, every 1 ft² of area conducts heat at 1 Btu/h. In practice, the units are usually dropped, and the thermal resistance is simply said to be "R-15." It's a good idea, though, to keep the units in mind. ◀

Example 16-6 Thermal Resistance of Wood.

A pine door is 4.8 cm thick. What is its thermal resistance in SI units?

   From Table 16-3, we see that the thermal conductivity of pine is 0.11 W/(m·C°). Using Eq. (16-2) and expressing the thickness as 0.048 m, we get

$$\text{Thermal resistance} = \frac{\text{thickness}}{\text{thermal conductivity}}$$

$$= \frac{0.048 \text{ m}}{0.11 \text{ W/(m} \cdot \text{C}°)}$$

$$= 0.44 \text{ m}^2 \cdot \text{C}°/\text{W}$$

This means that an area of 0.44 m² with a 1 C° temperature difference across it conducts heat at a rate of 1 W. But we cannot quote this result as the "R-value" since at present this expression still refers to old USCS units. This wood is actually R-2.4, not R-0.44. ◄

Notice that a high thermal resistance means that the material conducts heat very slowly. If we are looking for effective insulation, the higher the R-value, the better. Materials with low R-values, (thermal resistances) are poor insulators and good conductors.

Equation (16-1) may be rewritten in terms of thermal resistance:

$$\text{Rate of heat flow} = \frac{\text{(area)(temperature difference)}}{\text{(thermal resistance)}} \qquad (16\text{-}3)$$

Examples 16-3 and 16-4 can be solved by using this form of the equation. The student should verify that the same answers result.

Now the advantage of using thermal resistance and Eq. (16-3), rather than thermal conductivity and Eq. (16-1), is this:

rule for compound walls

Rule for Compound Walls: If a wall is built up of several layers of different materials, the total thermal resistance of the wall is just the sum of the thermal resistances of each layer.

In other words, if a wall is made up of R-2.0 sheathing, R-15 fiber glass insulation, and then a layer of R-1.4 plasterboard, then the total thermal resistance is better than R-18. This total value can be used in Eq. (16-3).

Example 16-7 Thermal Resistance of a Triple-Pane Window.

The triple-pane window in Fig. 16-20 has three 1/8-in panes of glass separated by two 1/4-in air spaces. What is the window's thermal resistance?

Because the airspaces are so narrow, convection currents are minimal. The air is effectively trapped, and it transfers heat by conduction. From Table 16-3, the thermal conductivity of air is 0.17 Btu·in/

(h·ft²·F°). One airspace is 0.25 in thick. Then from Eq. (16-2),

$$\text{Thermal resistance} = \frac{\text{thickness}}{\text{thermal conductivity}}$$

$$= \frac{0.25 \text{ in}}{0.17 \text{ Btu·in/(h·ft²·F°)}}$$

$$= \text{R-1.5}$$

For the glass, we have a thickness of 0.125 in and a thermal conductivity of 5.5 Btu·in/(h·ft²·F°). Then

$$\text{Thermal resistance} = \frac{0.125 \text{ in}}{5.5 \text{ Btu·in/(h·ft²·F°)}}$$

$$= \text{R-0.02}$$

If we add the R-values of three panes of glass and two thicknesses of air, we get a total thermal resistance of R-3.0 for the window.

Notice that the R-value of the glass itself contributes very little to the R-value of the window. Why, then, go to the expense of using three panes? Because the extra panes trap extra air, and it is the *air* that has a high thermal resistance. At the same time, the air gap cannot be too thick, because convection currents must be prevented. So three panes are better than two, which are much better than one. For a single pane, resistance to conduction and convection is less than the equivalent of R-1. ◄

Example 16-8 Heat Loss through a Triple-Pane Window.

A triple-pane window of the type described in Example 16-7 measures 21 in by 36 in. The inside surface temperature is 70°F while the outside surface temperature is 5°F. What is the rate of heat loss through the window?

We are using the old USCS units here, and the area has to be expressed in square feet:

$$\text{Area} = (21 \text{ in})(36 \text{ in}) \left(\frac{1 \text{ ft}^2}{144 \text{ in}^2}\right)$$

$$= 5.3 \text{ ft}^2$$

We have already estimated the thermal resistance as R-3.0. Then using

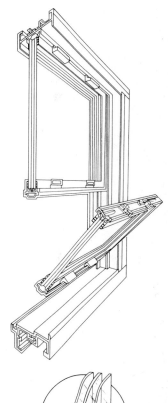

Fig. 16-20 A triple-pane window. Air trapped between the panes contributes to a high thermal resistance.

Application: Liquefied Natural Gas

Although oil has been shipped across the oceans for many years, there has always been a problem of what to do with the natural gas that is found with the oil. In the United States and western Europe, natural gas is certainly a valuable fuel. But because of the transportation problem, vast quantities have had to be burned off at wellheads in the Arab republics and South America.

One way to solve the transportation problem is to liquefy the natural gas. As a liquid, the fuel occupies a small enough volume to make it economical to transport in tankers. Unfortunately, the critical temperature of natural gas is the same as the critical temperature of the methane in it: −82.1°C [−114°F] at 45.8 atm pressure. Thus there is no way to keep natural gas liquefied at ordinary temperatures, even if the pressure is very high. At normal atmospheric pressure, liquefied natural gas (LNG) boils at −161.5°C [−259°F]. There are tremendous problems associated with carrying something this cold in a tanker.

To keep heat flow to a minimum, the LNG is stored in spherical tanks on the ship. This is done because heat flow depends on surface area, and a sphere is the shape which encloses the greatest volume with the least surface area. These tanks are made of aluminum, because steel and most other metals get brittle at such low temperatures while aluminum actually gets stronger.

Once on the ship, the LNG is not actually refrigerated. Instead, it is allowed to boil continuously, so evaporation cooling alone keeps a portion of the LNG in the liquid state. The same thing happens in boiling a pan of water: the liquid never rises beyond 100°C as the escaping steam carries away the heat. The evaporated LNG, which is now just natural gas, is piped into the ship's boilers to power the turbines. Thus none of the LNG is really lost in transport; part of it does the job of transporting the rest.

Eq. (16-3), we have

$$\text{Rate of heat flow} = \frac{(\text{area})(\text{temperature difference})}{(\text{thermal resistance})}$$

$$= \frac{5.3\ (65)}{3.0}\ \frac{\text{Btu}}{\text{h}}$$

$$= 110\ \text{Btu/h}$$

If a house has 12 such windows, the total heat loss through them is (12)(110 Btu/h), or 1300 Btu/h. But if the windows are single- rather than triple-pane, the heat loss is 3 to 4 times as much. ◀

As we just saw in Example 16-7, the insulating value of trapped air is very good. In fact, this is the idea behind using fiber glass in walls and attics: the fibers trap the air and prevent convection currents. There is nothing great about the fiber glass itself; it is the trapped air pockets that provide the insulation. This same principle is used in thermal underwear and down-filled jackets and sleeping bags. One thing to remember: compressing air-filled insulation drastically reduces its thermal resistance.

Exercises

14. Plasterboard has a thermal conductivity of about 3.0 Btu·in/(h·ft²·F°).
    (a) What is the R-value of a piece of plasterboard 3/8 in thick?
    (b) What is the R-value if it is 1/2 in thick?
    Answers: (a) R-0.13, (b) R-0.17
15. Sawdust is treated to make it fireproof, then is poured into an attic as insulation. The joists are 19 cm wide, so this is the thickness of the insulation. What is the thermal resistance in (a) USCS units, (b) SI units?
    Answers: (a) 18 ft²·F°/(Btu/h), or R-18, (b) 3.2 m²·C°/W
16. What thickness of concrete is needed to give the same thermal resistance as $1\overline{0}$ cm of fiber glass?
    Answer: 340 cm
17. What thickness of brick is needed to give the same thermal resistance as 3.5 in of rock wool?
    Answer: 62 in
18. A home is 72°F on the inside when the outside air temperature is 5°F. Its walls have a thermal resistance of R-19. What is the rate of heat loss through a room wall 8.0 ft high and 16 ft long?
    Answer: Less than 450 Btu/h

19. A certain home is built on-grade by pouring a slab of concrete 4.0 in thick. The floor is insulated with 1.7 in of fiber glass between pine spacers 16 in apart. On top of this is 5/8-in plywood, which has about the same insulating properties as pine. The floor surface is oak, 0.50 in thick. (*a*) What is the total thermal resistance of the floor? (*b*) What is the rate of heat loss through the floor if it measures 30 ft by 30 ft and the upper surface is at 72°F while the soil is at 43°F?
Answers: (*a*) R-7.5, (*b*) Less than 3500 Btu/h

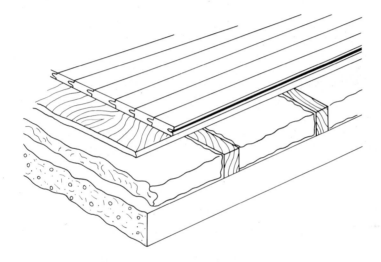

The methods of Sec. 16-9 can be used to estimate the rate of heat loss through a building's windows, walls, doors, and so on. Adding these gives the total rate of heat loss in British thermal units per hour or watts. There is also some additional loss resulting from the opening of doors, leakage around windows, and heat rising up chimney flues. Generally, this amounts to only an additional 5 to 10 percent (although in old and leaky homes it can be more).

16-10 Heating Requirements in Buildings

Suppose that a building conducts away a total of 24 kW/h [82 000 Btu/h]. This means that 24 000 kJ of heat needs to be supplied each second to keep the building from cooling off. In a home, we expect the furnace, or **heat pump**, to do most of the job. But in commercial buildings, where there is a lot of lighting and many people, other heat sources become important. Every 100-W electric light contributes 100 J each second, and every person contributes about 60 W [200 Btu/h] of sensible body heat. If the building has 50 people and 21 kW of lighting and other electric appliances operating, a total of 24 kW is supplied by these sources. No furnace or heat pump is needed at all. It is not unusual for a large office building or department store to get most of its heat from lighting and body heat.

Another potentially important source of heat is direct sunlight. Without any sophisticated collection devices, sunlight streaming through windows provides nearly 1 kW of heat for each 1 m² of window area perpendicular to the sun. (This is the same as about 300 Btu/h for each 1 ft².) These values are lower near sunrise or sunset, or on cloudy or smoggy days, but still they are significant. Even in a very cold climate, windows still admit much more heat than they lose if they are positioned facing the sun. So it makes sense (as the ancient Greeks discovered long ago) to build buildings with small North-facing windows and large South-facing windows. There is a two-story school in Liverpool, England, whose entire South-facing wall is double-pane glass. Enough heat is gained from the sun, the electric lighting, and student body heat that no furnace at all is needed. Unfortunately, this type of design leads to wider indoor temperature variations than are normally considered comfortable. Figure 16-21 summarizes the main sources of passive heat in a building.

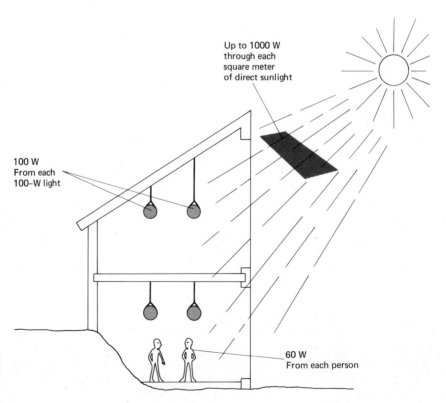

Up to 1000 W through each square meter of direct sunlight

100 W From each 100–W light

60 W From each person

Fig. 16-21 The important sources of passive heat in a building.

Solar heat can also be part of an *active* heating system, that is, a system where the heat is collected at one place and pumped somewhere else. In recent years, a vast number of collector designs have been developed, some using air to transfer the heat, others using water, and still others using other substances. The heat is stored so that it can be recovered during overcast days. Storage systems often use large water tanks or bins of gravel, since the internal energy of these substances can be changed a great deal with just a small change in temperature. In some cases, large quantities of wax or low-melting salts are used for heat storage; by exploiting the latent heat of fusion, a large amount of heat can be stored with practically no temperature change.

A great deal can be said about solar heating systems, but innovations are coming so rapidly that any details we supplied here would quickly be out of date. The most up-to-date developments can usually be found in the popular technical magazines on the newsstands.

16-11 Cooling

It is always more difficult to cool things than to heat them. This is because, as we mentioned earlier, every energy transformation produces at least some heat. So even when we cool things, heat has to be produced *somewhere*.

As we saw in Chap. 15, one way to lower an object's temperature is to cause a phase change. When ice is mixed with salt, the ice melts and absorbs its latent heat of fusion—which lowers the temperature of the mixture. And when alcohol or water is forced to evaporate quickly, perhaps by blowing air over the liquid, the latent heat of vaporization is carried away by the vapor, and this lowers the temperature of the liquid left behind. We have also seen that the temperature of a gas drops if the gas is allowed to expand rapidly. Most refrigeration and cooling devices use these principles.

Since refrigerators and air conditioners pump the refrigerant (usually one of the Freons) through a closed cycle, the cooling effect generated on expansion and vaporization is more than offset by the heating produced in condensation and compression. The net result is the flow of heat shown in Fig. 16-22. The heat is drawn in from the cooling compartment, where the temperature is being lowered (or at least a low temperature is being maintained). Work is done to operate the compressor that circulates the refrigerant. The heat output is the total of these two. Thus a home refrigerator actually heats the kitchen!

From Fig. 16-22 we see why a refrigeration unit may be called a "heat pump." By doing work, we can cause heat to flow from a cold region to a warm region; we have effectively pumped the heat uphill. Large heat pumps may be used to cool a building in the summer; then they may be reversed to heat the same building in the winter.

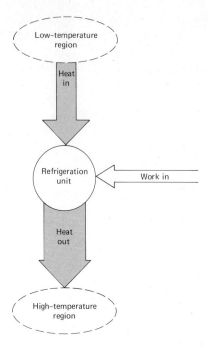

Fig. 16-22 Heat flow in a refrigerator, air conditioner, or
heat pump.

Air-conditioning units are rated according to the maximum rate at
which they can pump heat from the low-temperature side. Thus a room
air conditioner rated at 1.8 kW [6000 Btu/h] pumps 1.8 kJ of heat from
the room to the outside each second.

How big an air conditioner does a building need? If the outside is
warmer than the inside, heat is flowing *in*. The air conditioner must be
big enough to pump it back out at the same rate. We can calculate the
heat flow through the building's walls, doors, windows, and so on, by the
methods used earlier in this chapter. To this we must add the 60 W
generated by each person as well as the "wattage" of all the electric
appliances. Air-conditioning kitchens is particularly difficult because of
the additional heat generated in cooking and the use of hot water in the
dishwasher.

Exercises

20. On a typical summer day, a small store gains 15 kW by conduc-
     tion through its walls when the inside temperature is maintained
     at 21°C. As many as 20 people are expected to be in the store
     at one time, and 2800 W of lighting and other appliances is
     usually in use. There is no direct sunlight. What size air conditioner
     does the store need?
     Answer: 19 kW

21. A certain home needs 105 000 Btu/h of heat when the inside temperature is 70°F and the outside temperature is −5°F. If the summer temperature goes as high as 90°F, what is the approximate air conditioning requirement? Neglect body heat, sunlight, and heat from appliances.
Answer: 28 000 Btu/h

Summary

Heat flows from one place to another by the processes of conduction, convection, radiation, or a combination of these. If we understand these processes and the factors that influence them, we can control the flow of heat to our advantage. In buildings, for instance, we need to minimize the heat flow through the exterior walls. But in the condensers and radiators on large engines, we need to maximize the heat transferred to the surrounding air.

Heat may also be transferred through energy conversions. If a high-pressure gas is allowed to expand, it cools and absorbs heat from its surroundings. If this gas is pumped somewhere else and is compressed, then its temperature goes up and it gives off heat to its surroundings. The net effect is to transfer heat from one place to another and to generate some extra heat in the process.

Heat may also be transferred through chemical means. The internal energy in fossil fuels originated millions of years ago in sunlight absorbed by plants. Burning these fuels converts this stored energy back to heat. The amount of heat released by burning a substance is called its heat of combustion.

The first law of thermodynamics states that heat is a form of energy, and its transfer must be consistent with the law of conservation of energy. In fact, these two physical laws really amount to the same thing. It is often useful to draw a little sketch to see where the heat is coming from and where it is going in a heat-transfer process.

Terms You Should Know

internal energy	natural convection
first law of thermodynamics	wind-chill index
expansion cooling	thermal radiation
compression heating	greenhouse effect
heat of combustion	Newton's law of cooling
thermal conduction	thermal resistance
thermal conductivity	rule for compound walls
forced convection	heat pump

Problems

1. How many kilograms of fuel oil must be burned to heat 250 kg of water from 20 to 100°C? Assume that the heat is transferred to the water with an efficiency of 71 percent but that other heat losses are negligible.

2. A certain diesel-powered generator has an efficiency of 36 percent. How many liters of fuel oil does it consume in generating 240 kWh of electric energy?

3. A home is a 22°C [72°F] when the furnace stops working. The outside air temperature is −12°C [10°F]. In the first hour, the home's temperature drops to 18°C. (a) Notice that the temperature difference is originally 34C°. What percentage of this temperature difference remains after 1 h? (b) What is the inside temperature after another hour passes? (c) What is the home's temperature 8.0 h after the furnace stops?

4. A gas-fired hot-water heater has a cylindrical tank 31 cm in diameter and 120 cm tall. The sides and top are insulated with 3.0 cm of fiber glass. If the water is at 65°C [149°F] and the surrounding air is at 20°C [68°F], what is the rate of conductive heat loss from the tank? Neglect heat loss from the bottom and from the connecting pipes.

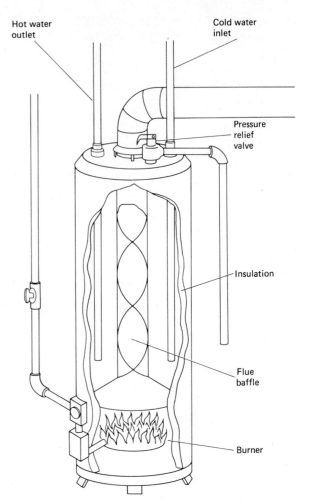

Hot water outlet

Cold water inlet

Pressure relief valve

Insulation

Flue baffle

Burner

17

Heat Engines

Introduction

An engine is a device that produces a useful and continuous mechanical energy output. Water turbines, wind machines, electric motors, and steam turbines are all engines. Engines that convert *heat* to mechanical energy are called, appropriately enough, *heat engines.* Diesel, gasoline, and steam engines are examples. Although most heat engines depend on a burning fuel to produce the heat that is converted to mechanical energy, combustion is not always necessary. Some heat engines get their heat from nuclear fission, geothermal sources, or, now in the experimental stage, direct sunlight or warm ocean currents.

17-1 The Second Law of Thermodynamics

We have already talked a little about this law without calling it by this name. When heat flows from warm places to cooler places, it is following the second law of thermodynamics. The fact that pressurized air leaks *out* of a tire, but never *in,* is a result of this same principle.

It seems that it is always easier to mix things than it is to separate them again. You can mix a bucket of hot water with a bucket of cold water to get two buckets of lukewarm water, but you can never take the lukewarm water and separate it back into the buckets of hot water and cold water (at least not without doing a great deal of work). A fresh pack of playing cards is arranged according to suit. Shuffling the cards mixes them up, but it is very unlikely that further shuffling will arrange them according to suit again. A house is a very ordered arrangement of wood, glass, nails, shingles, and so on. If it is struck by a tornado, it may turn into a pile of rubble. If the pile of rubble is hit by another tornado, we would be very surprised to see the materials shifted around to form a house again.

Technology is the act of imposing some order on nature that isn't already there. Nature itself behaves like the card shuffler—constantly trying to mix things again. For instance, it takes a long and complicated technological process to refine iron ore into steel, then to convert the steel into an automobile. But to convert the car body back into iron ore (which is the same as rust) requires no effort at all. If we wait long enough, it happens automatically.

This natural behavior is summarized in the second law of thermodynamics.

second law of thermodynamics

Second Law of Thermodynamics: The natural tendency of any system is to go from a state of order to a state of disorder. It takes an energy input to put a disordered system back into order.

Notice that a system has *order* if part of it is hot while part of it is cold. It is *disordered* (mixed up) if the hot part is mixed with the cold part so everything is lukewarm. In all cases, *disorder* increases if a system is left alone as time passes. Figure 17-1 shows a decaying house that is reverting back to nature according to the second law of thermodynamics.

17-2 Heat Flow in Engines

A heat engine can produce work only if part of the engine is hot while part of it is cooler. This sets up a natural flow of heat, some of which can then be converted to work. Figure 17-2 shows the flow of heat in a heat engine.

In a conventional car engine, for instance, the fuel mixture reaches temperatures of up to 2200°C [4000°F] at the beginning of the power stroke. Since the outside of the engine is at a temperature of perhaps 100°C [212°F], a great deal of heat flows. The cooling system carries away some of this heat, and much of it also leaves in the exhaust. By allowing the hot gases to expand against a piston, we also get some work out of the process. But the engine produces much more heat than it does work.

It would be nice to build an engine that changed all the heat into useful work. Unfortunately, this is impossible. The flow of heat through a

Fig. 17-1 If a house is not kept in constant repair (energy input), it deteriorates and reverts to nature (order to disorder). The principle is described by the second law of thermodynamics.

heat engine is like the flow of water through a water turbine. In Fig. 17-3 we see a water turbine with water streaming through. Since the expelled water is still moving, it still has energy; thus the turbine has changed only

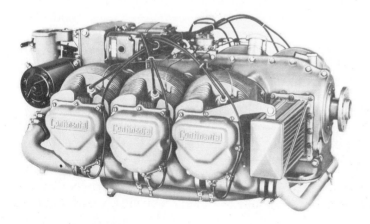

Fig. 17-2 Flow of heat in a heat engine.

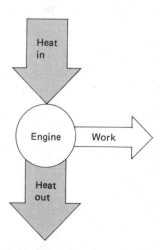

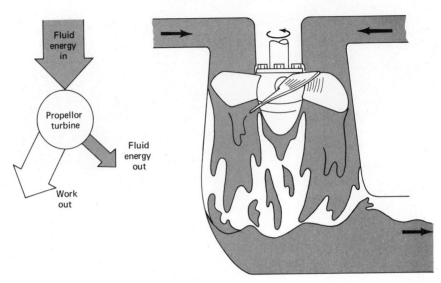

Fig. 17-3 A water turbine converts energy in a moving stream of water into work. If the stream does not move out of the exit pipe, no work can be produced. In the same way, a heat engine must have a flow of exhaust if it is to function.

part of the energy input to work. But if we cut off the output flow to stop the waste of energy, the wheel can no longer turn and there is no energy transformation at all. In the same way, a heat engine needs a flow of exhaust heat if it is to produce any work.

Furthermore, the second law of thermodynamics tells us that an engine's work output will never be large enough to pump all the exhaust heat back to the engine's input. Figure 17-4 shows this idea in a diagram. Suppose that we use a steam engine to operate a heat pump. The heat pump takes heat from the engine's low-temperature exhaust (condenser) and pumps it at high temperature back to the boiler. Here the heat boils more water, which drives the engine, which operates the heat pump, and so on. The device is a perpetual-motion machine. But for this to be successful, the boiler must have no natural tendency to cool off on its own, and the condenser must have no natural tendency to warm up. The second law of thermodynamics says that these tendencies are always there. The heat pump will not be capable of supplying enough heat to keep the engine turning, let alone to produce an excess work output. Stated another way, a heat pump supplying a 1-hp engine would need more than 1 hp to drive it.

17-3 Thermal Efficiency

We said that a heat engine can transform only part of the heat flow into work. The fraction transformed is called the engine's thermal efficiency.

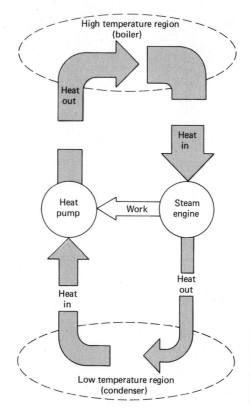

Fig 17-4 An impossible process. If an engine's work output was large enough to pump all its exhaust heat back to its input, it could be used to produce perpetual motion.

Definition: An engine's *thermal efficiency* is the work output expressed as a fraction (or a percentage) of the heat input:

thermal efficiency

$$\text{Thermal efficiency} = \frac{\text{work output}}{\text{heat input}} \qquad (17\text{-}1)$$

An engine's thermal efficiency depends on many design factors. But more than anything else, it depends on the input temperature and the exhaust temperature. The maximum thermal efficiency possible can be calculated if we know these temperatures:

$$\begin{matrix}\text{Maximum possible} \\ \text{thermal efficiency}\end{matrix} = 1 - \frac{\text{absolute exhaust temperature}}{\text{absolute input temperature}} \qquad (17\text{-}2)$$

Notice that this formula requires that the temperature be absolute, expressed on either the kelvin or Rankine temperature scales.

Example 17-1 Thermal Efficiency of a Gasoline Engine.

The air-fuel mixture in a certain engine reaches a peak temperature of 2200°C [4000°F] during combustion. The gases are at 1200°C [2200°F] just as the exhaust valve opens. What is the maximum possible thermal efficiency of the engine?

To use Eq. (17-2), we first have to express the temperatures on the kelvin scale. This amounts to adding 273° to the Celsius temperatures:

$$2200°C = 2473 \ K$$

$$1200°C = 1473 \ K$$

We wait until the final step to round off to two-digit precision. Then

$$\text{Maximum possible thermal efficiency} = 1 - \frac{1473 \ K}{2473 \ K}$$

$$= 1 - 0.596$$

$$= 0.40$$

So at best, this engine can convert only 40 percent of the heat input into work. In practice, because of some factors we will discuss shortly, the actual thermal efficiency is less than 30 percent. The engine puts out considerably more heat than work. ◄

Exercises

1. A steam engine has an input temperature of 520°C and an exhaust temperature of 100°C. What is its maximum possible thermal efficiency?
    Answer: 53 percent
2. Using ceramic materials, it may someday be possible to build gas-turbine engines with input temperatures of 1650°C. If the exhaust temperature is 480°C, what would be the maximum thermal efficiency of such an engine?
    Answer: 61 percent
3. In tropical climates, the ocean's surface may be at a temperature of 70°F while deeper water may be at only 55°F. Suppose that someone invented an engine to run on this small temperature difference. What would be the maximum thermal efficiency of such an engine?
    Answer: 2.8 percent

4. A certain steam turbine has a thermal efficiency of 35 percent and a mechanical power output of 7500 kW. (a) At what rate must heat be supplied to the engine? (b) At what rate must the condenser carry away heat?
Answers: (a) 21 000 kW, (b) 14 000 kW

Work, as we saw earlier, is the product of a force and the distance through which the force moves something. In most heat engines, the force is produced by an expanding gas (steam, hot air, oil smoke, etc.). The work done by such a gas is most conveniently expressed in terms of the gas's pressure and its volume change:

17-4 Thermodynamic Work

Work done by expanding gas =
(average pressure*) (increase in volume)   (17-3)

We are not particularly interested in calculating anything from this formula. The point is that the gas does more work when its average pressure is high, or when it gets to expand a great deal in an engine. Thus a steam engine can do more work if the steam pressure is high rather than low. It can also do more work if the piston displacement is large.
On the other hand, it also takes work to compress a gas:

Work needed to compress a gas =
(average pressure) (decrease in volume)   (17-4)

So the compression stroke on a gasoline engine requires more work when the volume decrease is large.
If an engine is to be based on the compression and expansion of a gas, more work must be done on expansion than on compression. Thus the expansion must take place at a higher average pressure than the compression. The change in pressure is achieved when a flow of heat changes the gas's temperature (Charles' law).

In Chap. 6 we told the story of Thomas Newcomen and his atmospheric engine. (You may want to read the account again before continuing here.) The engine's operation can be diagramed on a graph of pressure versus volume, as shown in Fig. 17-5. At point *a*, the piston is at the bottom of its stroke, and the small volume enclosed by the cylinder is at a partial vacuum. Then the steam valve is opened, and the pressure suddenly jumps to slightly above atmospheric (point *b*). This pressure lifts the

17-5 Indicator Diagrams: The Atmospheric Engine

*Sometimes called the *mean effective pressure*.

piston and increases the cylinder volume to point *c*, where the piston is at the top of its stroke. Now the steam valve is closed, and a spray of water is injected to condense the steam. This decreases the pressure to point *d*. The piston then moves into the partial vacuum in the cylinder, producing the power stroke as it returns to point *a*. Now the engine is ready to repeat the cycle.

The graph in Fig. 17-5 is called an indicator diagram.

indicator diagram

Definition: An *indicator diagram* is a graph showing changes in the pressure and volume of the working substance as an engine goes through one complete cycle.

We said that an engine's working substance has to expand at a higher pressure than it contracts at, if the engine is to have a work output. This means that the engine's cycle must be clockwise on the indicator diagram. Counterclockwise indicator diagrams may represent pumps or refrigerators, but not engines.

Furthermore, the net work done by an engine in one cycle is indicated by the *area* enclosed by the indicator diagram:

Net work per cycle = area enclosed by indicator diagram   (17-5)

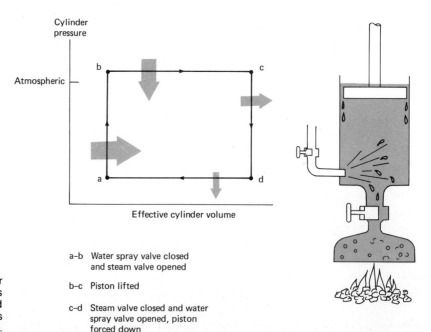

Fig. 17-5 Indicator diagram for an atmospheric engine. Arrows show where heat flows in and out of the engine during its working cycle.

a–b Water spray valve closed and steam valve opened

b–c Piston lifted

c–d Steam valve closed and water spray valve opened, piston forced down

So an engine modification that causes this area to be doubled also doubles the work output. Such modifications might include increasing the working pressure or increasing the cylinder volume.

But what about the engine's *power* output? Since the engine produces a certain amount of work each cycle, the power output depends on how fast the engine turns. The formula is

Power output of engine =
(net work per cycle)(operating frequency)   (17-6)

Application of this formula usually requires some unit conversions. Let's look at an example.

Example 17-2 Power Output of an Atmospheric Engine.

The engine has a total piston displacement of about 1.6 m³ [56 ft³]. The pressure difference between expansion and condensation is about 1.0 atm. The engine makes 8 cycles per minute. What is the power output?

The indicator diagram is rectangular, with a height of 1.0 atm and a width of 1.6 m³. The work done each cycle, then, is the area enclosed by this rectangle, or

$$\text{Net work per cycle} = (1.0 \text{ atm})(1.6 \text{ m}^3)$$

Using the fact that 1.0 atm is 101 325 Pa, or 101 325 N/m² (Table 5-2), we have

$$\text{Net work per cycle} = (1.6 \text{ atm·m}^3)\left(\frac{101\ 325 \text{ N/m}^2}{1 \text{ atm}}\right)$$

$$= 160\ 000 \text{ N·m}$$

$$= 160\ 000 \text{ J}$$

where we have used the fact that 1 N·m = 1 J.

The frequency of this engine is 8 c/min, or just 0.13 c/s, or 0.13 Hz. Then using Eq. (17-6) gives

$$\text{Power output} = (\text{net work per cycle})(\text{operating frequency})$$

$$= (160\ 000 \text{ J})(0.13 \text{ Hz})$$

$$= 21\ 000 \text{ J·Hz}$$

So far so good, but what is a joule-hertz? Since 1 Hz = 1 c/s and a cycle

Application: Perpetual-Motion Machines of the Second Kind

In Chap. 9 we saw that many people have tried to build machines which continuously put out more work than it takes to operate them. Such devices (which never do actually work) are called *perpetual-motion machines of the first kind*. The term "perpetual motion" stems from the idea that part of the work output could be used to supply the necessary work input—so the motion would continue indefinitely. The "first kind" refers to the fact that (if successful) such a machine would have to violate the first law of thermodynamics, or the law of conservation of energy.

A *perpetual-motion machine of the second kind* would be one that violated the second law of thermodynamics. Although few inventors have found this challenge very promising, at least one interesting attempt was made in the 1880s.

The steam engines of those days had tremendous appetites for coal. When the U.S. Navy gave up sails in favor of steam, it soon found that it was very expensive and complicated to maintain the required system of coaling stations across the globe. B. F. Isherwood, the Navy's chief engineer, was wide open to any helpful ideas.

A professor named John Gamgee thought he had the answer: a steam engine that needed no coal, or fuel of any kind. Gamgee reasoned that the steam engine's poor fuel economy lay in the tremendous amount of heat needed to boil the water into steam. Why not use a fluid that normally boils at a much lower temperature—ammonia, for instance? Since the boiling point of ammonia is less than 0°C, no fuel at all should be needed in the boiler as long as the surroundings are above freezing. In fact, if ammonia is boiled at 0°C, it will generate about 4 atm of pressure to drive the engine. Gamgee named his invention, appropriately enough, the "zeromotor."

When Isherwood heard about the idea, he was overjoyed. He wrote a favorable report to the Secretary of the Navy, and President Garfield himself expressed enthusiasm. Operating a ship's boilers from the energy in the water the ship floated in—it seemed too good to be true! And it was. When built, the zeromotor refused to work.

What went wrong? One requirement of the engine was that it had to take the ammonia through a closed cycle: boiler to engine to condenser back to boiler. This is what existing engines did with their water. Ammonia was too expensive (and poisonous) to blow the exhaust into the atmosphere. Besides, the original coal supply problem would just be replaced by an ammonia supply problem if the ammonia didn't cycle. And this was the problem—the ammonia *wouldn't* cycle. There was no way to condense the exhaust without refrigerating it. This refrigeration, it turned out, took more work than the engine could put out.

But we really don't have to go into the details to see why the idea failed. The boiler would be extracting heat from the ocean and leaving the water cooler. This would be possible if the water were much warmer than the air around the condenser (still, the efficiency would be very low). But if the water, the engine, and the air are all about the same temperature, there is no natural flow of heat and nothing to push the ammonia around a *closed* cycle. If the ammonia did cycle under these conditions, it would be a violation of the second law of thermodynamics.

Okay, so the condenser is out. What if we used a low-boiling but nontoxic substance like liquid air and exhausted it to the atmosphere? Would this make the idea work? Not really. The engine would turn and the boat might move, but the efficiency would be terrible. And the work involved in liquefying the air on the mainland (it still has to be done *somewhere*) would be greater than the work returned by the engine.

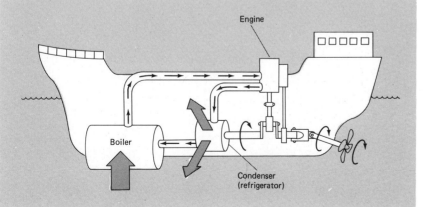

is not actually a unit (just a way to keep count), we can say that

$$1 \text{ J·Hz} = 1 \text{ J/s}$$

$$1 \text{ J·Hz} = 1 \text{ W}$$

So the engine's power output is 21 000 W, or 21 kW. Using the conversion in Table 9-4 of 1 kW = 1.341 hp, we may also write our engine's power output as 28 hp. Notice that this is an awfully large engine to be producing this relatively small amount of power. And this says nothing about the size of the boiler needed to keep it running. ◀

17-6 The Reciprocating Steam Engine

In 1763 the University of Glasgow in Scotland had a problem with its model atmospheric engine. A 27-year-old technician named James Watt was given the job of repairing it. Although the immediate adjustments were no problem, Watt recognized that the real difficulty with the engine was beyond repair. The alternate heating and cooling of the cylinder each cycle required so much steam that the boiler simply could not keep up with the task—especially since the brass cylinder was such a good conductor of heat.

Watt decided to improve on the engine's design. In his first attempt he used a wood cylinder, but he found it deteriorated too quickly and would not hold a very good vacuum in any case. But he kept thinking, and several years later a practical solution finally struck him. If the vacuum were created in a container *separate* from the cylinder, the steam would rush from the cylinder to fill the vacuum, and the engine itself would not have to be cooled. This invention was called the *separate condenser,* and Watt got a patent on it.

But there were further improvements to be made. Watt surrounded the cylinder with a steam jacket so no wasteful condensation at all could take place in the cylinder itself. He closed over the cylinder's open top so steam could be alternately admitted to the top and bottom sides of the piston, making the engine double-acting. Thus one cylinder could do the work of two. And gradually he increased the steam pressure so that the force of the expanding steam contributed to the work output. Now the device was no longer an atmospheric engine.

By 1773 Watt's first large engine was reliably pumping water to supply a waterwheel; this wheel, in turn, powered the machine shop of one Matthew Boulton, Watt's business partner. But the engine's real success dates from 1776. The American Revolution had begun, and John Wilkinson's ironworks in England was being flooded with orders for armament. One of Watt's engines was installed to pump the bellows to keep the forges glowing. And from miles around, people came to see this technological wonder in action. Boulton wrote, "I sell here . . . what all the world desires to have: power!" The Industrial Revolution had begun, with the *reciprocating steam engine.*

Although Watt himself never drew an indicator diagram for his engine, he was well aware that the device worked because the steam pressure produced a change in the effective cylinder volume. Many years later, an engineer named William Rankine devised a quantitative description of the cycle that now bears his name.

      Figure 17-6 shows a double-acting steam engine similar to Watt's design. The *Rankine cycle* for this engine is shown on the indicator diagrams in Fig. 17-7. The pressure and volume are graphed for only one

17-7 The Rankine Cycle

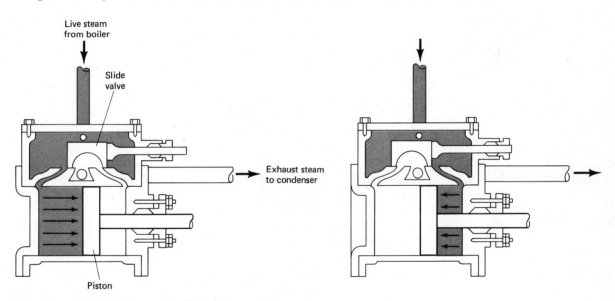

Fig. 17-6 A double-acting reciprocating steam engine.

side of the double-acting piston. At point *a* in the top diagram, the steam valve opens and closes off the exhaust port. Immediately, the pressure rises to the boiler pressure (point *b*). The piston then starts to move, increasing the cylinder volume as more steam flows in at the same pressure. At point *c*, the piston has reached the end of its stroke, and the slide valve closes the steam port and opens the exhaust port. The pressure quickly drops to the condenser pressure, which is a partial vacuum (point *d*). The piston is then returned to the starting point, and the cycle is repeated. The work done in one cycle is the area enclosed by this indicator diagram.

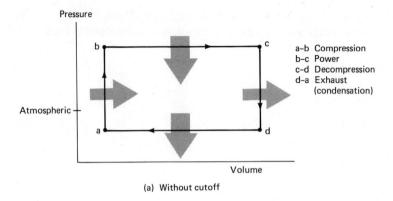

Pressure

b ●————————————→● c

a-b Compression
b-c Power
c-d Decompression
d-a Exhaust
      (condensation)

Atmospheric ┼

a ●←————————————● d

Volume

(a) Without cutoff

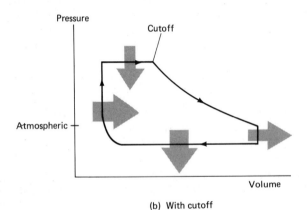

Pressure

Cutoff

Atmospheric ┼

Volume

(b) With cutoff

Fig. 17-7 The Rankine cycle: indicator diagram for a reciprocating steam engine.

Shortly after Watt invented this engine, he realized that its efficiency could be improved by cutting off the flow of steam partway
through the stroke. This allows the trapped steam to *expand* against the
piston, so much less steam (and fuel) is needed. Of course, less work is
also produced. But if the steam is cut off when the stroke is only one-
quarter completed, the engine still produces 60 percent of its maximum
power with only 25 percent of the fuel consumption. Thus the efficiency
is more than doubled.

By using properly designed valve gear, the engine's power output
may be varied by changing the cutoff point. But at full throttle, the engine's fuel consumption can be enormous. At the same time, such poor
efficiency means a great deal of exhaust heat, which in turn means that
the condenser has to be very large. If the condenser is not big enough to
remove the latent heat in the exhaust steam, the exhaust pressure rises
and shrinks the area enclosed by the indicator diagram. Then the engine
loses power, even though it is at full throttle.

Many people who have heard of the Stanley Steamer wonder why steam engines aren't used to power cars today. Using modern materials, a powerful steam engine could certainly be built that is both small and lightweight. But the problem is in the condenser. With no condenser, the engine is inefficient because the exhaust pressure and temperature are so high. (There is also the annoying problem of having to add water fairly often.) If a condenser is used, it has to be air-cooled in a car. And since air convection is such a poor way to transfer heat, the condenser may have to be quite a bit larger than the engine.

17-8 The Steam Turbine

Powerful *steam-turbine engines* are used to drive the dynamos at coal-fired and nuclear power plants. They are also used on many large ships. The idea is to let high-pressure steam expand as it travels at high speed through a series of curved vanes attached to a drive shaft. Generally, there are several stages. The exhaust from the first stage travels into a larger second stage (larger because the steam has expanded). The exhaust from the last stage travels into the condenser. The engine would operate without a condenser, but not nearly as efficiently. And then a continuous souce of very pure water would be needed to supply the boiler (any impurities build up deposits on the turbine blades, throwing the engine off balance). With a condenser, the same water is pumped back to the boiler to repeat the cycle again and again. The plumes at cooling towers at power plants (Fig. 17-8) are from the water that has been used to cool the condenser, not from steam that has actually traveled through the engine.

Figure 17-9 shows the Rankine cycle for this continuously circulated water in a steam turbine. From point *a* to *b,* high-pressure steam from the boiler expands through the engine, and its pressure drops as the steam does work. The steam then cools further and condenses into water in the condenser. This condensation causes a tremendous decrease in volume ($b - c$). The water is then pumped back into the boiler, where it is heated so its pressure increases back to point *a*. The cycle is then repeated.

Comparing Fig. 17-9 with Fig. 17-7, we see that the steam turbine has the effective cutoff occurring as soon as possible. It also lets the expansion go on as long as possible, rather than exhausting bursts of pressurized steam into the condenser. By fully exploiting the expansive property of the hot steam, the turbine can operate at a much higher efficiency than the reciprocating steam engine.

Steam turbines operate best at frequencies of 3000 to 18 000 rpm, which is quite high for their large size. In the engine, there may be hundreds of thousands of rotating vanes, failure of any one of which will quickly destroy the engine (Fig. 17-10). When the engine is started,

Fig. 17-8 The "plume" from this cooling tower contains water that has been used to cool the condensers in a large power plant. The heat energy exhausted to the atmosphere is actually greater than the electric energy produced.

thermal expansion of the parts can never allow things to get the slightest bit lopsided. For all their vast power, these engines are really quite delicate, and great care must be exercised during start-up.

In marine applications, the turbine usually turns dynamos which, in turn, supply electric motors driving the screws. There are several

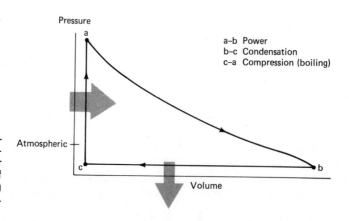

Fig. 17-9 The Rankine cycle: indicator diagram for a steam engine. The turbine makes better use of the expansive power of steam than the reciprocating steam engine does.

a–b Power
b–c Condensation
c–a Compression (boiling)

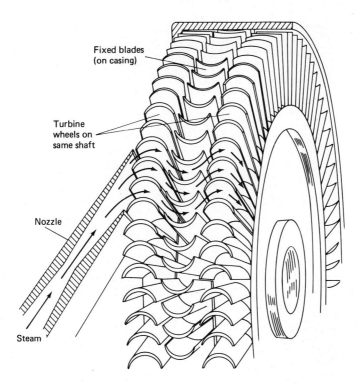

Fixed blades
(on casing)

Turbine
wheels on
same shaft

Nozzle

Steam

Fig. 17-10 Flow of steam in a
steam-turbine engine. The same
arrangement is used in gas-
turbine engines.

reasons for such an arrangement. For one thing, ship propellers do not
work very well at frequencies higher than a few hundred revolutions per
minute. Electric motors eliminate the need for a large, complicated gear
reduction system. For another thing, turbine engines are not reversible,
while electric motors are. Finally, there are many other good uses for the
electricity produced by the dynamo.

17-9 The Otto Cycle

Reciprocating steam engines and steam turbines are examples of
external-combustion engines. In such engines, the working substance is
separate from the fuel that provides the heat.

Internal-combustion engines burn the fuel in the engine itself,
which permits very high temperatures to be reached. They are exhausted
directly to the environment, which eliminates the need for a condenser.
Unfortunately, high exhaust temperatures have the same detrimental
effect on efficiency as low input temperatures, so internal-combustion
engines usually turn out to be less efficient than their external-
combustion cousins. But because of their smaller size, internal-
combustion engines are still better suited to mobile power plants.

Application: Materials Technology and Gas Turbines

The gas-turbine engine is the workhorse of commercial aviation. In an aircraft, the turbine's expanding exhaust gases usually provide the thrust directly; such an engine is called a *turbojet*. Even in turbojets, the output of the turbine shaft is still needed to drive the compressor, pumps, electric generator, and other essential moving parts. All gas turbines, including turbojets, operate on the cycle in Fig. 17-19.

Efficiency is extremely important in an aircraft engine. For one thing, an inefficient engine has a large fuel requirement, and the weight of the extra fuel reduces the plane's payload. For another thing, inefficient engines must be very large (and therefore very heavy) if they are to produce an appreciable power output. And, of course, jet fuel is very expensive. How do we achieve high efficiency in a gas turbine? The single most important factor is high combustion temperature. This is predicted by Eq. (17-2): to get a significant fraction of the heat changed into work, there must be a big difference between the combustion and exhaust temperatures.

Now it isn't difficult to boost the combustion temperature. The problems begin when the hot gases go roaring through the vanes of the high-pressure turbine. These parts are usually made of nickel-based superalloys that melt at around 1300°C [2400°F] and that rapidly lose their strength beyond 1070°C [1950°F]. Air-cooling these critical parts helps some, but even then the combustion temperature must be limited to around 1400°C [2600°F].

The problem is more than a matter of melting point. Turbine blades are subjected to stresses as high as 1400 kg_f/cm^2 [20 000 psi, or 140 MPa] because of the centrifugal force on the rim of the fast-turning turbine wheel. These stresses, coupled with the high temperature, cause the metal to "creep" slightly. Turbine materials must have high elastic limits at high temperatures, as well as high creep strengths.

Another problem is caused by thermal cycling. Engine parts are rapidly heated when a cold engine is started and when a pilot calls for more thrust. The outside surface of each part expands more quickly than the mass of metal inside.

When the engine is shut off, the surfaces contract more quickly than the insides. Turbine blades and vanes go through these heating-cooling cycles thousands of times in a year. Eventually, the resulting thermal stresses cause microscopic internal flaws that weaken the metal. The phenomenon is called *thermal fatigue.* So another requirement of turbine materials is that they have high resistance to this thermal fatigue.

At present, turbine parts must already be replaced periodically during the lifetime of a jet engine. To go to higher temperatures, new materials must be developed which will not aggravate the maintenance problem. A process called *single-crystal casting* offers promise, but a real breakthrough would come if ceramic materials could be used. So far, the main difficulty with ceramics is their low elastic limit in tension.

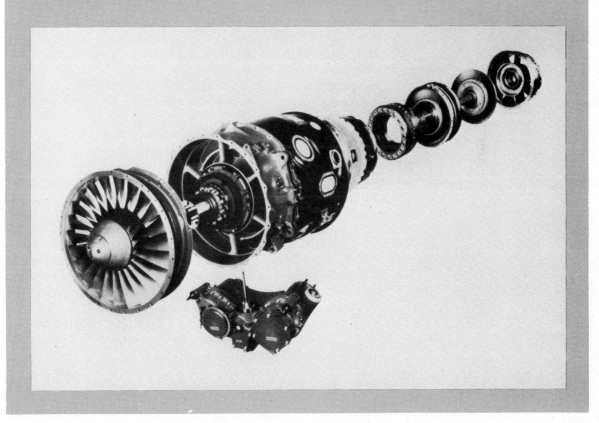

In 1680, Dutch scientist Christian Huygens suggested a gunpowder engine. A small charge would be placed in the bottom of a closed vessel containing a movable piston. When the gunpowder was ignited, the piston would be driven upward to a point where it opened an exhaust pipe, released the pressure, and fell again. But then came the problem: how to remove the ashes and quickly introduce a new charge in time to repeat the cycle. The engine was never successful.

It was not until the early 1800s that experiments with internal combustion really got off the ground. In 1807 two Frenchmen built an engine that burned lycopodium dust (the main ingredient in flea powder). It has since been referred to as the "flea-powder engine," and although it worked, no one ever took it very seriously. In 1863 another Frenchman (Jean Lenoir) successfully used a small, gasoline-powered engine to drive a carriage. But it was not an impressive demonstration, especially when compared to the powerful railroad locomotives then in use. Lenoir gave up.

Fortunately, the idea itself did not die. By 1876 a German, Nikolaus Otto, built a gasoline engine with a four-stroke cycle that worked quite well. In the United States, George B. Seldon applied for a patent on a gasoline-powered automobile in 1879. The patent was granted in 1895. By 1904 there were 35 companies selling their gasoline-powered cars in the United States. Within a few years, the Otto-cycle engine revolutionized transportation (Fig. 17-11).

Figure 17-12 shows the indicator diagram for the ideal, or theoretical, *Otto cycle*. This cycle has been the basis of operation of virtually all gasoline engines since the 1870s. Even designs like the Wankel rotary engine and the aircraft radial engines operate on this cycle.

At point *a* the piston is at the top of its stroke, the intake valve is open, and the exhaust valve is closed. The piston is drawn back to the bottom of its stroke (point *b*), and a fuel-air mixture (mostly air) flows in at atmospheric pressure. Now the intake valve is closed, and the piston is pushed back to the top of its stroke, compressing the mixture. At point *c* a spark initiates the combustion, which takes place very quickly and raises the pressure to point *d*. Now the high-pressure gases expand against the piston and drive it back to the bottom of its stroke (point *e*). Here the exhaust valve is opened, causing the pressure to suddenly drop to atmospheric, point *b* again. (A loud "pop" is heard during this decompression.) The piston is then pushed back to the top of its stroke (point *a*), to get rid of the remaining exhaust gases. Here the exhaust valve closes, the intake valve opens, and the cycle is repeated. This is called a "four-stroke cycle" because the piston moves 4 times in taking the engine through one complete cycle. It is also possible to have a two-stroke Otto cycle by eliminating the *b-a-b* section of the indicator diagram.

Fig. 17-11 The Otto-cycle engine revolutionized personal transportation. Both photographs were taken in 1908.

We said earlier that the work done by an engine in one cycle is the *area* enclosed by the indicator diagram. So any engine modification that increases this area also increases the engine's power output. Suppose that we increase the engine's stroke, or increase the cylinder diameter, so more volume is swept out by the piston. Figure 17-13 shows how this affects the indicator diagram. Since the new diagram encloses more area, the engine's power output is increased.

The power output may also be increased by increasing the pressure. This may be done by reducing the clearance volume (perhaps by planing the head) so the fuel-air mixture is compressed into a smaller volume before ignition. Or it may be done by *pumping* the fuel-air mixture

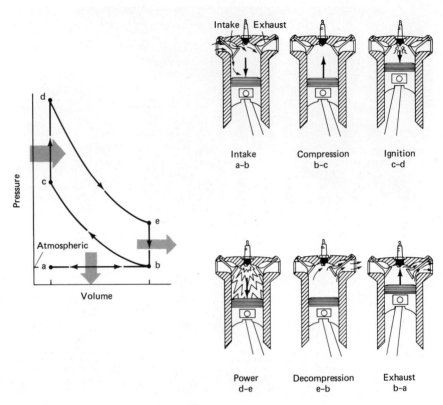

Intake   Exhaust

Intake
a–b

Compression
b–c

Ignition
c–d

Power
d–e

Decompression
e–b

Exhaust
b–a

Fig. 17-12 The Otto cycle: ideal indicator diagram for a spark-ignition internal-combustion engine.

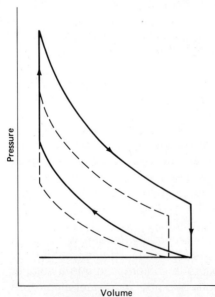

Fig. 17-13 Increasing the volume swept out by the piston increases the area enclosed by the indicator diagram of an Otto-cycle engine. This results in an increase in the engine power output.

into the cylinder at intake, rather than relying on atmospheric pressure to do the job. The pumping process is called *supercharging* if the pump is driven off the drive shaft, or *turbocharging* if the pump is driven by the engine's exhaust gases. Figure 17-14 shows how such modifications change the indicator diagram. Again, the gain in area is a measure of the gain in the engine's power output.

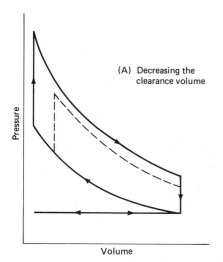

(A) Decreasing the clearance volume

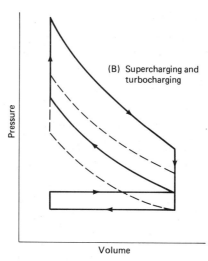

(B) Supercharging and turbocharging

Fig. 17-14 Increasing the pressure in an Otto-cycle engine increases the power output.

The Otto cycle is a theoretical way to describe the operation of a conventional gasoline engine. But when measurements are made on the test stand (and these measurements aren't easy), the actual indicator diagram looks more like Fig. 17-15. Here the engine is running at full load at a high frequency of rotation. The piston moves a bit during ignition, and it also moves a bit during decompression. But the biggest difference between Fig. 17-15 and Fig. 17-12 is the extra loop at the bottom. Since the cycle on this lower loop is counterclockwise, its area represents work done by the engine in moving the fuel-air mixture in and out. The exhaust gases must be forced out at a pressure higher than atmospheric to get them to travel through all the plumbing: the exhaust manifold, muffler, and the tail pipe. And the fresh charge is brought in at a pressure below atmospheric because of the constrictions in the carburetor and intake manifold and the fluid inertia of the fuel-air mixture. This parasitic loop is a drain on the engine's power output. To reduce its size, racing car engines use straight pipes with no mufflers (a noisy solution, but effective), as well as multibarrel carburetors and large manifolds. In fact, anything that lowers the exhaust pressure or raises the intake pressure increases the engine's power output.

17-10 The Spark-Ignition Engine in Operation

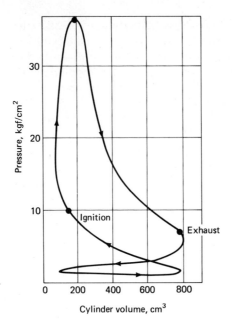

Fig. 17-15 Indicator diagram for
an Otto-cycle engine at full load
and high speed.

Figure 17-15 shows the engine's cycle when it is running at full throttle and full load. Most of the time, the engine runs at only 10 to 25 percent of full load. A car may have a 100-kW [140-hp] engine, but when it is cruising at interstate highway speeds, it is probably developing only around 25 kW [35 hp]. You know this is true, because when you pull out to pass a truck, you step on the accelerator and get an added burst of power. This additional power would not be available if the engine were already developing its maximum output.

At partial loads, less than a full charge of fuel and air is admitted to the cylinder. As a result, the upper part of the indicator diagram is considerably smaller. But the lower part still encloses about the same area (although the shape may be different). Now a larger fraction of the engine's power goes into the parasitic loop, and the engine's efficiency drops. Otto-cycle engines tend to be quite inefficient at partial loads.

17-11 Compression Ignition: The Diesel Cycle

We have already talked about compression heating—the fact that a gas's temperature increases if it is quickly compressed. In the early 1800s, before matches were in common use, a device known as the fire pump was often used for starting fires. This was simply a small cylinder with a tightly fitting piston. Some tinder was dropped into the cylinder, the piston was fitted in place, and with a quick snap of the wrist the trapped air was quickly compressed. The tinder burst into flame. This is known as *compression ignition.* Some suppliers of camping equipment still market the device.

It occurred to Rudolph Diesel, a German engineer, that the same principle could be used in an internal-combustion engine. This would eliminate the need for an electric ignition system. Diesel worked out his idea mathematically, then had some machinists build a working model to his specifications. The year was 1893. On the first trial, the engine promptly exploded, and Diesel wound up in the hospital. But he had proved his point: it was possible to compress air in a large metal cylinder to the point where the heat of combustion ignites a fuel spray.

The diesel engine developed rapidly: by 1906, engines of 200 to 400 kW [300 to 500 hp] were being produced. These engines were more efficient than gasoline engines, and they ran on a cheaper grade of fuel. Unfortunately they were very heavy for their power output, and their maximum frequency was relatively low. They were well adapted to boats, locomotives, small electric power plants, and heavy trucks and earth-moving equipment. With recent advances in materials technology, they have also begun to appear in automobiles.

A four-stroke *diesel cycle* is shown in Fig. 17-16. Air is drawn in

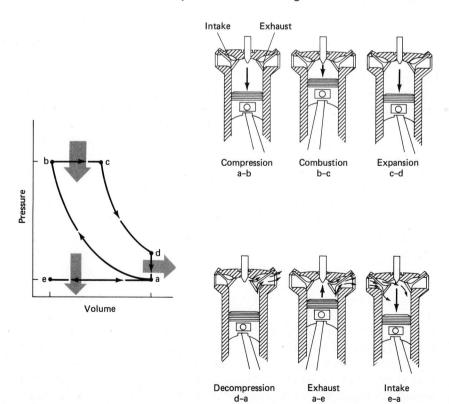

Fig. 17-16 The four-stroke diesel cycle. Ideal indicator diagram for a diesel engine.

from the atmosphere and compressed into the cylinder's small clearance volume, where it reaches a very high temperature ($a - b$). Fuel oil is then injected, and it burns as the piston begins to descend ($b - c$). For the remainder of the power stroke ($c - d$), the hot gases expand and begin to cool. At point d, the exhaust valve opens, and the pressure drops to atmospheric. The piston rises to expel the exhaust gases ($a - e$) and then descends and draws in fresh air through the intake valve ($e - a$). The cycle is then repeated.

During the compression stroke, the compressed air must reach a temperature of 500°C [900°F] or higher. This requires a compression ratio in the neighborhood of 20:1 (the compression ratio of a modern gasoline engine is typically 8:1). Such large compression ratios make diesels difficult to start, especially in cold weather. To start the engine, it must be turned over fast enough that the heat of compression is not conducted away as the air is compressed. Large multicylinder diesels are often started by turning them over with a small gasoline engine, and the exhaust valves on all but one cylinder may be propped open until the one cylinder starts. It is also common practice to preheat the cylinders electrically before attempting to start the engine. Because of the starting difficulties, large diesels are used primarily in applications where they run continuously for long periods.

17-12 The Stirling Engine

There are two serious practical problems with steam engines (including steam turbines). The first is the fact that steam becomes corrosive at temperatures above 600°C, causing steels to absorb hydrogen and become brittle. This limits the engine's working temperature, and because of Eq. (17-2) the maximum efficiency is also limited. The second problem is that all the waste heat must be carried away by the condenser, which therefore has to be very large.

In spite of these design problems, external-combustion engines offer many advantages, too. One is that they can be made essentially nonpolluting, and their emissions do not get worse as they age. They can also be adapted to a wide variety of fuels: coal, oil, nuclear, geothermal, solar, etc. And the type of fuel can even change from one day to the next. If it weren't for the temperature limit and the condenser problem, external-combustion engines would be very attractive as small, mobile power plants.

In fact, both problems are reduced in an external-combustion engine called the *Stirling engine*. The original device was invented in 1827 by a Scottish minister (he thought of the idea some 11 years earlier). It used no boiler, no water, and no valves. A continuous flame heated a sealed charge of air that drove the engine. Today it is still sometimes called the "hot-air engine," although modern versions usually use hydrogen gas as the working substance.

The indicator diagram for the Stirling engine is shown in Fig. 17-17. Starting at point *a*, the high-pressure gas expands against a piston, giving a power stroke. At point *b* the gas is still fairly hot. It is then exhausted (or transferred) to a cold cylinder without changing its volume, and on the way much of its heat is stored in a heat-absorbing material called the *regenerator.* This causes the pressure to drop to point *c*. Now a piston in the cold cylinder compresses the gas into a smaller volume, which also increases its pressure (point *d*). The gas then flows back through the regenerator into the hot cylinder, and along the way it picks

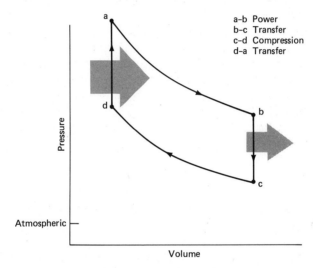

a–b Power
b–c Transfer
c–d Compression
d–a Transfer

Fig. 17-17 The Stirling cycle: indicator diagram for a Stirling engine.

up the heat that was previously stored. All this time, a flame has been heating the hot cylinder; so as the gas flows in, its pressure quickly rises to point *a*. The cycle is then repeated.

Figure 17-18 shows a schematic of the device. The actual engine can be built in many possible configurations; this diagram was chosen to clarify the principles of its operation. Here the two pistons slide in the same cylinder. During the power stroke (*a − b*), both pistons are driven downward. Then only the top piston moves back up, transferring the hot gas to the space between the pistons. Notice that this does not change the gas volume. During this transfer, heat is stored in the regenerators. The gas is further cooled by the cooling coils (or fins) around the lower part of the cylinder. The cool gas is now easy to compress (*c − d*), and as it travels through the regenerators and into the hot cylinder, its pressure quickly rises further (*d − a*).

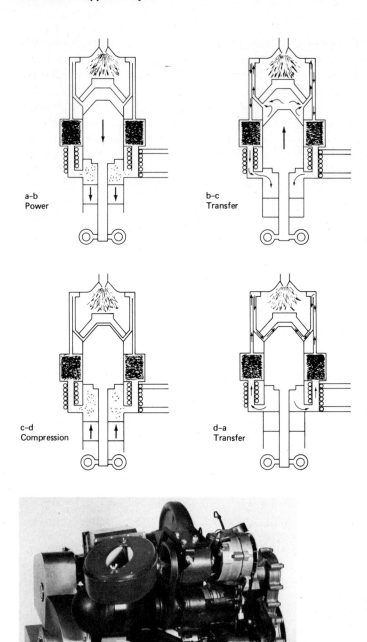

a–b
Power

b–c
Transfer

c–d
Compression

d–a
Transfer

Fig. 17-18 Operation of the
          Stirling engine.

Now this arrangement wouldn't work at all if the pistons weren't somehow connected. The two piston shafts run to a special crank-and-gear arrangement called a rhombic drive, which keeps them moving in proper relation to one another.

Because the Stirling engine recycles part of the exhaust heat, the cooling requirement for the engine is much less than that of steam engines. It can also operate at very high temperatures, which means that high thermal efficiencies can be achieved. And it turns out that the efficiency does not drop off drastically at partial loads, as with steam and gasoline engines. At present, Stirling engines have efficiencies of 30 percent with loads ranging down to 25 percent of full load, dropping to 28 percent at 10 percent load. This low-load efficiency is about double that of the Otto-cycle engine under the same conditions.

Stirling engines are being used in buses in some parts of Europe, and U.S. manufacturers are now experimenting with them. The major drawback at present seems to be high manufacturing cost. One other interesting thing about this engine: if you put work in, cranking the device through its cycle, it becomes a refrigerator and pumps heat from the cool area of the engine to the hot area.

17-13 The Brayton Cycle: Gas Turbines

A *gas-turbine engine* is shown in Fig. 17-19. Most commercial aircraft engines are of this type. Air is drawn in from the atmosphere and compressed by a rotating vane compressor. As it passes into the combustion

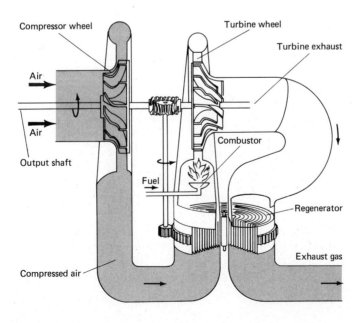

Fig. 17-19 Operation of the gas-turbine engine.

chamber, this high-pressure air is heated first by the engine's exhaust, then by the burning fuel. The air then expands through the turbine and is exhausted through the regenerator and into the atmosphere. The engine's cycle is called the *Brayton,* or *Joule,* cycle.

Figure 17-20 shows the ideal indicator diagram for this engine. Air is brought in at atmospheric pressure (point *a*) and then compressed so its pressure rises to point *b*. At first, expansion takes place with no drop in pressure (*b* − *c*) because heat is being added. As the air completes its expansion (*c* − *d*), its pressure drops, finally reaching atmospheric pressure in the exhaust. To get from *d* back to *a*, we have to figure that the surrounding atmosphere is effectively part of the engine's working substance.

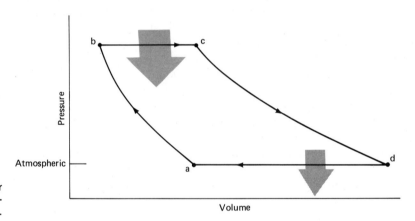

Fig. 17-20 Brayton cycle (or joule cycle): ideal indicator diagram for a gas-turbine engine.

Present-day gas turbines have full-load efficiencies of around 25 percent, but this drops drastically under partial-load conditions. If attempts at using ceramic materials are successful, gas turbines may achieve full-load thermal efficiencies as high as 55 percent. Ceramics would also probably bring down the cost, which is now rather high.

17-14 Engine Power and Torque

Earlier in this chapter, we saw how to calculate an engine's power output from its indicator diagram. Power calculated this way is called indicated power.

indicated power

Definition: An engine's *indicated power* output is the power developed by cycling the working substance within an engine. It can be calculated from the engine's indicator diagram.

Notice that the indicated power is the gross power developed by the working substance acting on the engine's piston or turbine. It does not account for the additional power needed to operate fuel pumps, valves, the ignition system, coolant pumps, oil pumps, or other essential accessories.

When the engine's power is measured by a brake or dynamometer at its output shaft, the actual power is always found to be less than the indicated power. This power is called the brake power.

> Definition: An engine's *brake power* is the power actually available at its output shaft. The brake power is always less than the indicated power.

brake power

The difference between an engine's indicated power and its brake power is the power needed to overcome friction and to operate the valve train, pumps, generator, and the other accessories we mentioned. This is called the friction power.

> Definition: An engine's *friction power* is the power needed to overcome friction and to perform all the external tasks necessary to the engine's operation.

friction power

The relationship between these three quantities is

Brake power = (indicated power) − (friction power)   (17-7)

In Fig. 17-21 we see how these quantities change with operating frequency. The graphs show the power developed by an Otto-cycle engine at full load, but the same general behavior is found in all types of engines. At low revolutions per minute, the indicated power increases in proportion to the frequency: doubling the frequency doubles the power. But as higher frequencies are reached, the indicated power no longer increases as rapidly. This is because the piston is now moving too rapidly to allow a full charge of fuel and air to flow in, and the exhaust is also being forced out rather than simply flowing out. As a result, the area inside the indicator diagram shrinks. (Engineers say that the engine's *volumetric efficiency* has decreased.) Eventually, at very high frequencies, the indicated power levels off at some maximum value. Meanwhile, the friction power starts off low but keeps increasing as the frequency is increased. At each frequency, the brake power is the difference between the indicated power and the friction power. When the brake power is calculated and graphed, we find that it rises at first, levels off, then actually decreases at high revolutions per minute. There is a single frequency where the brake power is a maximum for any given engine.

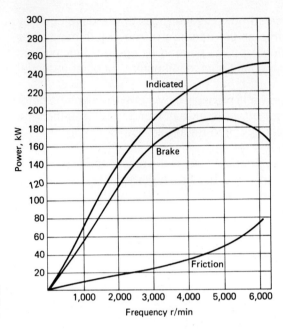

Fig. 17-21 Friction, brake, and indicated power at full load in an Otto-cycle engine.

When an engine is used to drive rotating machinery, it is often more useful to know the engine's torque than its power. As you might guess, the two quantities are closely related. If you know the brake power at a certain frequency, the torque is

$$\text{Torque, in N·m} = \frac{\text{power, in W}}{(2\pi)(\text{frequency, in rps})} \qquad (17\text{-}8)$$

Or if you know the output torque at a certain frequency, the brake power is

$$\text{Power, in W} = (2\pi)(\text{torque, in N·m})(\text{frequency, in rps}) \qquad (17\text{-}9)$$

The use of these formulas often requires conversion factors.

Example 17-3 Torque Output of a Car Engine.

A certain car engine has a brake power of 104 kW [139 hp] at 4800 rpm. What is its torque at this frequency?

A frequency of 4800 rpm is the same as 80.0 revolutions per second (rps). Then using Eq. (17.5),

$$\text{Torque, in N·m} = \frac{\text{power, in W}}{(2\pi)(\text{frequency, in rps})}$$

$$= \frac{104\ 000\ \text{W}}{(2\pi)(80.0\ \text{rps})}$$

$$= 207\ \text{N·m}$$

This may also be expressed as 21.1 kg_f·m, or 153 lb·ft. ◀

Exercises

5. A large turbine produces 325 MW of brake power at a frequency of 3550 rpm. What is the torque output?
   Answer: 874 000 N·m, or 89 100 kg_f·m
6. A certain machine requires 2200 N·m of torque at a frequency of 950 rpm. What is the brake power requirement of the engine that drives the machine?
   Answer: 220 kW

Summary

With fuel shortages threatening to get worse before they get better, engineers have taken a great interest in designing more efficient heat engines. There are really only a few possible approaches. One is to increase the engine's operating temperature and solve all the technical problems that this creates (problems of expansion and thermal stress in the materials, production of nitrogen oxides and other pollutants in the exhaust, etc.). Another way is to reduce the exhaust temperature and pressure, perhaps transferring some of the exhaust heat back to the engine's input, or perhaps by staging (using the hot exhaust to operate another engine). A third possibility is to reduce the engine's friction power by eliminating as many moving parts as possible. These ideas can be applied in a wide variety of ways, the best of which are probably yet to come.

Terms You Should Know

heat engine

second law of thermodynamics

thermal efficiency

indicator diagram

reciprocating steam engine

separate condenser

Rankine cycle

steam-turbine engine

Otto cycle

compression ignition

diesel cycle

Stirling engine

gas-turbine engine

Brayton, or Joule, cycle

indicated power

brake power

friction power

Problems

1. A certain reciprocating steam engine operates at a boiler pressure of 1300 kPa and an exhaust (condenser) pressure of 15 kPa. The engine has a single double-action piston. The indicator diagram is shown in Fig. 17-7a. The piston's clearance volume is 150 cm³ while the maximum cylinder volume is 590 cm³. (a) Calculate the work output of one side of the piston in one Rankine cycle. (b) Calculate the indicated power if the engine is turning at 1000 rpm. (c) If the friction power at 1000 rpm is 5.0 kW, what is the engine's brake power at this frequency? (d) What is the engine's torque output at this frequency?

2. A small turbine operates at a boiler pressure of 1030 kPa, absolute, and a condenser pressure of 12 kPa, absolute. The indicator diagram is shown in Fig. 17-8. In the boiler, each 1 g of water has a volume of 1.1 cm³. As it enters the condenser after passing through the engine, each 1 g of steam occupies 2500 cm³. Estimate the work produced by 1 g of water as it circulates through the engine, condenser, and back to the boiler.

3. The Otto-cycle engine in Fig. 17-15 is turning at 3000 rpm. Since this is a four-stroke cycle, each cylinder completes 1500 Otto cycles each minute. (a) How much work is done by one cylinder in completing one cycle? (b) What is the indicated power of one cylinder? (c) If this is an eight-cylinder engine, what is its total power output at this frequency?

LESS THAN HALF THE COVER PRICE!

OF 1981

(m will be served.)
ns and Canada
es. This rate includes
for two issues com-

1 year of People for $49.00
9 1 year of People for $49.00 (P33020).

Zip/Post Code

Apt. No.

S74650

! That's less than HALF SI's $1.50 cover price!

☐ 100 issues* of SPORTS ILLUSTRATED
just 69¢ an issue. That's less than half the
s-court, on-base and up-front. In SPORTS
at games of 1981. Use this card to catch the

C E R T I F I C A T E

Illustrated

18 Electricity

Introduction

ectrons have a property called electric
kert a very strong electric force on one
gether into atoms, and it allows atoms
oms and molecules may also hold one
pieces of solid matter. When you break
forces that you are overcoming.
nt for the impenetrability of matter. We
have seen that atoms themselves are mostly empty space. Why, then, can
you sit on a chair without passing right through? The answer is that the
outermost parts of all atoms and molecules are electrons. When two
objects are brought into contact, repulsion between these electrons prevents the one object from penetrating the other. In fact, elastic forces,
friction, and all other contact forces are actually electrical in origin.
Indirectly, then, electric forces affect everything we do.

In this chapter we see how electric forces may be used to produce
a flow of electric energy.

18-1 Electric Current

In any sizable chunk of matter, the number of protons and electrons is
normally equal. Since protons are found only in the nucleus of atoms,
they are seldom free to move off on their own. But under certain conditions, electrons may be separated from their atoms. The matter left be-

hind then has an unbalanced positive charge, and the electric force tries to pull the electrons back.

You may have had the experience of shuffling your feet across a carpet, then touching a doorknob and getting a mild electric shock. Now this doesn't always work; it depends on the carpet material, the material in the soles of your shoes, and the humidity of the air. But when it happens, the friction has been successful in tearing electrons loose from atoms in the carpet. The tiny spark is a flow of electrons trying to find their way back. The same effect happens on a larger scale when strong updrafts and downdrafts in thunderclouds rub against large numbers of raindrops. Instead of a small spark, in this case there is a bolt of lightning.

When charged particles (usually electrons) flow from one place to another, we say we have an *electric current*. The SI unit for electric current is the *ampere* (*amp,* for short). Using the standard SI prefixes, multiples of this unit can be formed; thus 1 mA = 0.001 A, and 1 μA = 0.000 001 A. The doorknob spark amounts to no more than a few milliamperes. The current flowing into an electric circular saw is about 10 A. The lightning bolt may have a current as high as 100 kA. In terms of electrons, a current of 1.0 A amounts to 6.25×10^{18} electrons moving past a point in 1 s.

Because there are so many electrons moving in typical electric currents, it is justifiable to think of the current as a continuous flow of charge rather than as a dribble of separate particles. In fact, an electric current behaves in many respects like a moving fluid.

18-2 Voltage

Electric currents flow because of the force of attraction between unlike charges and the repulsion between like charges. Figure 18-1 shows a device known as a Van de Graaff generator. A continuous rubber belt is turned by a crank or by a small motor. The belt rubs electrons off a glass cylinder at the bottom, leaving the base of the machine positively charged. The electrons are carried to the top of the device, where they leak onto a metal plate connected to the dome. Soon the outer surface of the dome becomes covered with electrons struggling to get away from one another. At the same time, these electrons are attracted back toward the positive base. Together, these forces combine to produce an "electrical pressure" that tries to get the electrons moving in a miniature lightning bolt. This electrical pressure is called *voltage*.

Now we can also look at this in another way. To carry the electrons away from the base of the Van de Graaff generator, an outside force had to act through some distance. This means that some work was done, and electric potential energy was stored. Just as a stretched spring with elastic potential energy tries to contract, or objects lifted above the earth

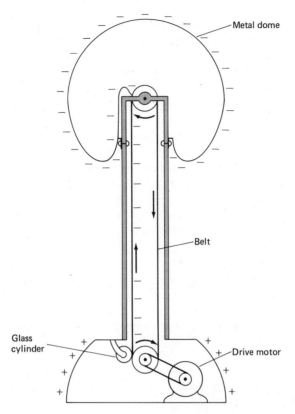

Metal dome

Belt

Glass
cylinder

Drive motor

Fig. 18-1 A Van de Graaff
generator.

tend to fall back to the ground, so the electrons are pulled back to the
base as they give up their potential energy. The voltage is actually a
measure of the potential energy stored per unit charge.

> Definition: *Voltage* is a measure of the potential energy   *voltage*
> or work per unit charge. It can also be thought of as the
> "electrical pressure" acting to force separated charges
> back together.

Voltage is measured in volts, which is an SI unit. Depending on its
size and construction, the Van de Graaff generator can develop voltages
of perhaps 100 kV up to several megavolts.

Notice that it is quite possible to have a voltage without a current.
This is the case with a flashlight that is shut off or a wall outlet that is not
being used. But we cannot have a current without a voltage (at least not
at ordinary temperatures). A spark cannot be drawn from a Van de Graaff
generator unless the voltage has grown large enough. And, of course, a

flashlight does not work without batteries. A voltage is always needed to produce a current flow.

18-3 Resistance   The atoms in some materials hold their electrons very tightly. Plastic, rubber, wood, and air fall into this category. It takes a very high voltage to push even a small current through these materials, which are called *insulators*.

On the other extreme, one or two electrons can easily be removed from most metal atoms. As a result, it takes only a small voltage to push a sizable current through copper, silver, aluminum, steel, and so on. Such materials are called *conductors*. Conductors and insulators both play important roles in electrical applications (Fig. 18-2).

But there is not a definite dividing line between conductors and insulators. Some materials (carbon, nichrome, and tap water, for instance) aren't really good conductors, but then they aren't insulators, either. And among the metals, some are much better conductors than others. It takes less voltage to push a 1-A current through a copper wire than through an aluminum wire the same size, for instance.

Fig. 18-2 Conductors and insulators both play important roles in electrical applications.

The ratio of the voltage to the current it produces is called the resistance.

Definition: An object's *resistance* is the ratio of an applied voltage to the current this voltage produces in the object: *resistance*

$$\text{Resistance} = \frac{\text{voltage}}{\text{current}} \qquad (18\text{-}1)$$

Ordinarily, we measure voltage in volts and current in amperes. Notice that these are both SI units. The unit of resistance, then, is volts per ampere. But since this is a bit cumbersome, it has been given a name of its own: the *ohm,* abbreviated Ω. Thus,

$$1\ \Omega = \frac{1\ \text{volt}}{1\ \text{amp}} \qquad (18\text{-}2)$$

Example 18-1 Resistance of a Lamp.

A courtesy lamp in a car operates on a voltage of 12 v. If this voltage produces a current of 1.5 A through the lamp, what is the lamp's resistance?

Using Eq. (18-1), we have

$$\text{Resistance} = \frac{\text{voltage}}{\text{current}}$$

$$= \frac{12\text{V}}{1.5\ \text{A}}$$

$$= 8.0\ \text{V/A}$$

$$= 8.0\ \Omega \qquad \blacktriangleleft$$

Now a device's electric resistance may not remain constant if the applied voltage changes. For resistance to be constant, a graph of voltage versus current would have to be a straight line. (In other words, doubling the voltage would double the current, and the ratio of voltage to current would not change.) In fact, a graph of voltage versus current is seldom exactly a straight line. Figure 18-3 shows the more common types of behavior. Lamp filaments, toaster elements, and spaceheater filaments have relatively low resistances at low voltage. But at higher voltages, the current causes a temperature increase, which also increases

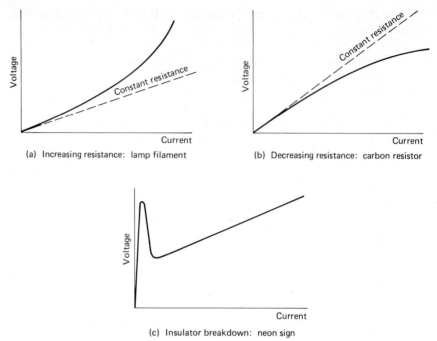

(a) Increasing resistance: lamp filament

(b) Decreasing resistance: carbon resistor

(c) Insulator breakdown: neon sign

Fig. 18-3 Resistance often changes when the voltage and current are changed.

the resistance. All metallic conductors behave to some extent like Fig. 18-3a. The carbon resistors common in electronic circuits have resistance that decreases slightly at high voltage. This is shown in Fig. 18-3b. Figure 18-3c shows the behavior of a neon sign. At low voltages, the sign's resistance is so high that practically no current flows. But at several thousand volts, the gas ionizes and begins to conduct; the result is a rapid increase in current. As the neon goes from the nonconducting to the conducting state, its resistance drops. The same general behavior occurs with all insulators: air, rubber, plastics, and so on. At a high enough voltage, any insulator "breaks down" and becomes a conductor. All electric devices carry a voltage rating. This should be considered the maximum safe voltage that can be used without taking a chance that the insulation might break down.

Now that we know about resistance, we can give a better definition of what we mean by conductors and insulators:

conductor

Definition: A *conductor* is something that has a low resistance. The lower the resistance, the better the conductor.

Definition: An *insulator* is something that has a very high resistance, usually greater than 10^8 Ω. The higher the resistance, the better the insulator.   *insulator*

Exercises

1. What is the resistance if a voltage of 110 V produces a current of 1.8 A?
    Answer: 61 Ω
2. A current of 0.52 A flows when the voltage is 12 V. What is the resistance?
    Answer: 23 Ω
3. A current of 125 mA results from a voltage of 9.0 V. What is the resistance?
    Answer: 72 Ω
4. A current of 35 mA is needed when the voltage is 120 V. What resistance must be used?
    Answer: 3.4 kΩ

The current in a device depends on the voltage that is pushing it and on the resistance that is holding it back. If we know the voltage and resistance, we can predict the current by rewriting Eq. (18-1):   **18-4 Ohm's Law**

$$\text{Current} = \frac{\text{voltage}}{\text{resistance}} \qquad (18\text{-}3)$$

Again, this formula is based on SI units: voltage in volts, resistance in ohms (Ω), and current in amperes. This formula is often referred to as *Ohm's law*.

For instance, if a 2-Ω resistance is connected to a 12-V battery, the current is 12 V/(2 Ω), or 6 A.

Ohm's law may also be written in terms of the voltage. If we know how much current we want to flow through a certain resistance, the voltage needed may be found by rewriting Eq. (18-3) as

$$\text{Voltage} = (\text{current})(\text{resistance}) \qquad (18\text{-}4)$$

Thus if a 3.0-A current is needed in a 12-Ω resistance, the voltage required is 36 V.

Application: Batteries

All batteries produce electricity through chemical reactions. A battery consists of one or more voltage-producing cells.

Primary cells use chemical reactions that cannot easily be reversed; as a result, such cells cannot be recharged. Since primary cells can be manufactured fairly cheaply, they are found in most flashlight and radio batteries.

Secondary cells use chemical reactions that can be reversed by reversing the direction of the current, that is, by forcing a current through the cell from the positive terminal to the negative terminal. Secondary cells are found in all rechargeable batteries.

A battery's voltage is determined by the chemical reaction in its cells and by the way the cells are connected in the battery. As a result, it is impossible to charge a 12-V battery to 18 V, for instance. The battery's voltage will change, however, if the temperature changes.

An important rating is a battery's ampere-hour (Ah) capacity. This is a measure of the total charge the battery will hold. A battery rated at 20 Ah can deliver an average current of 20 A for 1 h before becoming completely discharged. The same battery could deliver 10 A for 2 h or 5 A for 4 h.

CHARACTERISTICS OF SOME COMMON BATTERIES

Battery	Voltage	Ampere-hour rating	Mass, g
Primary			
"pen-light"	1.5	0.58	20
D-cell	1.5	3.0	90
#6 dry cell	1.5	30	860
#2U6 battery	9.0	0.325	30
#V-60 battery	90	0.47	580
#1 mercury cell	1.4	1.0	12
#42 mercury cell	1.4	14	166
Secondary (rechargeable)			
#CD-21 nickel-cadmium	6.0	0.15	52
#CD-29 nickel-cadmium	12.0	0.45	340
lead-acid car battery	6.0	84	13 000
lead-acid car battery	12.0	96	25 000

Exercises

5. What voltage is needed to produce a 65-A current when the resistance is 0.60 Ω?
    Answer: 39 V
6. What current flows when a voltage of 110 V is connected to a resistance of (a) 250 Ω, (b) 500 Ω, (c) 1000 Ω?
    Answers: (a) 0.44 A, or 440 mA; *(b)* 220 mA; (c) 110 mA
7. What current flows when a constant resistance of 3.3 kΩ is connected to a voltage of (a) 24 V, (b) 48 V, (c) 100 V?
    Answers: (a) 7.3 mA, (b) 15 mA, (c) 30 mA
8. A current of 2.5 A flows through a 16-Ω resistance. (a) What is the voltage? (b) What voltage is needed to double the current if the resistance remains constant?
    Answers: (a) 40 V, (b) 80 V

18-5 Circuits

Positive and negative electric charges may be separated and kept apart for long periods with no current flow. This happens in a thundercloud, a Van de Graaff generator, or the capacitor in a photoflash unit. Electricity stored like this is sometimes called *static electricity*. Static electricity becomes a current flow when a conducting path is provided between the positive charges and the negative charges. This current flow happens in a quick burst, which neutralizes the separated charges and depletes the voltage. The voltage must then have time to build up again before another current burst is possible.

Most electrical applications use current electricity rather than static electricity. With current electricity, a voltage source like a battery or a generator pumps the current continuously around a closed circuit.

Figure 18-4 shows a simple *circuit*. Here the current is pumped through a light bulb, which converts the electric energy to light and heat.

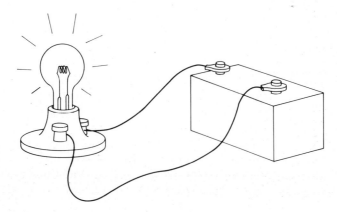

Fig. 18-4 A simple circuit. The battery pumps the current around a complete path.

Notice that the charges do not pile up in the bulb; for the light to work, there must be a return path to the battery. The current flowing out of the battery is exactly the same as the current flowing into the bulb, which is exactly the same as the current flowing out of the bulb, which is the same as the current flowing back into the battery. This same current flows inside the battery as well.

Obviously, the wires connecting the battery and bulb must be conductors. These conducting wires carry the current where we want it to go. The wires must be surrounded with insulation, to prevent the current from flowing where we *don't* want it to go. When the current goes someplace we don't want it to, we say we have a *short circuit*.

If the conducting path is broken, the current stops and the light goes out. Circuits are usually broken with switches or circuit breakers that have been wired into the circuit.

Since it is clumsy to sketch circuits by drawing things like batteries and light bulbs, we usually use *schematic symbols* instead. Figure 18-5 shows a few of the schematic symbols in common use. Figure 18-6 shows an example of how these symbols can be combined to show a particular circuit. Although the schematic circuit diagram doesn't look much like the actual arrangement of light bulbs, switches, and wires, if you follow the path of the current in each diagram, you will find that they really are the same.

18-6 Conventional Current

We have seen that an electric current is a flow of electrons, which are negative charges. They flow from the negative terminal of a battery (or other voltage source) toward the positive terminal. This is simply a result of the attraction between opposite charges.

But when Benjamin Franklin studied electricity in the late 1700s, no one knew about electrons. Franklin figured that the current flow was from positive to negative; and since everyone assumed he knew what he was talking about, they all followed his choice. Today it is still customary to say that the current flows from positive to negative, even though we know the electrons are doing just the opposite. This scheme is called conventional current.

conventional current

Definition: The *conventional current* in a circuit is the current assumed to flow from the positive side of the source through the circuit to the negative side of the source.

Why do we continue to use conventional current? Because it makes sense to have a current flowing from the high-voltage side of the source toward the low-voltage side. The scheme is then the same as

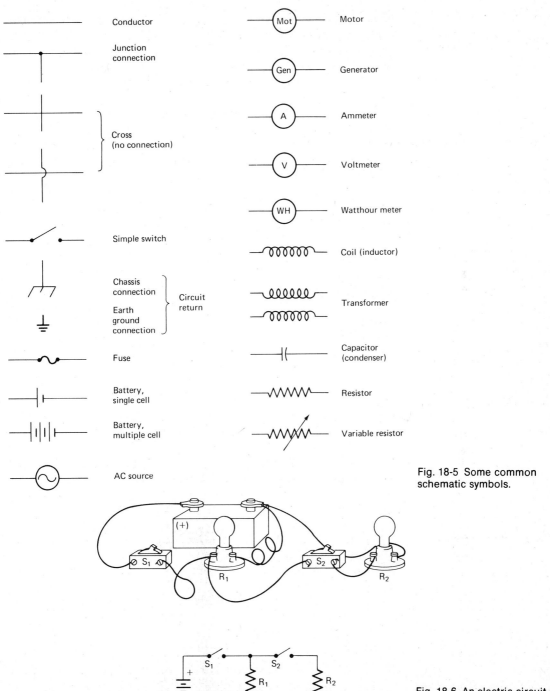

Conductor

Junction
connection

Cross
(no connection)

Simple switch

Chassis
connection   Circuit
                    return
Earth
ground
connection

Fuse

Battery,
single cell

Battery,
multiple cell

AC source

Motor

Generator

Ammeter

Voltmeter

Watthour meter

Coil (inductor)

Transformer

Capacitor
(condenser)

Resistor

Variable resistor

Fig. 18-5 Some common
schematic symbols.

Fig. 18-6 An electric circuit and
its schematic diagram.

having fluids flow from high pressure to low, and heat flow from high temperature to low. Besides, there *are* a few cases where the current contains a flow of positive charges. This happens in electrolytic solutions, in ionized gases, and in transistors, for instance.

Even so, *electron current* is sometimes used in electronics applications. Figure 18-7 summarizes the difference between the direction of electron current and conventional current.

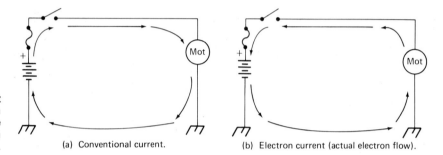

Fig. 18-7 Conventional current and electron current in a winch motor circuit. In vehicles, the circuit return is usually through the chassis.

(a) Conventional current.   (b) Electron current (actual electron flow).

18-7 Measurement of Voltage and Current

Voltage is usually measured with a *voltmeter,* and current with an *ammeter*. Since these two instruments measure different things, they have to be used in different ways.

Since an ammeter measures a kind of flow rate, the current must flow *through* the meter. This means breaking the circuit and actually wiring in the meter.* The positive terminal on the meter should be closer to the positive side of the source. Figure 18-8a shows the proper use of an ammeter.

Notice that the ammeter itself is part of the circuit, and therefore the ammeter affects the current it is supposed to measure. Manufacturers try to reduce this effect by making the meter's resistance as low as possible. This leads to problems when the meter is used incorrectly. In Fig. 18-8b, the ammeter provides a low-resistance path between the high- and low-voltage source terminals. We say that the meter has *short-circuited* the source. The resulting current will be very large, and the meter will probably be damaged. An ammeter should never be wired *across* an electric device.

A voltmeter, on the other hand, measures the electrical pressure difference between two points in a circuit. Figure 18-9a shows the right way to use a voltmeter. Notice that the circuit is not broken to make the voltage measurement. Instead, the voltmeter is wired *across* the source or across the circuit element of interest.

---

*There is one important exception: very large alternating currents are sometimes measured with an ammeter that has a coil which clamps around the conductor of interest.

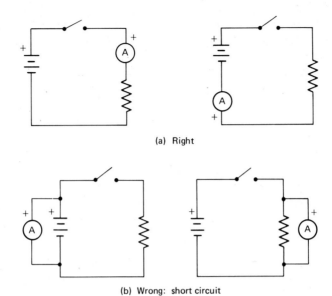

(a) Right

(b) Wrong: short circuit   Fig. 18-8 Use of an ammeter.

We said that ammeters have very low resistance. If this were the case with voltmeters, a large current would flow out of the circuit and through the meter. To keep this from happening, manufacturers build voltmeters with very high resistance. Since practically no current flows through a voltmeter, a circuit will not function properly if a voltmeter is

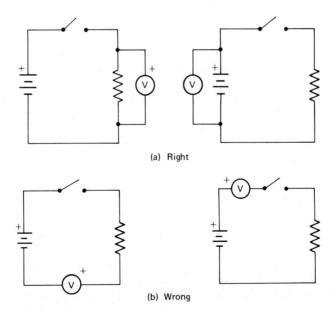

(a) Right

(b) Wrong   Fig. 18-9 Use of a voltmeter.

inserted as in Fig. 18-9b. A voltage will still be read, but no current will flow. Table 18-1 summarizes how voltmeters and ammeters are used.

TABLE 18-1   CHARACTERISTICS OF AMMETERS AND VOLTMETERS

	Ammeter	Voltmeter
Quantity measured	Current	Voltage
Meter resistance	Very small; often less than 1 Ω.	Very large; often greater than 1 MΩ.
Method of connecting to circuit	Circuit is broken; meter is wired in so circuit current flows through instrument.	Circuit is not broken; meter is connected across points where voltage difference is of interest.

Exercises

9. What should the meters read in the circuit shown?
   Answer: Voltmeter reads 110 V, and ammeter reads 2.0 A.

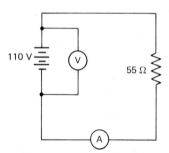

10. What should the meters read in the circuit shown?
    Answer: Voltmeter reads 18 V, and ammeter reads 9.0 A.

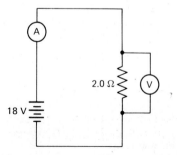

A voltage source sends energy into a circuit when it produces a current flow. This electric energy is transformed to other forms of energy by the circuit elements through which the current flows. In most cases of interest, energy does not pile up in the circuit. That is, the energy supplied by the source each second is the same as the energy converted to other forms each second.

What other forms are we talking about? To begin with, any resistance produces heat. Some resistances, like lamp filaments, produce light as well. Motors produce mechanical work. Speakers produce sound. Electrolytic cells produce chemical potential energy. There are other possibilities as well.

The *electric power* supplied by a source is very simply related to the voltage and current it puts out:

$$\text{Power, in W} = (\text{voltage, in V})(\text{current, in A}) \qquad (18\text{-}5)$$

Example 18-2 Power in a Home Circuit.

A certain circuit in a home is fused at 20 A. This means that the circuit can supply no more than 20 A without blowing the fuse. Of course, it can supply currents lower than this figure with no problem. If the voltage is 110 V, what power can the circuit supply?

When the fuse is just at the point of blowing out, the current is 20 A. Using Eq. (18-5), the power is then

$$\text{Power} = (\text{voltage})(\text{current})$$

$$= (110 \text{ V})(20 \text{ A})$$

$$= 2200 \text{ W}$$

We may also write this result as 2.2 kW. Most appliances have a tag that shows their power requirement, so we can quickly tell if they exceed the circuit's power limit. ◄

Table 18-2 lists typical power requirements for some common appliances. Most home appliances are designed to operate from a 110-V source. When the power requirement is very high, as with electric clothes dryers or air conditioners, a 220-V source may be specified. This allows the current to be kept to a reasonable value so large-gauge wire is not needed.

If several appliances are operated from the same source (or the same circuit), the total power needed is just the sum of the power requirements of all the appliances. So if a toaster and four 100-W bulbs are used at the same time, the total power needed is 400 W plus 1100 W, or 1500 W.

TABLE 18-2   *TYPICAL ELECTRIC POWER REQUIREMENTS OF SOME COMMON APPLIANCES*

Appliance	Power requirement, W
Air conditioner, portable	1 600
Circular saw, full load	900
Clock	3
Clothes dryer	5 000
Coffee pot	575
Dehumidifier	350
Dishwasher	1 500
Doorbell transformer	
No load	1
Full load	40
Electric blanket	200
Food mixer	130
Heat pump	12 000
Iron	1 000
Kitchen stove	12 000
Power drill, full load	230
Radio	60
Refrigerator	400
Television	330
Toaster	1 100
Water heater	2 500
Water pump	700
Vacuum cleaner	300

If we know the power and the voltage, we can find the total current by rewriting Eq. (18-5):

$$\text{Current, in A} = \frac{\text{power, in W}}{\text{voltage, in V}} \qquad (18\text{-}6)$$

Let's see how this equation is used.

Example 18-3 Electric Service for a Garage.

A certain garage is to be converted to a workshop. The question is whether one circuit fused at 30 A is sufficient to supply the electric power or whether two circuits are needed. If two circuits are needed, should they be 30 A each, or will two 20-A circuits be sufficient? (A 20-A circuit can be wired with smaller-gauge wire, which is not only cheaper but easier to work with.)

To answer this kind of question, we need to estimate the total power requirement. What kinds of appliances, in other words, can we expect to have in use at any one time? A typical list may be as follows:

Six 100-W light bulbs	600 W
Two 1200-W space heaters	2400 W
One 900-W power saw	900 W
One 575-W coffee pot	575 W
One 60-W radio	60 W

Of course, there may be other power tools, but only one can be used at a time if there is only one person.

Totaling the power requirement gives 4535 W. The voltage is standard 110 V. Then Eq. (18-6) gives the current:

$$\text{Current} = \frac{\text{power, in W}}{\text{voltage, in V}}$$

$$= \frac{4535 \text{ W}}{110 \text{ V}}$$

$$= 41 \text{ A}$$

With a single 30-A circuit, we'd constantly be blowing fuses and shutting off one thing to turn on another. Two 20-A circuits would come close to doing the job, but we would still have problems. A reasonable choice, then, is to go with two 30-A circuits. ◄

Exercises

11. What power is supplied by a 12-V battery when the current is (a) 2.0 A, (b) 10 A, (c) 40A?
    Answers: (a) 24 W, (b) 120 W, (c) 480 W
12. A certain 220-V circuit is fused at 30 A. What is the maximum power it can supply?
    Answer: 6.6 kW
13. A certain electric motor operates at 110 V. If a current of 14 A flows through the motor, what is its power input in (a) kilowatts, (b) horsepower?
    Answers: (a) 1.5 kW, (b) 2.1 hp
14. If a 110-V electric clock has power input of 3.0 W, what current flows through the clock?
    Answer: 27 mA
15. A kitchen stove has a power input of 12 kW at a voltage of 220 V. What current flows through the stove?
    Answer: 55 A
16. How many 50-W light bulbs can be operated from a 110-V circuit fused at 30 A?
    Answer: Up to 66 bulbs

Application: Electric Shock

The muscles of the body are controlled through tiny electric currents from the brain. If other currents are introduced into the body, muscle spasms and other unpleasant effects can result. Of course, severe electric shocks can even cause death (electrocution.)

Electric current, not voltage, causes shock. But since currents do not flow without voltage, many people mistakenly assume that it is the voltage which is felt. Actually, a person can be exposed to a very high voltage yet experience no shock if there is no conducting path through the body. For instance, birds may perch on bare high-tension wires with no bad effects because they touch only one wire.

If a person touches two conductors at different voltages, a current flows through the body. The amount of current depends on the resistance (Ohm's law). The resistance of the body varies from about 10 kΩ to several hundred kilohms, depending on where the contact points are, how far apart they are and how much perspiration is on the skin. The body's resistance can be measured easily and safely with an ohmmeter.

Although the effects of electric shock depend on a person's size and heatlh, the following gives a general idea of what happens. A current of 1 Ma is barely perceptible to most people. Current up to 8 Ma causes a noticeable tingling sensation. Current between 8 and 15 mA causes a severe jolt. Current over 15 Ma causes muscles to paralyze, and the person may not be able to let go if he or she has grabbed a

18-9 Electric Energy

When we use electricity to do a job for us, we have to be concerned about the electric power consumption. This tells us how much current must flow in the circuit, and from this we can tell whether the circuit can handle the load. But when it comes to paying for the electricity, we pay for *electric energy,* not power. A toaster consumes 370 times as much power as an electric clock, but if we use the toaster for 1 min in the morning while the clock runs all day, it costs almost 4 times as much to operate the clock.

As we saw in Eq. (9-7), energy is related to power by the relation

live wire. If the current passes through the base of the brain, breathing may stop. If it passes through the heart muscles, the heart may stop as well. It may be possible to resuscitate such a person if she or he is separated from the current source and first aid begins immediately. But currents over 75 mA are almost always fatal.

$$\text{energy} = (\text{power})(\text{time})$$

If the power is in watts and the time is in seconds, the energy is in units of joules. But the second is a rather small unit of time, and power is often expressed in kilowatts. To measure electric energy consumption over long periods, the unit kilowatthour (kWh) is generally used:

$$\text{Energy, in kWh} = (\text{power, in kW})(\text{time, in h}) \qquad (18\text{-}7)$$

Application: Electrolysis

Electrolysis is a process that converts electric energy to chemical energy. It is used in many industrial processes, including extracting aluminum from its ore; refining gold, silver, and other metals; manufacturing chlorine gas; and plating car bumpers, silverware, and other items.

A simple case is the electrolysis of water, which separates the water into hydrogen and oxygen gas. The idea is to pass an electric current through the water between two conductors (called electrodes). This causes positively charged hydrogen ions to move toward the negative electrode while negatively charged oxygen ions go to the positive electrode. At the negative electrode, the hydrogen ions pick up electrons and form neutral atoms. These atoms bind together in molecules that eventually form bubbles which float to the surface, where they can be collected. At the positive electrode, the oxygen ions give up electrons to form neutral oxygen atoms; oxygen gas can be collected above this electrode. The battery is needed to pump the electrons back to the negative electrode.

Since pure water does not conduct electricity very well, it is necessary to add some acid to lower its resistance. The acid pulls apart enough water molecules to start the process. This also speeds up the corrosion of the negative electrode, if it is made of a chemically active metal. Copper electrodes can be used for demonstration purposes, but for long-term operation the electrodes should be made of carbon or platinum.

Electrolysis of water may someday be used with wind- or wave-driven electric generators. Hydrogen fuel could then be produced during storms and stored until needed. As mentioned earlier, hydrogen is an attractive fuel for home

The electric meters on homes and other buildings measure the total electric energy supplied in kilowatthours. The cost per kilowatthour varies depending on locality (it is high in New York City and cheap in the Tennessee Valley, for instance). The cost of electric energy reflects the cost of the fuel used at the generating plant, as well as the depreciation of the equipment and machinery and profit. Oil-fired plants produce the

heating or for stationary heat engines. In liquid form, it has already been used in rockets.

Electrolysis also takes place when conventional lead-acid storage batteries are charged. Since hydrogen and oxygen gas escaping from the battery cells generate a potentially explosive mixture, care should be taken to keep open flames or sparks away when charging batteries.

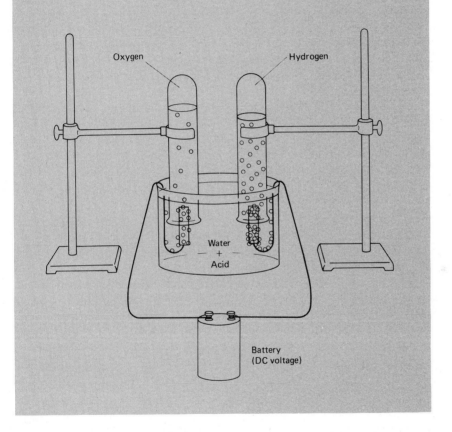

most expensive electricity, with nuclear and coal following in that order.*
Geothermal is next, followed by hydropower.

* Nuclear advocates claim that nuclear power is potentially cheaper than coal. This may eventually be the case as high-grade coal is mined out and environmental restrictions are tightened. Newly developing breeder technology also promises to lower the price of nuclear fuel.

Exercises

17. At a rate of $0.055 per kilowatthour, calculate the cost to do the following: (a) operate a color television for 10 h, (b) operate a 1200-W space heater continuously for 24 h, (c) operate an electric clock for 1 yr, (d) dry a load of clothes in 30 min, (e) operate a heat pump an average of 7 h per day for 1 month.
    Answers: (a) $0.18, (b) $1.58, (c) $1.45, (d) $0.14, (e) $138.60

18. A certain heat pump has an input rating of 12 kW. In moderate weather, this allows it to pump 86 kW of heat into a building. Suppose that the pump operates an average 3.8 h per day for a month and that the cost of electricity is $0.047 per kilowatthour. (a) What does it cost to operate the heat pump for a month? (b) What would it cost to get the equivalent amount of heat from electric resistance units?
    Answers: (a) $64.30, (b) $460.79

Summary   The principles of electricity are very much like the principles of fluid flow and heat flow. Although it is individual electrons traveling through conductors that transfer most electric energy, such large numbers of electrons are normally involved that we simply think of a continuous current flow. The current is pushed along by the voltage, and it is held back by the resistance. In all cases of practical importance, the voltage "pumps" the current around a closed loop, or circuit.

   Ohm's law tells us how to calculate the current if we know the voltage and the resistance. We may also calculate the current if we know the voltage and the power. The current is important because every circuit has a maximum current that it can safely handle. In homes and buildings, each circuit is fused for this maximum safe current; when this current is exceeded, the fuse blows or a circuit breaker trips.

Terms You Should Know

electric current	short circuit
ampere	conventional current
voltage	electron current
insulator	schematic symbols
conductor	ammeter
resistance	voltmeter
Ohm's law	electric power
circuit	electric energy

Problems

1. A hairdryer is rated at 775 W, 115 V. What current flows through this dryer when it is being used?

2. A 75-W bulb operates from a 110-V source. (a) What current flows through the bulb? (b) What is the bulb's resistance when it is lit?

3. An electric carburetor heater on a small airplane has a resistance of 3.5 Ω. (a) What current flows through the heater when the voltage is 12 V? (b) What electric power does the heater use?

4. Arc-welding a steel plate 2 cm thick requires a current of 200 A. To maintain this current takes a voltage of 22 V. What is the power requirement in kilowatts?

5. A certain electric motor has a power output of 2.25 kW [3.02 hp] at full load. If the efficiency is 96 percent and the voltage is 225 V, what current is being supplied to the motor?

6. A certain 12-V car battery is rated at 85 Ah. The headlamps are rated at 52 W each. Approximately how long can the fully charged battery operate the headlamps if the car's engine is stopped?

7. A small electric car could be driven by a 20-kW [27-hp] motor at a speed of 70 kmph [43 mph]. Estimate the ampere-hour capacity of the battery pack if the car is to have a range of 250 km [155 mi] at this speed starting with a full battery charge. Do this for a battery-pack voltage of (a) 12 V, (b) 36 V, (c) 96 V.

8. A sawmill is to be operated from a 22-kW motor. Electricity costs $0.085 per kilowatt-hour. Gasoline costs $0.28 per liter. An electric motor of this size has an efficiency of 96 percent, while a gasoline engine has an efficiency of 26 percent. Estimate the cost of one week's operation (40 working hours at full load) with each of these motors.

Magnetism and Alternating Currents

Introduction   In the last chapter, we discussed the principles of electricity and the relationship between four important quantities: voltage, current resistance and power. We saw that the conventional current in a circuit flows in the direction of high voltage to low voltage. This must be kept in mind when certain ammeters, voltmeters, electric motors, electroplating equipment, and radios and other electronic equipment are used. Batteries usually have their terminals labeled positive (+) and negative (−) so we can tell the direction of the current flow they produce.

But there are also voltage sources that periodically reverse their polarity. This is the case with the line voltages in our homes and most of our factories. Such periodically reversing voltages give rise to periodically reversing, or *alternating,* currents.

direct current   Definition: A *direct current* (dc) is a current that flows in only one direction. Such currents flow from batteries or dc power supplies, where the polarity of the source is always the same.

alternating current   Definition: An *alternating current* (ac) is a current that periodically reverses its direction of flow. Alternating currents originate in ac electric generators.

In this chapter, we discuss some of the special properties of alternating current, as well as the way ac is generated and transmitted. We begin with magnetism, since this is the basis for many of our later explanations.

Around 100 B.C. some unknown Chinese magician discovered an unusual property of a particular type of rock. Suspended from a string, the same end of the rock always swung around until it pointed north. The discovery came much later in Europe, where the rock was called a lodestone. By 1200 it was discovered that a steel needle stroked with a lodestone acquired this same north-seeking property. The idea of floating such a magnetized needle in a bowl of water led to the first ship's compass, which was a great boom to navigation.

We now know that compasses do not actually seek the North Pole of the earth; they align with a point called the north *magnetic pole,* which is some 1600 km from the North Pole and actually wanders about from time to time. In any case, there certainly is a force other than gravity between a compass needle and the earth. We may imagine the earth to be enveloped by invisible magnetic lines of force that originate near the earth's south magnetic pole and disappear into the ground near the north magnetic pole (Fig. 19-1). At any point on the earth's surface, a compass needle aligns with the local direction of these lines of force. The *magnetic lines of force* also extend into and above the atmosphere, layer upon layer like the skins of an onion. This entire pattern is called a *magnetic field.*

19-1 Permanent Magnets

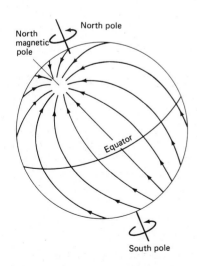

(A) At the earth's surface

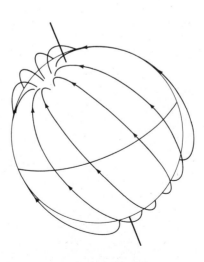

(B) Above the earth's surface

Fig. 19-1 Magnetic lines of force around the earth.

Suppose now that we lay a bar magnet (or an oversized compass needle) on a table. We then place a small compass nearby. An interesting thing happens: the north-seeking pole of the compass points toward the south-seeking pole of the bar magnet! Unlike magnetic poles attract, just like unlike charges attract. Since the term "north-seeking pole" is a bit cumbersome, we simply call this pole of a magnet or compass its "north pole." Then the other pole is the south pole.

By moving a small compass from point to point around a magnet, we can map the lines of force around the magnet. The direction of the line of force at any point is the same as the direction in which the compass points when placed there. Figure 19-2 shows the lines of force around a bar magnet and a U-magnet.

Permanent magnets are usually made of iron or steel or alnico (which is an alloy of aluminum, nickel, and cobalt). These materials are said to be *ferromagnetic*. Placed near a magnet, ferromagnetic materials at least temporarily become magnets themselves. This causes them to be attracted to magnets. Most other substances (copper, aluminum, plastics, and so on) have no magnetic properties except possibly at temperatures near absolute zero. Thus a magnet cannot be used to lift copper or aluminum.

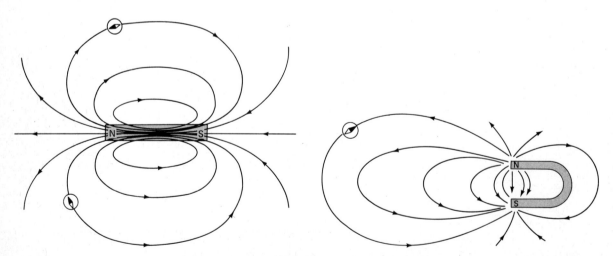

Fig. 19-2 Magentic lines of force around two types of permanent magnet. At any pont, the direction of the line of force is the same as the direction in which a compass needle would point if placed there.

In Fig. 19-3 we see a coil of wire which is carrying an electric current. If we place a small compass near the coil, we find that there is a magnetic field around it. And if we plot the lines of force, we find much the same pattern around the coil as we had with the bar magnet of Fig. 19-2. The coil acts like a magnet as long as there is a current flowing through it. For obvious reasons, we refer to such a magnet as an *electromagnet.* This magnet is stronger if a ferromagnetic material is placed inside the coil, but even without such a core we still have a magnet.

Since the lines of force in Fig. 19-3 form closed loops, it may not seem that there are north and south magnetic poles here. But if we compare Figs. 19-2 and 19-3, we have to conclude that the north pole of a coil is where the lines of force leave and the south pole of a coil is where the lines of force enter. Then the current-carrying coils behave just like permanent magnets: their unlike poles attract and their like poles repel. This similarity in behavior is shown in Fig. 19-4.

It is not necessary to have a coil of wire to get a magnetic effect from an electric current. *Any* electric current sets up magnetic lines of force. Figure 19-5 shows the lines of force around a straight current-carrying conductor. Again, this pattern can be discovered with the help of a compass. If two such conductors are brought near each other, they attract or repel each other depending on whether the currents are in the same or in opposite directions. This effect is actually the basis for defining the ampere as a unit of current. One ampere is the current that, if maintained in two parallel conductors 1 m apart in a vacuum, produces a force of attraction of 2×10^{-7} N for each meter of length. Although this is a small force, it can be measured with a current balance.

If a coil electromagnet is pivoted so it can rotate and is then placed near a permanent magnet, it turns until its lines of force line up with those of the magnet. If the current in the coil is then reversed, its lines of force change direction and the coil turns further. By periodically reversing the current flow, the coil can be kept in constant motion.

This simple idea is used in the *dc motor.* The rotating coil, which has a ferromagnetic core, is called the *armature.* The stationary permanent magnet is called the *field magnet.* On large motors, the field magnet may be wound over with field coils to make it extra strong. The armature winding has to stay connected to a voltage source as it rotates, and without getting the leads all twisted up. This is done with a *commutator* and *brushes.* The commutator is a metal cylinder that rotates concentric with the armature shaft; the brushes are spring-loaded contacts that brush against the commutator as it rotates. In the simplest dc motor, the commutator is split down the middle, with one half insulated from the other. The ends of the armature coil are wired to those two halves of

19-2 Electromagnets

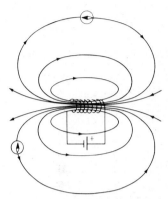

Fig. 19-3 A coil carrying an electric current produces much the same magnetic field pattern as a permanent bar magnet.

19-3 The dc Motor

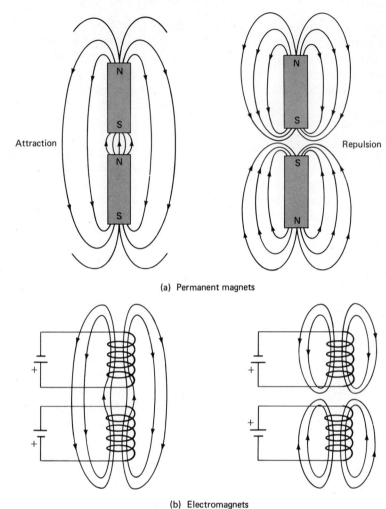

Fig. 19-4 Electromagnets follow the same rule of attraction and repulsion as permanent magnets.

the split commutator. While one brush makes contact with one half of the commutator, another brush makes contact with the other half. When the armature and commutator rotate 180°, each commutator half swings into contact with the opposite brush. This reverses the current in the armature coil. The principle is shown in Fig. 19-6.

    In practice, a dc motor may have more than one armature coil and more than one field magnet. This results in a larger and more continuous torque output without appreciably increasing the motor's outside dimensions. The details of such arrangements are beyond the scope of this book.

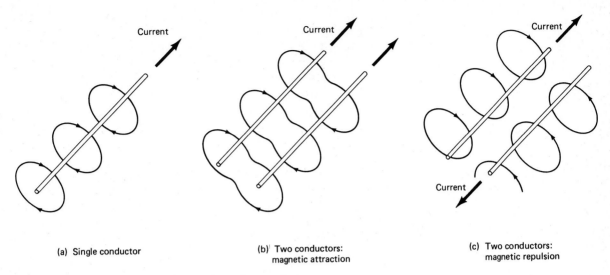

(a) Single conductor

(b) Two conductors: magnetic attraction

(c) Two conductors: magnetic repulsion

Fig. 19-5 Magnetic fields around current-carrying conductors.

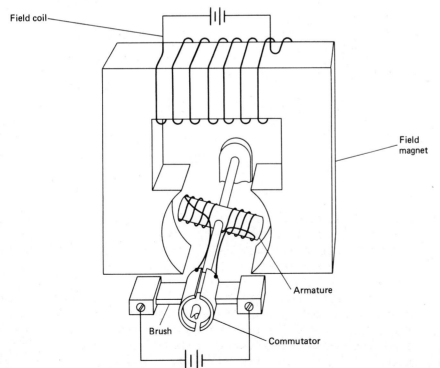

Field coil

Field magnet

Armature

Brush

Commutator

Fig. 19-6 Principle of the dc electric motor.

The rotational frequency of a dc motor is governed by two factors: the current through the coils and the load on the motor. Increasing the current or decreasing the mechanical load increases the frequency of rotation. Some small dc motors "run away" if the load is suddenly removed; they may turn faster and faster until centrifugal force and friction destroy them. In larger motors, special precautions are usually taken so this cannot happen.

19-4 Magnetic Induction

A magnet does not attract stationary electric charges. If a magnet is placed near the terminal of a battery, the battery just ignores it. In other words, the magnetic force and the electric force are two different things.

Still, there is certainly a relationship between these two forces. An electric force can produce a current, which in turn sets up a magnetic field. As we have seen, the resulting magnetic force attracts or repels another wire or coil carrying a current.

Can the process be reversed? Is it possible to use a magnetic force to produce an electric force? The answer is yes. This happens any time the magnetic force on a conductor is *changing*. The principle is referred to as Faraday's law.

Faraday's law of induction

Faraday's law of induction: A magnetic force field that changes in time will develop a voltage in any conductor placed in the field. A voltage produced in this way is called an *induced voltage*.

Figure 19-7 shows two coils of wire wound on the same iron bar. The coils are insulated from each other and from the bar. One coil is connected through a switch to a voltage source, while the other coil is connected to a light bulb. When the switch is first closed, a current starts to flow in the lower coil, and a magnetic field begins to grow in the bar.

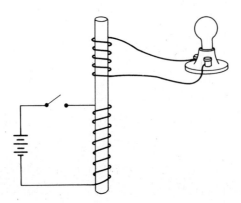

Fig. 19-7 Faraday's law: the light blinks when the switch is first closed or first opened.

Application: The D'Arsonval Meter Movement

This device is the heart of most ammeters and voltmeters. It consists of a small electromagnet which pivots in the field of a larger permanent magnet. The electromagnet rotates against a countertorque generated by a spiral spring. When a current flows in the electromagnet, or moving coil, the electromagnet rotates in an attempt to align its field with that of the permanent magnet. But the farther it rotates, the greater is the countertorque supplied by the spiral spring. The equilibrium point is indicated by a pointer on a scale. Properly calibrated, the scale will read the current in the coil.

      The zero position of the pointer can be adjusted by twisting a screw that varies the tension in the spring. The pointer itself moves through an arc of less than 180°. Meters that give a pointer deflection of more than 180° do not have D'Arsonval movements.

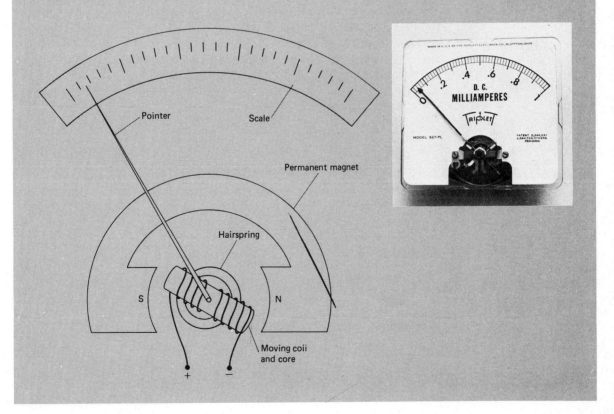

According to Faraday's law, this changing field induces a voltage in the upper coil, which causes the bulb to blink. Once the current from the source reaches a steady value, the magnetic field no longer changes and the bulb goes out. When the switch is opened, the magnetic field collapses, and a voltage is again induced in the upper coil. The bulb blinks again. The point is that the bulb blinks only when the magnetic field *changes*. Notice that a steady magnetic field does not induce a voltage in the upper coil and does not light the bulb.

Now any scheme that changes a magnetic field can be used to induce a voltage in a conductor. In an automobile's alternator, for instance, field magnets rotate around a stationary coil. The resulting voltage is used to keep the car's battery charged.

19-5 The dc Generator

The motor in Fig. 19-6 is a coil rotating in a magnetic field. As the coil rotates, the magnetic field within the coil changes. According to Faraday's law, this should induce a voltage in the coil. As a matter of fact, if we replace the battery with a voltmeter and spin the armature by hand, we find that the device does generate a voltage. In fact, any electric motor operated in reverse becomes a generator. In some early automobiles (the 1914 Hudson, for instance), the starter motor and generator were one and the same.

If we look closely at the voltage output of a *dc generator*, we find that it is not steady. The largest voltage is induced when the armature coil lines up with the permanent magnetic field. (This is also where the armature is most difficult to turn.) When the armature coil becomes perpendicular to the permanent field, the induced voltage momentarily drops to zero. A graph of the voltage output versus time is shown in Fig. 19-8. This particular generator is being rotated once every 0.50 s, or at a frequency of 2.0 rps. Notice that the output voltage occurs in a series of pulses. If very steady dc voltage is needed, special steps must be taken to smooth out these pulses.

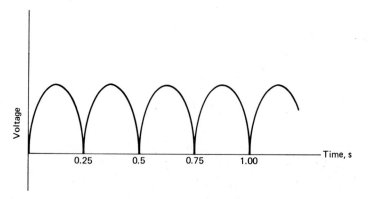

Fig. 19-8 Voltage output of a simple dc generator rotated at a constant rate of 2.0 rps.

Even though a dc generator's voltage comes in pulses, the polarity stays the same. This is because the commutator switches sides just as the magnetic field in the armature coil reverses. If we replace the commutator with two slip rings rubbing against the brushes (Fig. 19-9), the output voltage swings positive and negative as the field reverses. We now have an *ac generator.* The voltage is shown for an ac generator rotating at 2.0 rps.

   The stationary part of an ac generator is called the *stator* while the rotating part is called the *rotor.* In many cases, the coil is the stator, and the magnet or magnets make up the rotor. An ac generator built on this design is called an *alternator.* Its voltage output, however, is the same as that shown in Fig. 19-9.

19-6 The ac Generator and Motor

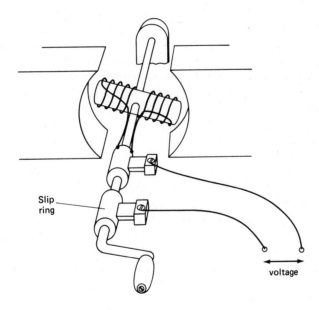

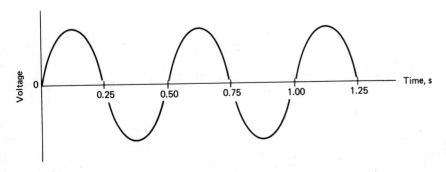

Fig. 19-9 The ac generator and its voltage output. This generator is being rotated at a constant rate of 2.0 rps.

One interesting and very simple ac generator is shown in Fig. 19-10. This device is an alternator design, since the magnet does the moving. A voltage is induced in the coil as the magnet sweeps past. The ignition timing can be advanced or retarded by moving the magnet forward or backward on the flywheel (with the engine stopped, of course).

An *ac motor* is an ac generator operated in reverse: by putting electric energy *in,* we get mechanical work *out.* Since ac motors have no split commutator, they depend on the voltage source's polarity changes to reverse the current in the field coil. An ac motor does not operate on dc. (There are, however, *universal motors* that run just as well on ac or dc.)

Regardless of the load, ac motors operate at a fairly constant speed. Most motors for power tools operate at either 1750 or 3500 rpm. Varying the speed of an ac electric motor is fairly difficult and always results in a reduction in efficiency. For this reason, lathes and other power machines must use mechanical transmissions or else dc motors to get a variable frequency of rotation.

Since the rotational frequency of an ac motor is relatively constant, increasing the load also increases the power output. This increased power must come from somewhere. Because the source voltage cannot

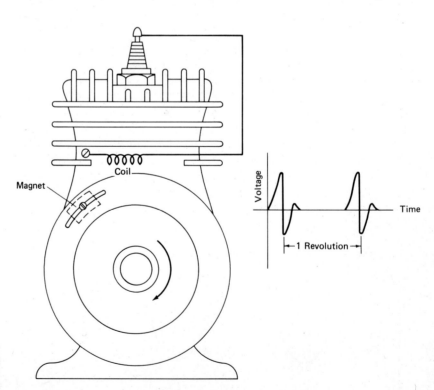

Fig. 19-10 Simple alternator used to fire a spark plug on a small gasoline engine.

change, the increased power comes from an increase in current. Thus *the current through an ac motor automatically changes with the load.*

Suppose that you switch on a power saw. The motor armature and the saw blade quickly accelerate to 1750 rpm. A relatively large current is drawn from the source to develop the power to do this. Now the motor and blade are idling at 1750 rpm. Very little power is required, and so very little current is drawn. You now begin to saw a piece of hardwood. This requires an increased power output, and so an increased current flows.

The motor's rated power output is the maximum power it can safely develop under load. You can exceed this rating by up to about 25 percent for very short periods. But this slows the motor slightly, resulting in a loss of efficiency and the generation of excess heat. Eventually the insulation will melt and the motor will burn up. And the increased current under overload can easily blow a fuse or breaker in the main circuit.

19-7 Characteristics of ac

The ac voltage used in the United States has a standard frequency of 60 Hz. This means that the voltage goes through 60 complete cycles, positive to negative and back again, each second. If you connect a dc voltmeter to a wall outlet, it reads zero. The voltage changes too rapidly to move the needle. (On some meters, you may notice the needle vibrating very rapidly about the zero position.) So a zero reading on a dc voltmeter does *not* mean the absence of any voltage.

An ac voltmeter is needed to measure ac voltage. Similarly, an ac current does not register on a dc ammeter, and a special ac ammeter must be used. People have suffered nasty electric shocks after using the wrong meter to verify that a wire was dead and then touching the wire.

Figure 19-11 shows how a 117-V ac line voltage varies in time. Notice that the voltage actually varies between a positive peak of 165 V and a negative peak of −165 V. Why, then, don't we say that the voltage is simply 165 V? Because if this source is used to light a lamp, we find that the lamp glows as brightly as if it were lit by a 117-V dc source. The *effective* value of this ac voltage is 117 V.

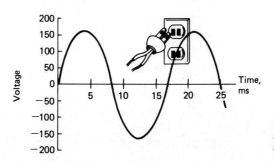

Fig. 19-11 A 117-V ac line voltage. The voltage actually peaks at 165 V.

effective (or rms) voltage

Definition: The *effective*, or *rms*, *voltage* of an ac source is the equivalent dc voltage that would produce the same power consumption in a resistance.

When we talk about "ac voltage," we always mean the effective voltage. When we mean the peak voltage, we have to be careful to say so explicitly. The peak voltage is related to the effective voltage in this way:

$$\text{Peak ac voltage} = (1.414)(\text{effective ac voltage}) \qquad (19\text{-}1)$$

$$\text{Effective ac voltage} = (0.707\ 1)(\text{peak ac voltage}) \qquad (19\text{-}2)$$

The current in an ac circuit varies at the same frequency as the voltage. It also has a peak value and an effective value. The two are again related by the factor 1.414.

Example 19-1 Insulation and ac Circuits

A certain electric wire has insulation rated at 600 V dc. What is the maximum ac voltage that can be used with this wire?

If the voltage across this insulation exceeds 600 V, the insulation breaks down and begins to conduct. This can lead to blown fuses or worse, a shock or fire hazard. If we use ac, then, the *peak voltage* should not exceed 600 V.

Using Eq. (19.2), this gives an effective voltage of

$$\text{Effective voltage} = (0.707)(\text{peak voltage})$$

$$= (0.707)(600\ \text{V})$$

$$= 424\ \text{V}$$

Of course, to be on the safe side, we should keep the voltage *below* this value. ◀

The nice thing about using effective ac voltage and current is that then all the equations in Chap. 18 work equally well for ac and dc. In particular, the power in an ac circuit is just the product of the current and the voltage.

Exercises

1. Find the peak voltage if the effective ac voltage is (*a*) 6.3 V, (*b*) 18 V, (*c*) 117 V, (*d*) 230 V, (*e*) 18 kV.
    Answers: (*a*) 8.9 V, (*b*) 25 V, (*c*) 165 V, (*d*) 325 V, (*e*) 25 kV

2. Calculate the effective ac voltage if the peak ac voltage is (*a*) 8.9 V, (*b*) 100 V, (*c*) 480 V, (*d*) 22 kV.
    Answers: (*a*) 6.3 V, (*b*) 70.7 V, (*c*) 339 V, (*d*) 16 kV

19-8 Rectification

To change dc to ac requires expensive devices like inverters or motor-generator sets. To go the other way—ac to dc—is fairly simple. This is called *rectification*. This is accomplished with a relatively cheap device called a *rectifier*.

   A rectifier acts like a fluid check valve, permitting current to flow in only one direction. Figure 19-12 shows a rectifier connected to an ac source and a load. Conventional current flows in the direction of the black arrowhead in the rectifier symbol. Since the source voltage appears at the load only half of the time, this circuit arrangement is called a *half-wave rectifier*.

   It is also possible to convert an ac voltage to a dc voltage like the dc generator output in Fig. 19-8. This is done by connecting four rectifiers in a bridge circuit, shown in Fig. 19-13, called a *full-wave rectifier.* You

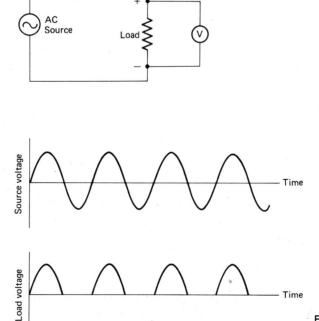

Fig. 19-12 Half-wave rectification using a single rectifier.

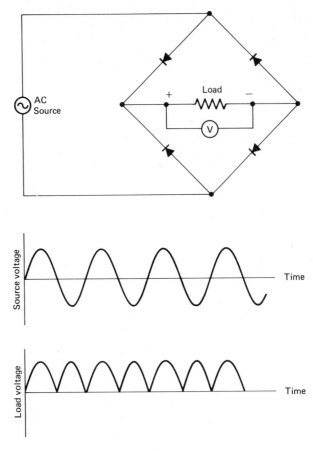

Fig. 19-13 Full-wave rectification using a bridge circut.

can follow the operation of this circuit by remembering that each rectifier
can conduct only in the direction of the arrowhead. Even though the ac
source changes polarity, the current through the load is always in the
same direction. The alternators found on car engines have a full-wave
bridge circuit built in so the output can be used to charge the car's
battery. In this case, the battery is the load.

19-9 Transformers

If an ac source is connected to a coil, the coil's magnetic field alternates
at the same frequency as the source. This alternating field will induce an
ac voltage in a second coil if it is close enough. The effect is shown in
Fig. 19-14, where the induced voltage produces a current that lights a
lamp. Notice that this does not work with a dc source. Although the lamp
will blink momentarily when the dc source is connected or disconnected,
an ac source is needed to light the lamp continuously.

The arrangement of two coils shown in Fig. 19-14 is a single *transformer*. The coil connected to the ac source is called the *primary coil,* while the coil connected to the load is the *secondary coil.* Usually the coils are wound on an iron core to keep the magnetic field aligned,

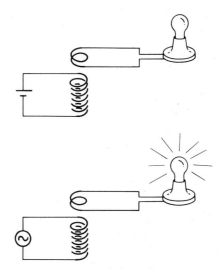

Fig. 19-14 A simple transformer. A voltage is induced in the upper coil when an alternating or changing current flows in the lower coil. The device does not work with steady dc.

and often the coils are wound right over one another, separated by a layer of insulation. In a large, well-designed transformer, as much as 99 percent of the electric power supplied to the primary coil eventually reaches the load. This is the same as saying that the transformer's efficiency is as much as 99 percent.

But even though there is little loss of power in a transformer, the secondary *voltage* can be considerably different from the primary voltage. Whether the voltage increases or decreases depends on whether the secondary coil has more or fewer turns than the primary coil. For an efficient transformer, the relationship is

$$\frac{\text{Secondary voltage}}{\text{Primary voltage}} = \frac{\text{number of turns on secondary coil}}{\text{number of turns on primary coil}} \qquad (19\text{-}3)$$

If the secondary coil has more turns than the primary coil, then the secondary voltage is higher than the primary voltage, and we have a *step-up transformer.* If the secondary coil has fewer turns than the primary, then the secondary voltage is lower than the primary voltage, and we have a *step-down transformer.* For instance, if the secondary has 3 times as many turns as the primary, then the secondary voltage is 3 times the primary voltage. But if the primary has 3 times as many turns as the secondary, then the secondary voltage is one-third the primary voltage. Transformers are built in many sizes and shapes; a few types are shown in Fig. 19-15.

Since a transformer's power output is nearly the same as its power input, the voltage can be stepped up only at the expense of current. The relationship may be written as

$$(\text{Secondary voltage})(\text{secondary current}) = \\ (\text{efficiency})(\text{primary voltage})(\text{primary current}) \qquad (19\text{-}4)$$

This formula applies whether we have a step-up transformer or a step-down transformer.

Example 19-2 Power and Current in ac Arc Welding.

An ac arc-welding power supply is basically a large step-down transformer. The primary voltage is typically 230 V. Let's say that the circuit supplying this voltage is fused at $3\bar{0}$ A. What is the maximum current we can draw from the secondary coil if the secondary voltage is $4\bar{0}$ V?

The maximum power supplied to the transformer is

$$\text{Power} = (\text{voltage})(\text{current})$$
$$= (230 \text{ V})(3\bar{0} \text{ A})$$
$$= 6900 \text{ W}$$

If the efficiency is 98 percent, the power leaving the secondary is about 6760 W. This value is the right-hand side of Eq. (19-4):

$$(\text{Secondary voltage})(\text{secondary current}) = 6760 \text{ W}$$

Since the secondary voltage is $4\bar{0}$ V, the secondary current is

$$\text{Secondary current} = \frac{6760 \text{ W}}{4\bar{0} \text{ V}}$$
$$= 170 \text{ A}$$

With this much current, we could weld a steel plate about 0.6 cm thick in a single pass. ◀

Fig. 19-15 Transformers come in many sizes and shapes.

Application: The Eddy-Current Brake

The eddy-current brake is a simple device sometimes used on rotating machinery. A metal disk (usually aluminum) rotates with the shaft of the motor or machine to be stopped. An electromagnet is mounted so the disk passes between its poles. When the magnet is switched on with the disk rotating, eddy currents are generated in the vicinity of the magnet. These currents produce their own magnetic fields which interact with the electromagnet and place a drag on the disk. This brings the disk and machine to a stop much more quickly than one might guess.

Since there is no friction in the eddy-current brake, there is nothing to wear out but the switch. As in most other dc applications, the switch is shunted with a condenser (capacitor) to prevent arcing and to lengthen its life.

In some cases where no dc is available, small eddy-current brakes are operated from permanent magnets. In such devices, the magnet must be kept clear of the disk to allow the disk to rotate freely. To apply the brake, the magnet is slid into position by a mechanical linkage.

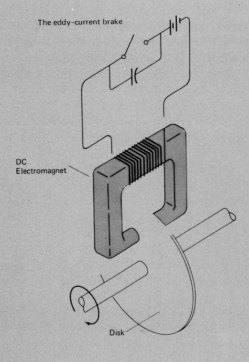

The eddy-current brake

DC Electromagnet

Disk

Exercises

3. A certain transformer has 250 turns on its secondary coil and 100 turns on its primary coil. Find the secondary voltage if the primary voltage is (a) 115 V, (b) 6.3 V, (c) 0.919 V.
   Answers: (a) 288 V, (b) 16 V, (c) 2.30 V

4. A transformer is to be wound with 12 turns on the primary coil. The primary voltage is 110 V. How many turns are needed on the secondary coil to produce a secondary voltage of 15 kV?
   Answer: About 1600 turns

5. A current of 30 mA flows through a neon light when the voltage is 15 kV. The transformer efficiency is 95 percent. (a) What power is supplied by the secondary coil? (b) What power is being supplied to the primary coil? (c) If the primary voltage is 110 V, what is the current in the primary circuit?
   Answers: (a) 450 W, (b) 470 W, (c) 4.3 A

6. A certain high-voltage power transformer steps up 18 kV to 275 kV. The efficiency is 98 percent. If the primary current is 8.2 A, what is the current in the secondary circuit?
   Answer: 0.53 A, or 530 mA

We use dc in our cars and other vehicles, while we use ac in our homes. Why? Wouldn't it make sense to use the same type of electricity for all applications?

19-10 Electric Power Transmission

We use dc in our vehicles because no one has yet invented an ac battery. By using dc, electric energy stored while the car is running can be made available to start the engine after it's been stopped. But this also means using all dc motors in the vehicle (starter motor, windshield wiper motor, electric fan motor, heater motor, seat-adjustment motors in luxury cars, etc.). To be compatible, the alternator must also have a dc output, which is why the rectifiers are built in.

Okay, if dc works so well in our cars, why don't we use it in our homes, too? The answer has to do with the transmission of electric power from the generating station. There may be many miles of wire between the plant and the home. Although copper (or aluminum) wire is a good conductor, a very long length of it has a sizable resistance. Since it takes a voltage to push a current through a resistance, the voltage in a wire drops at increasing distances from the source. This is similar to the drop in fluid pressure in pipes.

Most appliances are designed to operate on only a narrow range of voltages, say 110 to 120 V. Suppose our power plant supplied a voltage of 120 V dc. Then if we lived next door to the plant, we would get the full 120 V. Homes located farther from the plant (Fig. 19-16) would get

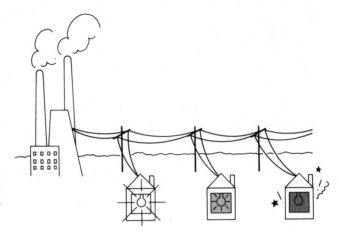

Fig. 19-16 Because of voltage drops in the power lines, the available voltage decreases at larger distances from the power station.

a reduced voltage. At distances of several kilometers from the power plant, the line voltage may drop considerably below the 110 V needed to operate the appliances. Thus, it would be necessary for the power companies to locate a complete power-generating station within a few kilometers of each of its customers.

Let's suppose that a power plant is called on to deliver 12 000 W of power to a certain consumer, at a voltage of 120 V. Let's suppose further, for the sake of illustration, that the power line between the station and the consumer has a resistance of 0.2 Ω. To deliver 12 000 W at 120 V, the current must be 100 A. With this current, the voltage lost in the lines is 20 V [100 A(0.2 Ω)], and so the consumer gets only 100 V instead of the 120 V desired. Moreover, the consumer gets only 10 000 W of power. The other 2000 W is lost as heat in the power lines.

As we mentioned, with alternating current there is an alternative. The power plant can use a transformer to step the 120 V up to 12 000 V ac. The 12 000 V is fed into the power line at a current of only 1 A. This still gives 12 000 W of power transmitted. With this 1-A current flowing through the 0.2-Ω wires, the voltage drop is only 0.2 V. At the consumer's home, the voltage is then 11 999.8 V. Another transformer (mounted on the utility pole) is then used to step this voltage down to 119.998 V to supply the consumer. With this arrangement, the consumer gets nearly the full voltage, and only 0.2 W of power is lost in the power line. The situation is shown in Fig. 19-17.

We see, then, that it is possible to transmit large amounts of electric power over large distances by using a high enough voltage. Within cities, power line voltages are usually within the range of 12 000 to 14 000 V. The high-tension lines crisscrossing the countryside commonly have voltages of 200 to 300 kV. At such high voltages, it is useless to even try to insulate the wires; they are usually bare aluminum, which also keeps

Fig. 19-17 Voltage losses are
minimized by using step-up and
step-down transformers.

the weight down and permits wider spacing of the very expensive supporting towers. In some long-distance transmission lines, voltages in
excess of 1 MV (1 million V) are being used.

19-11 Eddy Currents

You can make a simple transformer by winding two wire coils, one over
the top of the other, on a small bar of iron. But if you do, the efficiency is
terrible. The problem is that the changing magnetic field induces currents in the iron bar itself, and this uses up power which would otherwise
go into the external secondary circuit. What happens to this lost power?
It heats up the bar, which gets fairly hot fairly quickly. The currents in the
bar are called eddy currents.

> Definition: *Eddy currents* are loops of current induced in
> conducting materials placed in changing magnetic
> fields.

eddy currents

Eddy currents must be prevented if a transformer is to be efficient.
This can be done by using a *laminated core* of iron (Fig. 19-18). The core
is built up from flat iron plates insulated from one another by layers of

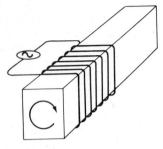

Solid iron bar

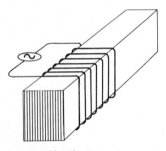

Laminated iron bar

Fig. 19-18 A laminated core is used to prevent eddy currents in transformers and motor and generator armatures.

lacquer. There are still induced voltages in the bar, but with much less internal current flow there is much less power loss. Laminated cores are used not only in transformers, but also in motor and generator armatures. In some cases, the core may be molded from powdered iron. This is even more effective than lamination in reducing eddy currents.

If eddy currents are allowed to occur, they set up their own magnetic field in opposition to the magnetic field that produces them. This can be seen in the simple demonstration in Fig. 19-19. Two aluminum rings are suspended by strings so they are free to swing. One ring has been cut to prevent eddy currents. If an end of a bar magnet is thrust in and out of each ring, one will begin to swing but the other one won't. Which ring swings? The one that hasn't been cut. The eddy current in this ring sets up a magnetic field that interacts with the bar magnet. The other ring has no eddy current (since there is no complete conducting path), and so it has no magnetic field of its own. It remains unaffected by the bar magnet.

The automobile speedometer works on this principle. A flexible shaft (connected to a front wheel or the transmission) rotates a magnet inside a hollow aluminum cylinder called the "speed cup" (Fig. 19-20). Eddy currents in the speed cup set up a magnetic field that tries to drag the cup in the direction of the rotating magnet. A spiral spring connected to the cup lets it rotate only until the induced magnetic torque balances the spring's elastic countertorque. This spring also returns the cup to the same zero point when the cable stops rotating. Also connected to the speed cup is the most important part of the instrument: the needle that indicates the car's speed on a scale.

Fig. 19-19 An eddy current produces a magnetic field. If one end of a bar magnet is thrust in and out of a suspended aluminum ring, the ring begins to swing. If the ring is cut so that no eddy currents can flow, it cannot be set into motion in this way.

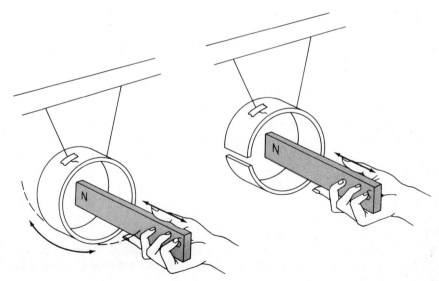

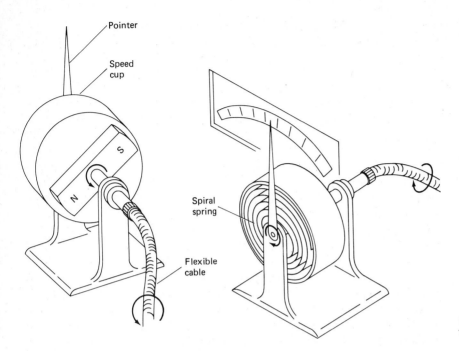

Fig. 19-20 Principle of the automobile speedometer.

We have seen that a magnet does not pick up metals like copper, aluminum, brass, lead, and so on. But a changing magnetic field induces eddy currents in these and all other conductors. This principle is sometimes used to separate nonferrous metals from trash in recycling plants. After the iron and steel are removed directly by a magnet, the trash is slid down a ramp where it passes between the poles of another very strong magnet (Fig. 19-21). This deflects the aluminum cans, copper wire, etc., to one side while letting paper and plastics pass straight through.

Summary

Electricity and magnetism are closely related. Electric currents produce magnetic fields, and changing magnetic fields can be used to generate electric currents. Batteries, solar cells, and thermocouples produce fairly steady voltages whose polarity does not change, and a current that flows in response to such a voltage is a direct current (dc). But when a voltage is generated by magnetic induction (using changing magnetic fields), the polarity periodically reverses. Such reversing, or alternating, voltages produce alternating currents (ac). Changing ac to dc is done simply and cheaply by using rectifiers. Changing dc to ac is fairly difficult and requires expensive inverters or motor-generator sets.

The major advantage of ac is that the voltage can easily be changed with little loss of power. A device that does this is called a transformer. By using step-up transformers at the generating station and step-down transformers near the user, electric power can be transmitted many kilometers with little loss.

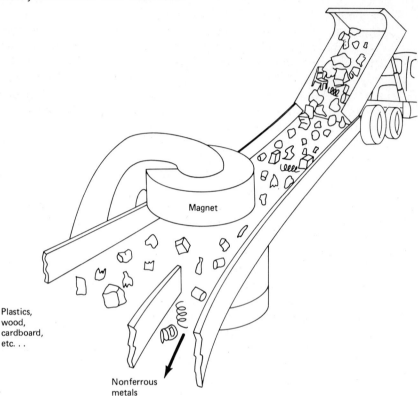

Fig. 19-21 Eddy currents permit the separation of nonferrous metals from other trash.

Magnet

Plastics, wood, cardboard, etc. . .

Nonferrous metals

Terms You Should Know

direct current	ac motor
alternating current	ac generator
permanent magnet	effective (or rms) voltage
magnetic pole	rectification
magnetic field	half-wave rectifier
magnetic lines of force	full-wave rectifier
ferromagnetic	transformer
electromagnet	primary coil
Faraday's law of induction	secondary coil
dc motor	eddy current
dc generator	laminated core

Problems

1. A doorbell transformer operates on 115 V and produces an output of 18 V. The transformer efficiency is 95 percent. The doorbell draws 1.9 A of current. (a) Approximately what power is used by the bell? (b) What power is supplied by the primary source? (c) What current flows in the primary circuit?

2. In industrial applications, pilot lights must often operate at 6.3 V rather than the full 115-V line voltage. This prevents a shock hazard if the light is accidentally broken. Generally, a small step-down transformer is built into the lamp assembly. The pilot light is rated at 15 W at 6.3 V. (a) What is the current through the lamp filament? (b) What is the current in the transformer's primary coil? Assume 98 percent transformer efficiency.

3. A certain electric saw is driven by a 110-V motor with an output rating of 0.85 kW. The motor's efficiency is 96 percent. (a) What effective ac current must the source supply when the motor operates at full load? (b) What effective ac current must the source supply when the motor operates at 25 percent overload? (c) What peak current flows at full load?

4. The secondary coil on some transformers is "tapped" at various points to provide a multiple-voltage source. One such transformer operates on a primary voltage of 230 V. The primary coil has 124 turns while the secondary has 163 turns. Starting from one end of the secondary coil, which is grounded, there are taps at 33, 65, 94, 120, 143, and the full 163 turns. What voltages can be obtained from the transformer if one end of the output is grounded?

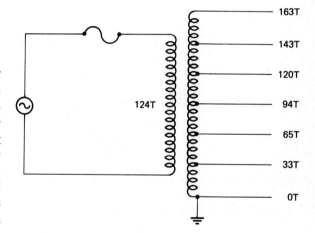

Appendix A

Scientific Notation (Powers-of-10 Notation)

Scientific notation is a scheme for writing very large and very small numbers without all the zeros. It is based on the fact that moving the decimal point in a number is the same as multiplying the number by a power of 10.

Some powers of 10 (actually, the *integer*, or whole number, powers of 10) are listed in Table A-1. The table results from the fact that

TABLE A-1   SOME INTEGER POWERS OF 10

10^9	1 000 000 000
10^8	100 000 000
10^7	10 000 000
10^6	1 000 000
10^5	100 000
10^4	10 000
10^3	1 000
10^2	100
10^1	10
10^0	1
10^{-1}	0.1
10^{-2}	0.01
10^{-3}	0.001
10^{-4}	0.000 1
10^{-5}	0.000 01
10^{-6}	0.000 001

$10^2 = 10 \cdot 10 = 100$, $10^3 = 10 \cdot 10 \cdot 10 = 1000$, and so on. The negative powers are defined as the reciprocals of the corresponding positive powers. Thus $10^{-2} = 1/10^2 = 1/100 = 0.01$, $10^{-3} = 1/10^3 = 1/1000 = 0.001$, and so on. The table can be extended in both directions indefinitely.

Now if we are given one of the numbers in this table, it is quite simple to tell what power of 10 we have. All we need to do is move the decimal point until we are left with 1.0. The number of places we moved the decimal is the power of 10.

Example A-1   What Power of 10 Is 100 000 000 000?

Although the decimal point isn't written here, it is implied after the last zero. Moving the decimal to the left and counting places gives

$$
\underset{11\ 10\quad 9\ 8\ 7\quad\ 6\ 5\ 4\quad\ 3\ 2\ 1}{1\,0\,0\ \ 0\,0\,0\ \ 0\,0\,0\ \ 0\,0\,0}
$$

Thus the number can be written as 10^{11}. ◄

Example A-2   What Power of 10 Is 0.000 000 001?

Now the decimal has to be moved to the right:

$$
\underset{1\ 2\ 3\quad\ 4\ 5\ 6\quad\ 7\ 8\ 9}{0.\ \ 0\,0\,0\ \ 0\,0\,0\ \ 0\,0\,1}
$$

Remember that we want to be left with 1.0, so we had to make the last jump *past* the 1. This gives a total of nine jumps. Since the original number was smaller than 1 (*much* smaller), we write the power of 10 with a negative exponent: 10^{-9}. ◄

What if we are given a number that is not an integer power of 10 — 31 400, for instance? No problem. We still count the number of places that we move the decimal point and write this as a power of 10 multiplying the number that is left. Thus 31 400 becomes 3.14×10^4. Similarly, 49 000 000 would be 4.9×10^7, and 0.000 937 would be 9.37×10^{-4}.

How do we decide how far to move the decimal? Mathematically, it doesn't really matter. We can put the decimal wherever it is convenient, as long as we reflect this in the power of 10. Thus 0.000 051 can be written as 0.51×10^{-4}, or 5.1×10^{-5}, or 51×10^{-6}. These are all the same number.

Example A-3   Express 59 000 as 59 Times a Power of 10.

Since the decimal has to be moved three places to the left, the result is 59×10^3.   ◄

Sometimes, particularly when using SI multipliers, we know what power of 10 we want to end up with. In such cases, we simply move the decimal a number of places equal to the exponent on the 10.

Example A-4   Express 0.000 891 m in Micrometers.

Since 1 μm $= 10^{-6}$ m, we want -6 as the power of 10. We get this by moving the decimal six places to the right:

$$0.000\ 891 = 891 \times 10^{-6}$$

Thus the result may be written as 891 μm.   ◄

Of course, we always have to be careful not to get the positive and negative powers mixed up. The simplest way is to remember that the positive powers are always numbers bigger than 1, while the negative powers are smaller than 1.

There are also rules for multiplying and dividing numbers written in scientific notation. These rules can be found in most basic mathematics textbooks. The rules aren't needed if you have a calculator which accepts numbers in scientific notation. You can also get around this problem by writing the numbers out if you have to do arithmetic with them.

Appendix
B

The following table summarizes the units used to measure most of the physical quantities discussed in this book. The SI base units (the meter, second, kilogram, kelvin, and ampere) are defined according to the standards described in Table 1-1. All other SI units are defined in terms of these base units. Non-SI units are related to the SI units by the conversion factors tabulated within the chapters.

Physical Quantities and Their Units

Quantity	U.S. engineering unit	Other common USCS units	SI unit	Common non-SI metric units
Length, distance	foot (ft)	inch (in), yard (yd) mile (mi)	meter (m)	—
Time	second (s)	minute, hour, day, week, year, etc.	second (s)	minute, hour, day, week, year, etc.
Mass	slug	pound-mass (lb_m)	kilogram (kg)	metric ton (t)
Force, weight	pound (lb)	ounce, ton	newton (N)	kilogram-force (kg_f)
Area	square foot (ft²)	acre, square inch, square yard, etc.	square meter (m²)	hectare

Quantity	U.S. engineering unit	Other common USCS units	SI unit	Common non-SI metric units
Volume	cubic foot (ft^3)	gallon, fluid ounce, quart, fifth, etc.	cubic meter (m^3)	liter (L)
Speed, velocity	feet per second (ft/s)	miles per hour (mph), inches per second (ips), knot	meters per second (m/s)	kilometers per hour (km/h, or kmph)
Acceleration	$\dfrac{ft}{s^2}$	standard gravitational acceleration (g)	$\dfrac{m}{s^2}$	standard gravitational acceleration (g)
Frequency	cycles per second (cps)	revolutions per minute (rpm)	hertz (Hz)	rpm, rps
Mass density	$\dfrac{slug}{ft^3}$	$\dfrac{lb_m}{ft^3}$	$\dfrac{kg}{m^3}$	$\dfrac{g}{cm^3}$, $\dfrac{kg}{L}$
Weight density	$\dfrac{lb}{ft^3}$	—	$\dfrac{N}{m^3}$	$\dfrac{kg_f}{L}$
Pressure, stress	$\dfrac{lb}{ft^2}$	lb/in² (psi), in Hg, standard atmosphere (atm)	pascal (Pa)	kg_f/cm², mmHg, standard atmosphere (atm)
Flow rate	$\dfrac{ft^3}{s}$	gallons per minute (gpm)	$\dfrac{m^3}{s}$	L/s, L/min
Torque	lb·ft	lb·in	N·m	kg_f·m, kg_f·cm
Work, energy	ft·lb	horsepower-hour (hp·h), Btu	joule (J)	calorie, kg_f·m, kilowatthour (kWh)
Power	$\dfrac{ft·lb}{s}$	horsepower (hp), Btu/h	watt (W)	metric horsepower (cheval vapeur)
Temperature	Rankine degree (°R)	Fahrenheit degree (°F)	kelvin (K)	Celsius degrée (°C)
Heat of combustion, latent heat	—	$\dfrac{Btu}{lb_m}$	$\dfrac{J}{kg}$	$\dfrac{cal}{g}$
Specific heat capacity	—	$\dfrac{Btu}{lb_m·°F}$	$\dfrac{J}{kg·K}$	$\dfrac{cal}{g·C°}$, $\dfrac{J}{kg·C°}$
Thermal conductivity	—	$\dfrac{Btu·in}{h·ft^2·°F}$	$\dfrac{W}{m·K}$	$\dfrac{W}{m·C°}$
Electric current	—	—	ampere (A)	—
Voltage	—	—	volt (V)	—
Electric resistance	—	—	ohm (Ω)	—

Index